AF518337

Textbook of
BIOLOGY

THE AUTHORS

Dr. Amit Kumar Jain has more than fourteen years of professional experience of teaching, research and administration. Dr. Jain joined IGNOU as Assistant Regional Director (ARD) IGNOU Regional Centre Ahmedabad Gujarat in 2011, and presently working at IGNOU RC Karnal Haryana. Prior to this, he has served as Assistant Professor in Microbiology, J.V. College, Baraut, Uttar Pradesh. He obtained Ph.D in 2005 and Doctor of Science (D.Sc.) Degree (Botany) in 2015 from CCS University, Meerut. He has qualified ICAR-NET in Plant Physiology and Biochemistry. He has also acquired the degree of B.Ed, MA(Edu), MA in Distance Education (MADE), PG Diplomas i.e. PGDNBT, PGDDE, PGDESD, PGDRD, PGDIPR and PGDEMA through ODL Mode. He is the Fellow of International and National level academic societies i.e. Linnean Society of London (UK), Society of Innovative Educationalist & Scientific Research Professional (FSIESRP) and Botanical Society (India) as well as Life member of the Indian Science Congress Association (ISCA), Kolkata and Centre of Education Growth and Research (CEGR), New Delhi. He has earned various national and international awards. In 2018, he awarded Innovative Researcher and Dedicated Academician Award 2018 by INSRPM Malaysia. Dr. Jain has published 05 Books, 50 Research Articles in national and international journals, and published 8 Book chapters edited by various authors. Dr. Jain has participated in a large number of national and international seminars, conferences, workshops, training and added programmes. He is the member of various committees of academic bodies of other reputed organizations.

Prof. J.D.S. Panwar got Ph.D. in Plant Physiology from Indian Agricultural Research Institute (IARI), New Delhi. Joined IARI, Plant Physiology in 1971 and served the Institute in various capacities as Scientist, Senior Scientist, Principal Scientist and as Head, Division of CPBM at IIPR, Kanpur. Retired as Head, Division of Plant Physiology, IARI, New Delhi. He was engaged in research for 37 years, teaching 19 years and guided to M.Sc. and Ph.D. students. He worked on several projects as PI and published 06 books, more than hundred research articles in journals of national and international repute. He has international exposure at University of Illinois, Urbana Champagin, USA as UNDP/FAO fellow. Prof. Panwar is a fellow of ISPP and ISPRD 1993 onward, awarded as ISPP, G.S. Sirohi best paper publication award (2003). He served as expert/members of academic and administrative bodies at national level. He leaded as Vice Chancellor, Mahatma Gandhi University, Meghalaya as well as Venketashwara Open University, Arunachal Pradesh. He has actively participated in the Vice Chancellor`s conferences organized by Association of Indian Universities, New Delhi in collaboration of different universities. In 2016, he has been awarded as Excellence Education Award by Confederation of Indian University at Kolkata. He has also participated in round table discussion on GST organized by "Education Promotion Society for India (EPSI)" in 2017.

Textbook of
BIOLOGY

Amit Kumar Jain
D.Sc. Ph.D, NET, FLS, FBS, FINSRPM, FAPS
Assistant Regional Director
Indira Gandhi National Open University (IGNOU)
Regional Centre Karnal, Haryana
Former Assistant Professor in Microbiology, J.V. College, Baraut, UP

J.D.S. Panwar
Ph.D, FISPP, FISPRD
Former Vice Chancellor, Venketashwara Open University, Arunachal Pradesh
Former Vice Chancellor, Mahatma Gandhi University, Meghalaya
Former Director, N.K. School of Nursing, Rampur, Uttar Pradesh
Former Head, Division of CPBM, IIPR, Kalyanpur, Kanpur
Former Head, Division of Plant Physiology, IARI, New Delhi

2022

Daya Publishing House®
A Division of
Astral International Pvt. Ltd.
New Delhi – 110 002

ISBN: 978-93-5461-222-0

Published by : **Daya Publishing House®**
A Division of
Astral International Pvt. Ltd.
– ISO 9001:2015 Certified Company –
4736/23, Ansari Road, Darya Ganj,
New Delhi-110 002
Ph. 011-43549197, 23278134
E-mail: info@astralint.com
Website: www.astralint.com

Digitally Printed at : **Replika Press Pvt. Ltd.**

Acknowledgment

First and foremost praise thanks to the God, the Almighty for never ending grace, and blessings to pursue our thoughts.

This entire tenure made us to compromise with many of our family commitments and whole heartedly thanks to our beloved parents and family members for imparting unstinted support and encouragement. Sacrifice of our family kids especially Aarna Jain and Prakhar Panwar can't be expressed in words, they have compromised with their almanacs during writing this book.

The completion of this task could not have been possible without motivation and encouragement of many peoples. We are thankful to our teachers, colleagues, friends and well-wishers spanning all over the sphere, inspiring us for writing this book. We extend our thanks to the authors of various books who helped us in upcoming of the subject material to the present status.

We express our heartfelt thanks to our publisher for liberal use of the recent available expertise in the field of printing and publication of the book which enabled us for timely publication.

We shall welcome the constructive suggestions from the readers that will be a source of light for further improvement in the next edition.

"Dedicated to our much-loved parents and family"

(Authors)

Preface

Education is the key to development. A progressively improved education system largely determines the pace and the quality of national development. Biology is a natural science concerned with the study of life and living organisms. It provides detailed information about the zoological as well as botanical aspects of life.

The subject material of this book is divided into seven chapters. Chapter 1 covers the world of living and life processes. Chapter 2 present the control and co-ordination in animals as well as plants. Chapter 3 explains the reproduction in animals and plants. Chapter 4 covers the basic concepts of evolution, sex determination and origin of life. Chapter 5 deals with sources, forms and utilization of energy. Chapter 6 and 7 covers the environment and management of natural resources. The most important feature of this book is that we have included remember sockets with a different types of questions in the last of each chapter from core material for evaluating the learning skill of the students.

- Remember sockets
- Very short answer type questions
- Shorty answer type questions
- Long answer type questions
- Multiple choice questions

The language of book is very simple and diagrams are keep to the simple and line drawing that a might produce under examination conditions. Thus, this book will be helpful to the students from a variety of biological discipline and can be used as a basic subject material for the students and teachers in various academics as well as technical institutes. Each part of the book is cautiously intended to make it students-friendly and to make learning of Biology an enjoyable involvement. There is always ample scope of improvement in the concerned subject material.

In spite of sincere efforts in translation, editing and printing some inadvertent errors and omissions may be found in the book. However, our efforts to make it more refined and impeccable will continue. Any constructive suggestion towards its further improvement will be gratefully considered. We invite constructive comment and valued suggestions from the students and teachers to further improve the book in subsequent editions.

We hope that the book will be useful for the students for whom it is written.

Authors

Contents

1

CHAPTER

WORLD OF LIVING

Life Processes: 'Living Being'. Basic concept of nutrition, respiration, transport and excretion in plants and animals

Something which is living (not dead) is said to be alive, means having life. Thus, those things which are alive are called living things. All the plants and animals including human beings are living things. The most important criterion to decide whether something is alive (or not) is the movement. Movement is one of the most important signs of life in an organism. All the living moves by themselves without any external help. In some cases, the movements of living things are quite fast which can be easily observed by eye, when the movements are very slow and hence observed with difficulty. For example, the movements in most of the animals are fast, and observed easily but the movements in plants are usually slow and observed with difficulty. Animals and plants move in different ways. Animals can move their body parts and can move from one place to another i.e. a bird moves when it flies in the sky and a fish moves when it swims in water. All these movements show are alive (Fig 1.1A and B). The plants are fixed in the soil at a place, so they cannot move like animals from place to place. The plants can only move parts of their body such as leaves, flowers, shoots and roots. The plant parts move towards a stimulus such as sunlight, gravity or water, etc. For example, the shoot, the leaves and flower of a sunflower plant move by bending towards the sun so as to face the sunlight. The leaves of a *Mimosa pudica* plant (sensitive plant) move by folding up when touched with a finger (Fig 1.2). Plants also show movement by growing their roots and shoots bigger.

(A) Fish swimming

(B) Bird flying

Fig. 1.1 Living things move

Fig. 1.2 *Mimosa pudica* plant

Non-living things (which are not alive) cannot move by themselves. For example, a stone is a non-living thing which cannot move by itself from one place to another or show any other type of movement. We will have to move it by applying force from outside. All the living things (plants and animals) are made up of tiny living units called cells. The cells themselves are made up of still smaller particles called molecules. The movements over very small scale (as those in the molecules of living things) are invisible to the naked eye. The invisible molecular movement is, however, necessary for the existence of life. In fact, viruses do not show any molecular movement in them (until they infect some cell) and this has created controversy about whether they are truly alive or not. In addition to movement, the living things also show some other characteristics. All the living things have some common characteristics which make them different from non-living things such as:

- They can move by themselves and can grow.
- They can respond to changes around them.
- They need food, air and water.
- They respire (release energy from food).
- They excrete (get rid of waste materials from their body).
- They can reproduce. They can have young ones.

Table 1.1 Differences between living and non-living things

Sr. No.	Living things	Non-living things
1.	A living thing has a self-built organization, consisting of several components, that exists beyond the level of individual.	The organization is imposed limited.
2.	Living things obtain simple molecules from outside and convert them into complex protoplasmic constituents.	Such activity is not seen.
3.	Living things perform growth, development and differentiation.	No such activity occurs.
4.	Living beings have property of self-repair.	Non-living things have no such property. The repairs can only be done by outside agencies.
5.	Living organisms have ability to sense the surroundings and protect themselves.	Protection of non-living things is imposed.

6.	Living things have a definite life span.	There is no definite life span.
7.	Living beings reproduce and multiply. They produce offspring's and pass on genes to next generation.	Multiplication is imposed.
8.	Living organisms have ability to evolve in time.	Non-living things do not have ability to evolve.

1.1 WHAT ARE LIFE PROCESSES?

All living organisms perform some basic functions or processes in order to remain alive. These processes are needed for the maintenance of their bodies and catching to their daily activities. Some of these processes continue to occur even when the organisms are not doing any activities. The maintenance processes continue to occur even when we are just sitting or sleeping. This is because when we are asleep, a large number of biological processes continue to occur which require energy. All these processes which together keep the living organisms alive and perform the job of body maintenance are called life processes. All the organisms perform some basic functions to keep themselves alive. The basic functions performed by living organisms to maintain their life on this earth are called life processes. The basic life processes common to all the living organisms are such as (i) Nutrition and Respiration (ii) Transport and Excretion (iii) Control and Coordination (Response to stimuli) (iv) Growth (v) Movement and Reproduction

1.1.1 Nutrition:

It is a process by which the organisms obtain and utilize the nutrients (food). These nutrients are the source of energy which comes from outside the body of the individual organism. Since the life on earth depends mainly on carbon-based molecules (e.g. carbohydrates, proteins, fats, etc.). Nutrition of different organisms involves different processes. Plant nutrition involves the process of photosynthesis in which the green plants synthesize carbohydrates with the help of carbon-di-oxide and sunlight. Animal nutrition involves taking of food inside the body and its digestion.

1.1.2 Respiration:

It is a process that involves breakdown of nutrients through oxidation and release of energy for cellular needs. For this, many organisms take oxygen from outside the body. Respiration is the process which releases energy from the food absorbed by the body.

1.1.3 Transport:

It varies in single cell and multi cell organism. In the case of single-celled organisms (e.g. *Amoeba*), no specific organs for getting nutrients, oxygen, water or removal of wastes are needed because the entire body surface of such organisms remains in contact with the environment. They enter into the cell by simple diffusion. However, in case of multicellular organisms like humans and higher plants, simple diffusion is not sufficient to meet the oxygen and nutritional requirement of all the cells. Such organisms develop specific organs and tissues for transportation of materials from one part of the body to another part. This process is called transport. Transport is the process in which a substance absorbed or made in one part of the body is moved to other parts of the body.

1.1.5 Control and Coordination (or Response to Stimuli):

It is a process which helps the living organisms to survive in the changing environment around them. The process of growth involves the change from a small organism to a big organism or an adult organism. In movement, the organism either moves from one place to another or moves its body parts while remaining at the same place. The process of reproduction involves the making of more organisms from the existing ones, so that organisms could live on this earth forever.

> The basic functions performed by living organisms to maintain their life on this earth are called life processes. The basic life processes common to all the living organisms are such as (i) Nutrition and Respiration (ii) Transport and Excretion (iii) Control and Coordination (Response to stimuli) (iv) Growth (v) Movement and Reproduction

1.2 ENERGY IS NEEDED FOR THE LIFE PROCESSES:

All the living organisms need energy to perform various life processes. They get this energy from food. Food is a kind of fuel which provides energy to all the living organisms. The living organisms use the chemical energy for carrying out various life processes. They get this chemical energy from food through chemical reactions. Actually, living organisms continuously need energy for their various life processes and other activities which they perform. For example, energy is required by an organism even during sleep. This is because when we are asleep, a number of biological processes keep on occurring in the body which requires energy. Our heart beats non-stop even when are asleep to pump blood throughout the body and this beating of heart requires energy. Thus, the working of heart requires a continuous supply of energy. The energy required by an organism comes from the food that the organism eats. Thus, food is the basic requirement of all the living organisms for obtaining energy. Several processes are given in detailed here.

1.3 NUTRITION

All living organisms need nutrients (food) to build up most of their own body molecules and to get energy for doing work. Nutrients are inorganic as well as organic substances which the organisms obtain from their surroundings in order to synthesize their body constituents and use them as a source of energy. Food is an organic substance. The simplest food is glucose (called simple sugar). A more complex food is starch. Starch is made from glucose. The general name of substances like glucose (sugar) and starch is 'carbohydrates'. Carbohydrates are the common foods for getting energy. Fats and proteins are also foods (Fig 1.3). A wider definition of food, however, also includes mineral, salts, vitamins and water which are essential for the normal growth and development of an organism). The process of consuming or taking the food and utilizing it is called nutrition. It is a process in which food is obtained in order to utilize it to provide energy for performing various metabolic activities of the organism. Actually, the term 'nutrition' comes from the word 'nutrient'. A nutrient is an organic or inorganic substance required for the maintenance of life and survival of a living organism. In most simple terms, a nutrient can be said to be a particular type of food. A nutrient can be defined as a substance which an organism obtains from its surroundings and uses it as a source of energy or for the biosynthesis of its body constituents (like tissues and organs). For example, carbohydrates and fats are the nutrients which are used by an organism mainly as a source of energy whereas proteins and mineral salts are nutrients used by an organism for the biosynthesis of its body constituents like skin, blood, etc. The food taken in by an organism contains a large number of nutrients like carbohydrates, fats, proteins, minerals, vitamins and water, etc. Thus, nutrition is a process of intake of nutrients (like carbohydrates, fats, proteins, minerals, vitamins and water) by an organism as well as the utilization of these nutrients by the organism. We will now describe the various ways of procuring food (or obtaining food) by the different organisms. In other words, we will now describe the different modes of nutrition of the various organisms.

(A)

(B)

(C)

(D)

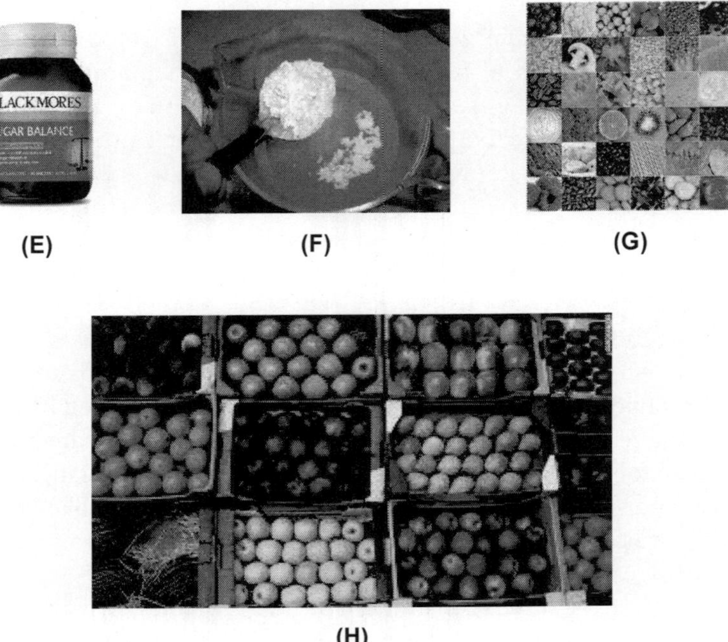

(E) (F) (G)

(H)

Fig. 1.3 Energy yielding products/substance

> Nutrients are inorganic as well as organic substances which the organisms obtain from their surroundings in order to synthesize their body constituents and use them as a source of energy. Food is an organic substance. The simplest food is glucose. It is also called simple sugar. A more complex food is starch. Starch is made from glucose. The general name of substances like glucose (sugar) and starch is 'carbohydrates'. Carbohydrates are the common foods for getting energy. Fats and proteins are also foods.

1.3.1 Modes of Nutrition:

A mode of nutrition means methods of procuring food or obtaining food by an organism. Different organisms have different methods for obtaining the food. Depending on the mode of nutrition all the organisms can be categorized into two groups such as (i) autotrophic and (ii) heterotrophic.

1.3.1.1 Autotrophic Nutrition:

The word 'auto' means 'self' and 'trophe' means 'nutrition'. Thus, autotrophic means self-nutrition. In autotrophic nutrition, the organism makes (or synthesizes) its own food from the inorganic raw materials like carbon-di-oxide and water present in the surroundings by using the sunlight energy. The green plants have an autotrophic mode of nutrition. The autotrophic bacteria also obtain their food by the autotrophic mode of nutrition (Fig 1.4). The organisms having autotrophic mode of nutrition are called autotrophic organisms or autotrophs. The autotrophic organisms contain the green pigment called chlorophyll which is capable of trapping the sunlight energy. This trapped sunlight energy is utilized by the autotrophs to make food by combining inorganic materials like carbon-di-oxide and water present in the environment by the process of photosynthesis. Thus, autotrophs make their own food by photosynthesis. So, autotrophs are the producers of food. The food produced by autotrophs (green plants) is also used by human beings and other animals.

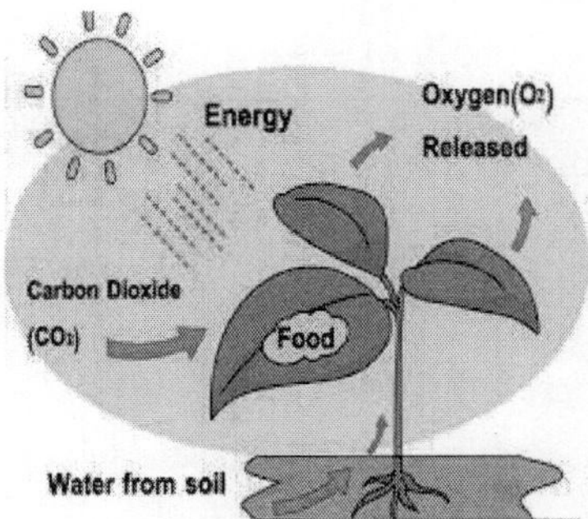

Fig. 1.4 Autotrophic Nutrition

1.3.1.2 Heterotrophic Nutrition:

The word 'heteros' means 'others' and 'trophe' refers to 'nutrition'. Thus, 'heterotrophic' means 'nutrition obtained from others'. A heterotrophic organism is a consumer which derives its nutrition from other organisms. That is, a heterotrophic organism has to eat other organisms for its nutrition(Fig 1.5). Those organisms which cannot make their own food from inorganic substances like carbon-di-oxide and water, and depend on other organisms for their food are called heterotrophs. All the animals are heterotrophs (because they cannot make their food from inorganic substances like carbon-di-oxide and water and obtain their food from other plants or animals). Thus, man, dog, cat, deer, tiger, bear, lion, cow, etc., are all heterotrophs. The non-green plants (like yeast) are also heterotrophs. Heterotrophs depend on autotrophs and other animals as heterotrophs for their food.

Fig. 1.5 Heterotrophic Nutrition in Animals

Depending on the mode of nutrition all the organisms can be categorized into two groups such as (i) autotrophic and (ii) heterotrophic. In autotrophic nutrition, the organism makes (or synthesizes) its own food from the inorganic raw materials like carbon-di-oxide and water present in the surroundings by using the sunlight energy. The green plants have an autotrophic mode of nutrition. In heterotrophic nutrition, the organism cannot make (or synthesize) its own food from the inorganic raw materials like carbon-di-oxide and water, and uses the food made by autotrophic organisms directly or indirectly.

1.3.1.2.1 Types of Heterotrophic Nutrition:

A heterotrophic organism can obtain its food from other organisms in various ways such as (i) Saprotrophic Nutrition, (ii) Parasitic Nutrition, and (iii) Holozoic Nutrition

(i) Saprotrophic Nutrition (Saprophytic Nutrition):

'Sapro' means 'rotten', so a saprotrophic organism draws its food from rotting wood of dead and decaying trees, rotten leaves, dead animals and household wastes like rotten bread, etc. the organisms having saprotrophic mode of nutrition are called saprophytes (Fig 1.6A). Thus, saprophytes are the organisms which obtain their food from dead plants like rotten leaves, dead and decaying animal bodies, and other decaying organic matter like rotten bread. Fungi like bread moulds, mushrooms, yeast, and many bacteria are saprophytes. The saprophytes break down the complex organic molecules present in dead and decaying matter and convert them into simpler substances outside their body. These simpler substances are then absorbed by saprophytes as their food. Saprotrophic nutrition is that nutrition in which an organism obtains its food from decaying organic matter of dead plants, dead animals and rotten bread, etc.

(ii) Parasitic Nutrition:

The parasitic nutrition is that nutrition in which an organism derives its food from the body of another living organism (called its host) without killing it. The organism which obtains the food is called a 'parasitic' and the organism from whose body food is obtained is called the 'host'. A parasite is an organism (plant or animal) which feeds on another living organism called its host (Fig 1.6B). A parasite receives its food from the host, but gives no benefit to the host in return. A parasite usually harms the host. The host many be a plant or an animal. Most of the diseases which affect mankind, his domestic animals (like dogs and cattle) and his crops are caused by parasites. Parasitic mode of nutrition is observed in several fungi, bacteria, a few plants like *amarbel* and some animals like *Plasmodium* and roundworms. Thus, the micro-organism '*Plasmodium*' (which causes malaria disease) is a parasite. Roundworm which causes diseases in man and domestic animals (like dogs and cattle) is also a parasite. Roundworms live inside the body of man and his domestic animals.

(iii) Holozoic Nutrition:

Holozoic nutrition means 'feeding on solid food' (which may be a plant product or an animal product), Most of the animals (*including human beings*) take the solid food into their body by the process of ingestion. The ingested food is then digested or broken down into simpler substances which are then absorbed into the cells of the body, and the undigested and unabsorbed waste materials are egested (thrown out) of the body (Fig 1.6C). The holozoic nutrition in which an organism takes the complex organic food materials into its body by the process of ingestion, the ingested food is digested and then absorbed into the body cells of the organism. The undigested and unabsorbed part of the food is thrown out of the body of the organism by the process of egestion. The human beings and most of the animals have a holozoic mode of nutrition. In other words, man, deer, cat, dog, cattle, bear, giraffe, tiger, lion, fish, frog and Ameoba, etc., have the holozoic mode of nutrition.

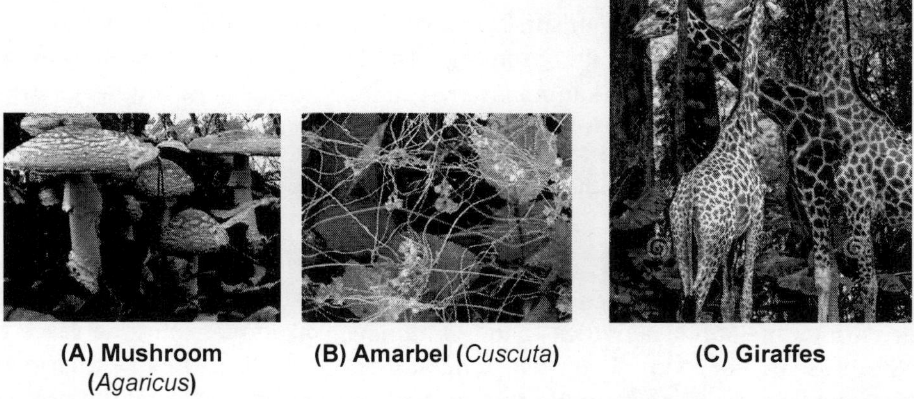

(A) Mushroom (*Agaricus*) **(B) Amarbel (*Cuscuta*)** **(C) Giraffes**

Fig. 1.6 Examples of Heterotrophic (A) Saprophytic (B) Parasitic (C) Holozoic

TABLE 1.2 DIFFERENCES BETWEEN AUTOTROPHIC AND HETEROTROPHIC NUTRITION

Sr. No.	Autotrophic Nutrition	Heterotrophic Nutrition
1.	It occurs in green plants, some bacteria and in some protists.	It occurs in animals and in plants which lack chlorophyll.
2.	Chlorophyll is necessary for trapping the sunlight.	Chlorophyll is absent; as such they do not trap the sunlight.
3.	Food is self-manufactured using CO_2 and water as raw materials.	Food is obtained directly or indirectly from autotrophs.

4.	Digestion of food does not occur.	Digestion is required to convert complex organic substances present in food into simpler and soluble forms.
5.	Autotrophs are placed at the bottom of the food chain as producers.	Heterotrophs are placed above autotrophs in the food chain as consumers.

Saprophytes are the organisms which obtain their food from dead plants like rotten leaves, dead and decaying animal bodies, and other decaying organic matter like rotten bread. Fungi like bread moulds, mushrooms, yeast, and many bacteria are saprophytes. The organism which obtains the food is called a 'parasitic' and the organism from whose body food is obtained is called the 'host'. A parasite is an organism (plant or animal) which feeds on another living organism called its host.

Parasitic mode of nutrition is observed in several fungi, bacteria, a few plants like amarbel and some animals like *Plasmodium* and Roundworms. The holozoic nutrition in which an organism takes the complex organic food materials into its body by the process of ingestion, the ingested food is digested and then absorbed into the body cells of the organism. The undigested and unabsorbed part of the food is thrown out of the body of the organism by the process of egestion. The human beings and most of the animals have a holozoic mode of nutrition. In other words, man, deer, cat, dog, cattle, bear, giraffe, tiger, lion, fish, frog and Ameoba, etc., have the holozoic mode of nutrition.

1.4 NUTRITION IN PLANTS:

Just like other organisms, plants also require food which can supply energy for their various metabolic activities. Though animals can move from one place to another in search of food, plants just stand still at one place and make their own food. Green plants are autotrophic and synthesize their own food by the process of photosynthesis. 'Photo' means 'light' and 'synthesis' means 'to build', thus 'photosynthesis' means 'building up by light'. The plants use the energy in sunlight to prepare food from carbon-di-oxide and water in the presence of chlorophyll. Chlorophyll is present in the green coloured bodies called 'chloroplasts' inside the plant cells. In fact, the leaves of a plant are green because they contain tiny green coloured organelles called chloroplasts (which contain chlorophyll).

The process, by which green plants make their own food (like glucose) from carbon-di-oxide and water by using sunlight energy in the presence of chlorophyll, is called photosynthesis. Oxygen gas is released during photosynthesis. The process of photosynthesis can be represented as:

$$\underset{\substack{\text{Carbon dioxide}\\\text{(From air)}}}{6CO_2} + \underset{\substack{\text{Water}\\\text{(From soil)}}}{6H_2O} \xrightarrow[\text{Light energy}]{\text{Chlorophyll}} \underset{\text{Glucose}}{C_6H_{12}O_6} + \underset{\text{Oxygen}}{6O_2}$$

The process of photosynthesis takes place in the green leaves of a plant. In other words, food is made in the green leaves of the plant. The green leaves of a plant make the food by combining carbon-di-oxide and water in the presence of sunlight and chlorophyll (Fig 1.7). The carbon-di-oxide gas required for making food is taken by the plant leaves from the air. This CO_2enters the leaves through tiny pores in them called stomata. Water required for making food is taken from the soil. This water is transported to the leaves from the soil through the roots and stem. The sunlight provides energy required to carry out the chemical reactions involved in the preparation of food. The green pigment called chlorophyll present in green leaves helps in absorbing energy from sunlight. Oxygen gas is produced as a by-product during the preparation of food by photosynthesis. This oxygen gas goes into the air. The food prepared by the green leaves of the plant is in the form of a simple sugar called glucose. This glucose food made in the leaves is then sent to the different parts of the plant The extra glucose is changes into another food called starch. This starch is stored in the leaves of the plant (Fig 1.8). Glucose and starch belong to a category of foods called carbohydrates. The foods like carbohydrates prepared by photosynthesis contain chemical energy stored in them. Thus, the green plants convert sunlight energy into chemical energy by making carbohydrates. The food prepared by photosynthesis provides all the energy to a plant which it needs to grow, and when we eat plant foods (like food grains, fruits and vegetables), the chemical energy stored in them is released in our body during respiration. The photosynthesis takes place in the following three steps such as:

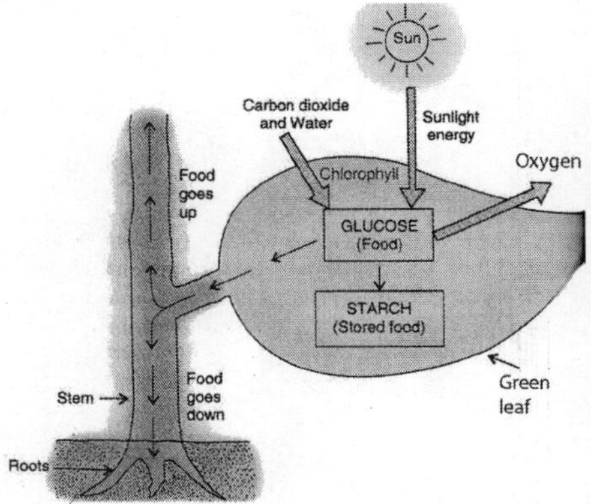

Fig. 1.7 Preparation of food in green plant through photosynthesis

Fig. 1.8 Food items made by tree plants through photosynthesis

(i) Absorption of sunlight energy by chlorophyll.

(ii) Conversion of light energy into chemical energy, and splitting of water into hydrogen and oxygen by light energy.

(iii) Reduction of carbon-di-oxide by hydrogen to form carbohydrate like glucose by utilizing the chemical energy (obtained by the transformation of light energy).

Please note that the three steps involved in photosynthesis need not take place one after the other immediately. They can take place at different times. For example, desert plants take up carbon-di-oxide at night and prepare an intermediate product which is acted upon by the sunlight energy absorbed by chlorophyll when the sun shines during the next day.

The process, by which green plants make their own food (like glucose) from carbon-di-oxide and water by using sunlight energy in the presence of chlorophyll, is called photosynthesis. Oxygen gas is released during photosynthesis. The photosynthesis takes place in the following three steps such as (i) Absorption of sunlight energy by chlorophyll (ii) Conversion of light energy into chemical energy, and splitting of water into hydrogen and oxygen by light energy (iii) Reduction of carbon-di-oxide by hydrogen to form carbohydrate like glucose by utilizing the chemical energy (obtained by the transformation of light energy), sunlight, chlorophyll and carbon-di-oxide are necessary for photosynthesis by green plants

1.4 NECESSARY CONDITIONS FOR PHOTOSYNTHESIS:

It has been found by experiments that the presence of sunlight, chlorophyll, carbon-di-oxide and water is necessary for the process of photosynthesis. Some of the experiments to show that sunlight, chlorophyll and carbon-di-oxide are necessary for photosynthesis by green plants are given here

1.4.1 Experiment to Show that Sunlight is Necessary for Photosynthesis

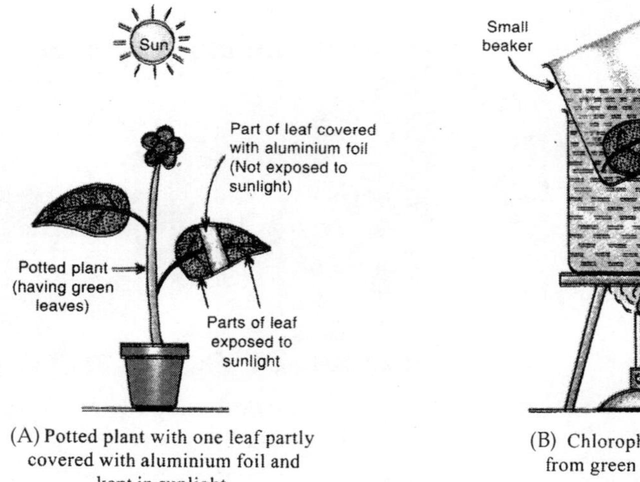

(A) Potted plant with one leaf partly covered with aluminium foil and kept in sunlight.

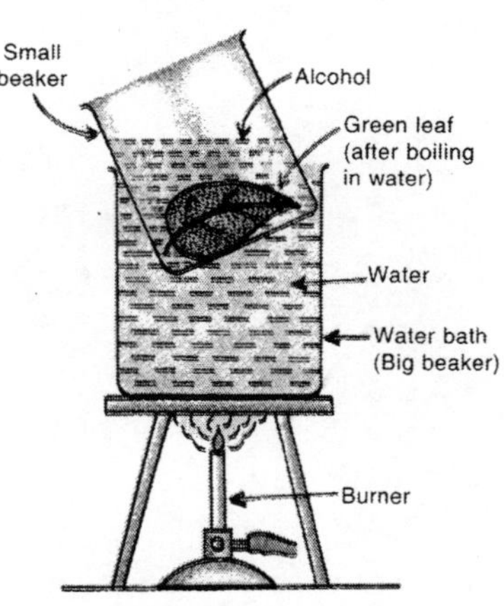

(B) Chlorophyll being removed from green leaf by boiling in alcohol.

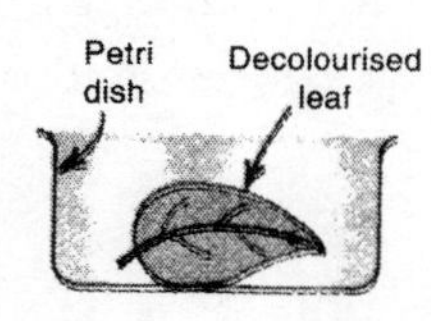

(C) Chlorophyll removed leaf.

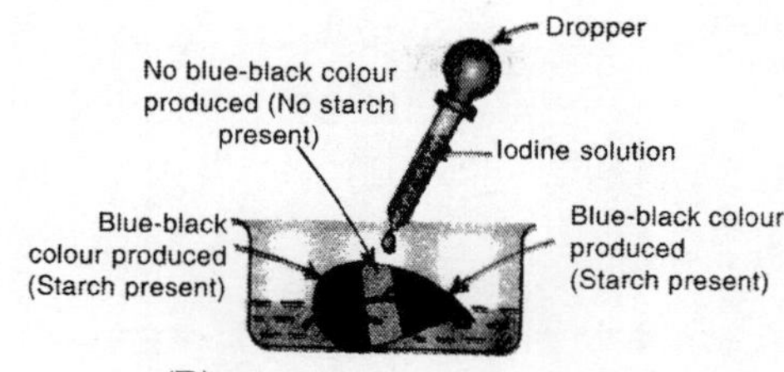

(D) Iodine solution being poured over decolourised leaf.

Fig. 1.9 Experiment to show that Sunlight is necessary for Photosynthesis

(a) Take a potted plant having green leaves and place it in a completely dark place for about 3 days to destarch its leaves. So, in the beginning of the experiment, the leaves do not have any starch in them.

(b) Take a thin strip of aluminum foil or black paper and wrap it in the centre of one leaf on both the sides (while the leaf is still attached to the plant). The aluminum foil should be fixed tightly to the leaf by using paper clips so that sunlight may not enter it from the sides. The aluminum foil should cover only a small part of the leaf so that the remaining part of the leaf remains uncovered and exposed to sunlight may not fall on this covered part of the leaf.

(c) Keep this potted plant (with partially covered leaf) in bright sunlight for 3-4 days.

(d) Pluck the partially covered leaf from the plant and remove its aluminum foil. Immerse this leaf in boiling water for a few minutes. This will break down the cell membranes of leaf cells and make the leaf more permeable to iodine solution (so that it may reach the starch present inside the leaf cells). This leaf is now to be tested for the presence of starch. But before testing for starch, chlorophyll has to be removed from the leaf. This is because chlorophyll interferes in the test for starch due to its green colour.

(e) Put the plucked leaf in a beaker containing some alcohol. Place the beaker containing alcohol and leaf in a water bath. A water bath is being used here for heating alcohol because alcohol is a highly inflammable liquid. So, if alcohol is heated directly over a flame, then it will catch fire at once. We may use bigger beaker containing water as a water bath.

(f) Heat the water in the bigger beaker or water bath. Then the alcohol in the smaller beaker will also get heated and start boiling soon. This boiling alcohol will extract or remove chlorophyll from the green leaf.

(g) Boil the green leaf in alcohol till all its green pigment 'chlorophyll' is removed. The leaf will now become almost colourless or pale (and the alcohol will turn green).

(h) Remove the colourless leaf from alcohol and wash it thoroughly with hot water to soften it and remove any chlorophyll which may be sticking to it.

(i) Place the colourless leaf in a petri-dish. Drop iodine solution over the decolourised leaf with the help of a dropper. Observe the change in colour of leaf.

Observation:

The middle part of leaf which was covered with aluminum foil does not turn blue-black on adding iodine solution showing that no starch is present in this middle part of the leaf. This is because sunlight could not reach the covered middle part of the leaf due to which the covered 'middle part' of leaf could not do photosynthesis to make starch. The uncovered part of leaf (on both sides of the aluminium foil) which was exposed to sunlight turns blue-black on adding iodine solution showing that starch is present in this part of leaf. This means that the part of leaf which was exposed to sunlight could be photosynthesis to make starch. Since the part of leaf which was covered and hidden from sunlight does not contain starch but the part of leaf which was exposed to sunlight contains starch, therefore, we conclude that sunlight is necessary for photosynthesis (to make food like starch).

Conclusion:

It is concluded from above experiment that (i) Sunlight is necessary for the process of photosynthesis, and (ii) Leaves make starch as food by photosynthesis. Most of the common plants have leaves which are totally green (because all the parts of such leave contain the green pigment called chlorophyll). But there are some plants whose leaves are partly green and partly white. The green part of such a leaf contains chlorophyll but the white part of such a leaf does not contain chlorophyll. The leaves which are partly green and partly white are called 'variegated leaves' (Fig 1.10). The plants such as Croton and Coleus have variegated leaves which are partly green and partly white. We will use a plant having variegated leaves in the next experiment to show that chlorophyll is necessary for the process of photosynthesis in plants.

(A) Croton Plant

(B) Coleus Plant

Fig. 1.10 Variegated Leaves

1.4.2 Experiment to show that Chlorophyll is necessary for Photosynthesis

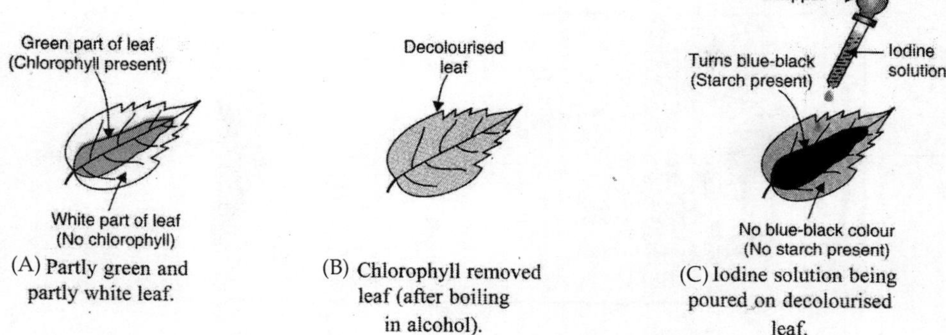

(A) Partly green and partly white leaf.

(B) Chlorophyll removed leaf (after boiling in alcohol).

(C) Iodine solution being poured on decolourised leaf.

Fig. 1.11 Experiment to show that Chlorophyll is necessary for Photosynthesis

(a) Take a potted plant like croton whose leaves are partly green and partly white. The green part of the leaf has chlorophyll, while the white part of the leaf does not have chlorophyll.

(b) Place this plant in a completely dark place for about 3 days to destarch its leaves.

(c) Take out the potted plant from the dark place and keep it in bright sunshine for 3-4 days.

(d) Pluck the variegated leaf from the plant, boil it in water for a few minutes and then remove its green colour chlorophyll by boiling it in alcohol. The green parts of the leaf get decolourised. So, we get decolorized leaf.

(e) Wash the decolourised leaf with hot water to soften it and remove any chlorophyll which may be sticking to it.

(f) Pour iodine solution over the colourless leaf and observe the change in colour of the leaf.

Observation:

It is observed that the outer part of leaf that was originally white (without chlorophyll) does not turn blue-black on adding iodine solution showing that no starch is present in this outer part of the leaf.

Conclusion:

It is concluded that the photosynthesis to make starch does not take place without chlorophyll. The inner part of leaf which was originally green (contained chlorophyll) turns blue-black on adding iodine solution showing that starch is present in this inner part of the leaf. From this observation we conclude that the photosynthesis to make starch takes place in the presence of chlorophyll. In other words, chlorophyll is necessary for the process of photosynthesis to take place.

1.4.3 Experiment to show that Carbon-di-oxide (CO_2) is necessary for Photosynthesis

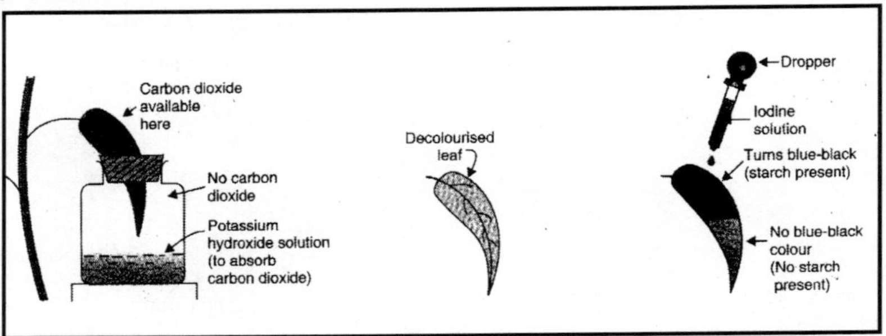

Fig. 1.12 Experiment to show that CO_2 is necessary for Photosynthesis

(a) Take a potted plant having long and narrow leaves and place it in a completely dark place for about 3 days to destarch its leaves.

(b) Take a glass bottle having a wide mouth and put some potassium hydroxide solution (KOH solution) in it. (This KOH solution is to absorb the carbon-di-oxide gas from the air present in the glass bottle so that no carbon-di-oxide is left in the air inside the glass bottle).

(c) Take a rubber cork which fits tightly into the mouth of the glass bottle and cut it into two halves.

(d) Put a destarched leaf of the potted plant (while it is still attached to the plant), in-between the two halves of the cut cork and then fit the cork in the mouth of the glass bottle. The upper half of the leaf should remain outside the glass bottle and only the lower half of the leaf should be inside the glass bottle.

(e) The potted plant (with its one destrached leaf half inside the glass bottle containing KOH solution) is kept in sunlight for 3 to 4 days. During this period, the upper half of the leaf (which is outside the glass bottle) gets carbon-di-oxide from the air but the lower half of the leaf (which is inside the glass bottle) does not get any carbon-di-oxide. This is because all the carbon-di-oxide of the air present in the glass bottle has been absorbed by potassium hydroxide solution, and no fresh air can come into the closed glass bottle.

(f) Pluck the leaf from the plant and take it out from the glass bottle. Remove the green coloured chlorophyll from the leaf by boiling it in alcohol. In this way, we get a decolourized leaf.

(g) Wash the decolourized leaf with water to remove any chlorophyll which may be sticking to it.

(h) Pour iodine solution over the colourless leaf and observe the change in colour of the leaf.

Observation:

It is observed that the lower half part of the leaf (which was inside the glass bottle having no carbon-di-oxide around it), does not turn blue-black on adding iodine solution showing that no starch is present in this lower half of the leaf. From this observation we conclude that the photosynthesis to make starch in the leaf does not take place without carbon-di-oxide.

The upper half part of the leaf (which was outside the glass bottle, having carbon-di-oxide around it) turns blue-black on adding iodine solution showing that starch is present in this upper half of the leaf. From this observation we conclude that photosynthesis (to make starch) takes place in the presence of carbon-di-oxide. In other words, carbon-di-oxide is necessary for the process of photosynthesis to take place.

1.5 RAW MATERIALS FOR PHOTOSYNTHESIS:

The preparation of carbohydrates (food) by plants by the process of photosynthesis requires two materials (substances) such as carbon-di-oxide, and water. Thus, the raw materials for photosynthesis are: (i) Carbon-di-oxide and (ii) Water.

1.5.1 How the Plants obtain Carbon-di-oxide:

There are a large number of tiny pores called stomata on the surface of the leaves of plants (The singular of stomata is stoma). The green plants take carbon-di-oxide from air for photosynthesis. The carbon-di-oxide gas enters the leaves of the plant through the stomata present on their surface. Each stomata is surrounded by a pair of guard cells. The opening and closing of stomatal pores is controlled by the guard cells. When water flows into the guard cells, they swell, become curved and cause the pore to open (Fig 1.13A). On the other hand, when the guard cells lose water, they shrink, become straight and close the stomatal pore. A large amount of water is also lost from the cells of the plant leaves through open stomatal pores. So, when the plant does not need carbon-di-oxide and wants to conserve water, the stomatal pores are closed(Fig 1.13B). The oxygen gas produced during photosynthesis also goes out through the stomatal pores of the leaves. Please note that in addition to leaves, the stomata are also present in the green stems (or shoots) of a plant. So, the green stems (or shoots) of a plant also carry out photosynthesis. It is clear from the above discussion that stomata allow the movement of gases in and out of plant cells. In other words, the gaseous exchange into plants takes place through the stomata in leaves (and other green parts).

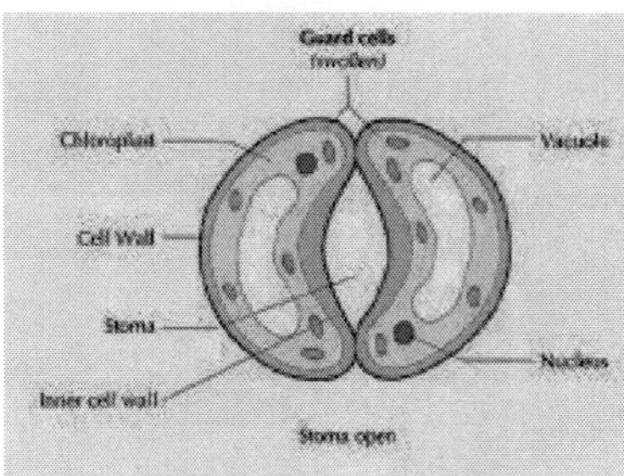

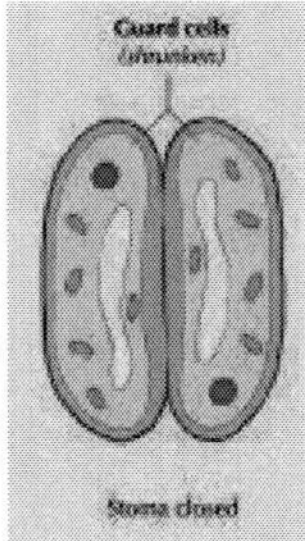

(A) Open stomata **(B) Closed stomata**

Fig. 1.13 Structure of Stomata

Please note that in most broad-leaved plants, the stomata occur only in the lower surface of the leaf but in narrow-leaved plants, the stomata are equally distributed on both the sides of the leaf. It is also to be noted that the aquatic plants (water plants) use the carbon- di-oxide gas dissolved in water for carrying out photosynthesis.

1.5.2 How the Plants obtain Water for Photosynthesis:

The water required by the plants for photosynthesis is absorbed by the roots of the plants from the soil through the process of osmosis. The water absorbed by the roots of the plants is transported upward through the xylem vessels to the leaves where it reaches the photosynthetic cells and utilized in photosynthesis.

The two raw materials, carbon-di-oxide and water, are required by the plants to prepare energy foods called carbohydrates (such as glucose and starch). But the plants also need other raw materials such as nitrogen, phosphorus, iron and magnesium, etc., for building their body. The plants take materials like nitrogen, phosphorus, iron and magnesium, etc., from the soil. For example, nitrogen is an essential element used by the plants to make proteins and other compounds. The plants take up nitrogen from the soil in the form of inorganic salts called nitrates (or nitrites), on in the form of organic compounds which are produced by bacteria from the atmospheric nitrogen.

1.6 SITE OF PHOTOSYNTHESIS: CHLOROPLASTS:

The absorption or light energy and carbon-di-oxide and ultimate conversion of CO_2 to starch with the evolution of O_2 all takes place with in illuminated chloroplast. The chloroplast have photosynthetic pigments and on analysis it has been found that chlorophyll contains 40-50% proteins, 23-25% phospholipids, 5 %RNA, 5-10% chlorophyll a and b and 1-2% carotenoids.

Chloroplasts can be seen easily by using a light microscope. In a cross-section of a leaf, chloroplasts can be seen as numerous disc-like organelles in the photosynthetic cells of the palisade tissue just below the upper epidermis. Each chloroplast is covered by a double-membrane. Inside the membrane, the centre of chloroplast is filled with matrix or stroma. A number of organized flattened membranous sacs called thylakoids are present in the stroma. Thylakoids are arranged in stacks like the piles of coins called grana (singular; granum) or the intergranal thylakoids. In addition, there are flat membranous tubules called the stroma lamellae connecting the thylakoids of the different grana. The stroma of the chloroplast contains enzymes required for the synthesis of carbohydrates and proteins. It also contains small, double-stranded circular DNA molecules and ribosomes. Chlorophyll pigments are present in the thylakoids.

In the structure of a leaf, the middle layers in the leaf (palisade layer and spongy layer) contain photosynthetic cells called mesophyll cells. These cells contain more chlorophyll than other plant cells(Fig 1.14). A typical photosynthetic cell of a green leaf may contain 100 or more tiny chloroplasts in it, and a whole leaf may contain thousands of photosynthetic cells. Carbon-di-oxide needed for photosynthesis enters from the air into the leaf through the stomata in its surface, and then diffuse into the mesophyll cells and reaches the chloroplasts. Water is carried to the leaf by xylem vessels and passes into the mesophyll cells by diffusion and reaches the chloroplasts. There is a thin, waxy protective layer called cuticle above and below a leaf which helps to reduce the loss of water from the leaf.

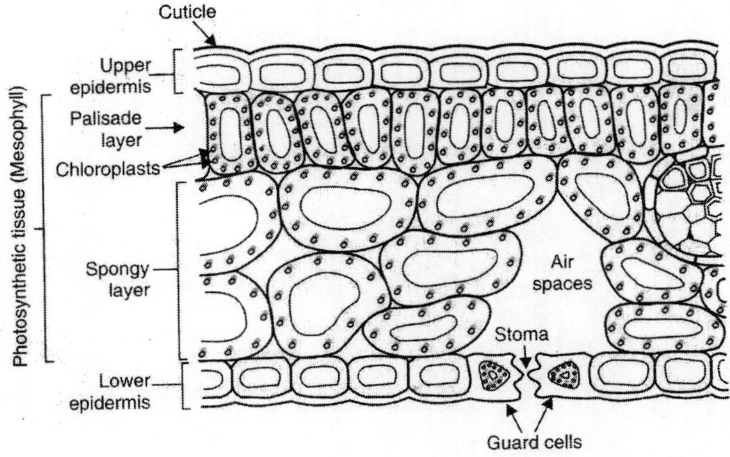

Fig. 1.14 Structure of a leaf show chloroplasts

1.6.1 Chlorophyll Pigments:

The chlorophylls are the primary pigments which take an active part in the process on photosynthesis (Fig 1.15). Without the presence of chlorophyll process of photosynthesis, is not possible. At present eight types of chlorophylls are known which are chlorophyll a, b, c, d, e, bacterio-chlorophyll a, b and chlorobium chlorophyll. The molecular formula of chlorophyll is as follows (Fig 1.16A and B):

(a) Chlorophyll a $C_{55}H_{72}O_5N_4Mg$

(b) Chlorophyll b $C_{56}H_{70}O_6N_5Mg$

The chlorophyll molecule consists of a ring structure composed of four Pyrol nuclei linked by four +CH bridges. Chlorophyll a and b have exactly the same structure excepting in Chlorophyll b, a $-CH_8$ side chain is replaced by –CHO group.

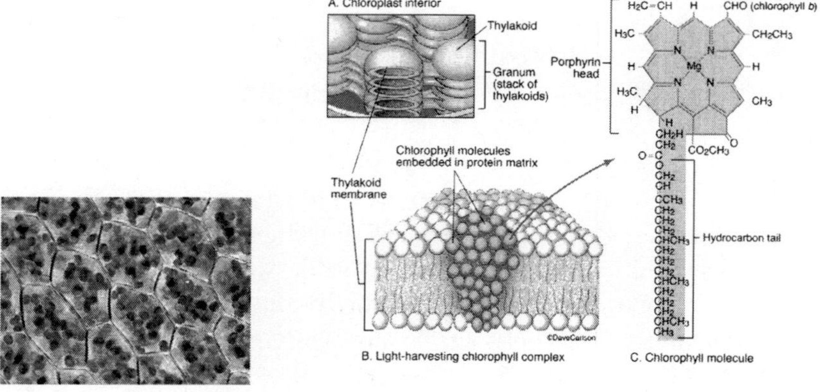

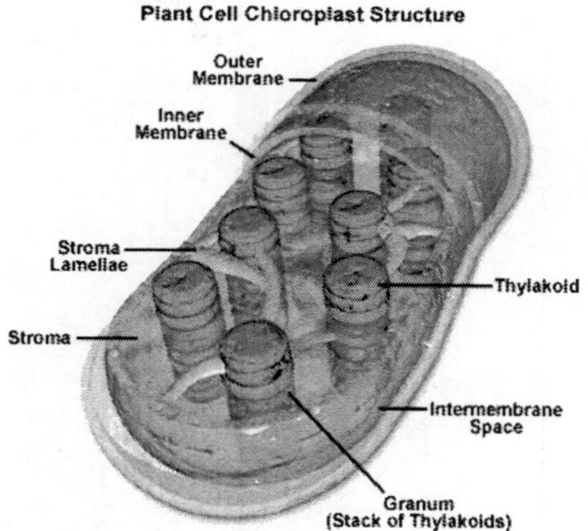

Fig. 1.15 Structure of Chloroplast

Chlorophyll a

Chlorophyll b

Where R= $CH_2CH_2CO_2CH_2CH{=}C(CH_2CH_2CH_2CH)_3CH_3$ (with CH_3 side groups)

Fig. 1.16 Chlorophyll 'a' and Chlorophyll 'b'

The important properties of Chlorophyll may be summarized as:

- It absorbs the radiant energy of sun
- It absorbs particular wavelength of light and single type of chlorophyll has particular absorption spectrum
- It brings about photoionization
- It has fluorescent property.

In a cross-section of a leaf, chloroplasts can be seen as numerous disc-like organelles in the photosynthetic cells of the palisade tissue just below the upper epidermis. Each chloroplast is covered by a double-membrane. Inside the membrane, the centre of chloroplast is filled with matrix or stroma. A number of organized flattened membranous sacs called thylakoids are present in the stroma. At present eight types of chlorophylls are known which are chlorophyll a, b, c, d, e, bacterio-chlorophyll a, b and chlorobium chlorophyll.

The chlorophyll molecule consists of a ring structure composed of four Pyrol nuclei linked by four +CH bridges. Chlorophyll a and b have exactly the same structure excepting in Chlorophyll b, a CH_3 side chain is replaced by –CHO group.

1.7 FACTORS AFFECTING THE PHOTOSYNTHESIS:

The rate of photosynthetic process is affected by several external factors (light, temperature, carbon-di-oxide and water) as well as internal factors (Chlorophyll, Protoplasm and Accumulation of Photosynthetic Products).

1.7.1 External Factors:

a. Light.

Light is the most essential and important factors to control the rate of photosynthesis as light supplies the energy necessary for this process. The ultimate source of light for photosynthesis in green plants is solar radiation. Out of the total solar energy reaching to the earth, only a very small portion about 2% is used in photosynthesis. **Light intensity, quality (wave length) and duration affect the process of photosynthesis.** Under low light intensity (e.g. during cloudy days), the rate of photosynthesis is also low. Increase in light intensity causes increase in the rate of photosynthesis up to a certain limit. At very high light intensity, the photosynthesis is decreased due to photo-oxidation of the constituents (solarization). The quality of light also affects photosynthesis. Green pigments chlorophylls absorb mostly the blue and the red regions of the spectrum. They reflect green light. Therefore, photosynthesis is high in blue and red light. Photosynthesis does not occur in green light. Duration of light also affects the rate of photosynthesis.

b. Temperature.

The rate of photosynthesis increases by increase in temperature upto 40°C. Above this temperature, there is a decrease in the photosynthesis. Similarly, low temperature also inhibits photosynthesis. The temperature affects photosynthesis by affecting the activity of enzymes. We know that the dark reaction of photosynthesis involves several enzymes. These enzymes function at a specific range of temperature. Low temperature lowers the activity of enzymes and high temperature cause inactivation of enzymes. The range of temperature at which photosynthesis takes place is quite variable. For most plants this is between 10 to 34°C. However some plants occurring in rigid zones as lichens, conifers etc. show photosynthesis at -35 to -20°C while some algae may photosynthesize at 75°C.

c. Carbon-di-oxide:

Carbon-di-oxide is present in the atmosphere in the concentration of 0.033% by volume. It is really a low concentration for photosynthesis. Thus, the rate of photosynthesis increases by increasing the concentration of CO_2 in the atmosphere if the light and temperature conditions are not limiting. At the same time, very high concentration of CO_2 becomes toxic to plants and inhibits photosynthesis. In normal conditions, it is observed that photosynthesis rate increase with increase of CO_2 concentration from 3 to 1%. If CO_2 supply is stopped, there would be no photosynthesis. It has also observed that rate of photosynthesis at a given light intensity is directly correlated with CO_2 supply.

d. Water:

Water is an essential raw material in photosynthesis. This rarely acts as a limiting factor because less than 1% of the water absorbed by a plant is used in photosynthesis. However, the rate of photosynthesis is lowered if the plants are inadequately supplied with water. Under water deficient conditions the stomatal apertures closed to reduce the loss of water by transpiration. As a consequence, the entry of CO_2 also stopped into the leaves. During photochemical reactions, water brings about various chemical changes. The water is necessary for turgidity of the cell, and controls opening and closing of stomata which is necessary for gaseous exchange. Thus water affect photosynthesis directly and indirectly.

Light intensity, quality (wave length) and duration affect the process of photosynthesis. Under low light intensity (e.g. during cloudy days), the rate of photosynthesis is also low. Increase in light intensity causes increase in the rate of photosynthesis up to a certain limit. At very high light intensity, the photosynthesis is decreased due to photo-oxidation of the constituents (solarization). The quality of light also affects photosynthesis. Green pigments chlorophylls absorb mostly the blue and the red regions of the spectrum. They reflect green light. Therefore, photosynthesis is high in blue and red light. Photosynthesis does not occur in green light. Duration of light also affects the rate of photosynthesis. The range of temperature at which photosynthesis takes place is quite variable. At the same time, very high concentration of CO_2 becomes toxic to plants and inhibits photosynthesis. However, the rate of photosynthesis is lowered if the plants are inadequately supplied with water. Under water deficient conditions the stomatal apertures closed to reduce the loss of water by transpiration.

1.7.2 Internal Factors:

a. Chlorophyll:

It is the most important component because only through this substance the light energy gets converted into chemical energy Plants which are devoid of chlorophyll are unable to synthesize food materials.

b. Protoplasm:

Apart from chlorophyll content and hydration of protoplasm, enzymes present in protoplasm also affect the rate of photosynthesis. Since the enzymes are numerous and in different proportions in the plant cells insignificant knowledge has been obtained regarding nature and properties.

c. Accumulation of Photosynthetic Products:

Accumulation of the carbohydrates usually tends to reduce the process. The carbohydrates formed during photosynthesis simultaneously being translated to other plants. During the photosynthetic period more food material is formed than translocated parts. Due to this food be accumulated in the assimilatory cells and results in lowering the process of photosynthesis.

1.8 PROCESS OF PHOTOSYNTHESIS:

The photosynthesis is a process which occurs only in green plants and by which these green plants synthesize energy rich carbohydrates from water and carbon-di-oxide with the process of consumption. It is a very complex process including a series of chemical reactions. These chemical reactions takes place in definite sequence and get influenced by specific enzyme or a group of enzymes. The study of photosynthesis demonstrated the existence of two phases, a light phase and a dark phase. The reactions of light phase require light and hence are also called photochemical reactions whereas reactions of dark phase require no light (dark) and are purely chemical reactions.

1.8.1 Light Phase or Light Reactions : Hill Reactions or Photochemical Reactions:

It takes place in grana region of chloroplast. It is completed in to two steps such as (i) Photolysis and (ii) Transfer of Hydrogen. The pigment chlorophyll present in the chloroplasts absorbs visible light and after absorption creates such condition that water breaks into protons (H+), electron (e-) and molecular oxygen (O_2). This is called photolysis. The molecular oxygen (O_2) goes into the atmosphere. The electrons and protons released by the photolysis of water are used up in the production of assimilatory power in the form of NADPH and ATP (reduced nicotinamide adenine dinucleotide phosphate and adenosine triphosphate).

1.8.2 Dark Phase Dark Reaction Blackmann's Reaction:

The assimilatory power, generated in the light phase of photosynthesis is used during this phase where carbon-di-oxide (CO_2) of atmosphere is utilized in the production of carbohydrate. This step was discovered in details by Calvin, Benson and Bassam in the year 1955. It catalyzes assimilation of CO_2 to carbohydrates. These reactions are called dark reactions. It occurs in stroma or matrix of chloroplasts. The reactions do not require light. **They are also known as Path of Carbon or CO_2 fixation.** Here CO_2 first of all combines with RuBP (Ribulose biphosphate). After a series of reactions, carbohydrate is synthesized and RuBP is regenerated.

Table 1.3 Difference between Light and Dark Phase of Photosynthesis

Sr. No.	Light Phase	Dark Phase
1.	It occurs over thylakoids.	It occurs in matrix of chloro-plasts.
2.	The phase is dependent on light.	The phase does not require light.
3.	Temperature has little effect over it.	It is influenced by temperature.
4.	It produces NADPH and ATP (assimilatory power).	It consumes NADPH and ATP.
5.	Oxygen is evolved.	Glucose is the end product.

Light Phase or Light Reactions : Hill Reactions or Photochemical Reactions:

It takes place in grana region of chloroplast. It is completed in to two steps such as (i) Photolysis and (ii) Transfer of Hydrogen. The pigment chlorophyll present in the chloroplasts absorbs visible light and after absorption creates such condition that water breaks into protons (H^+), electron (e^-) and molecular oxygen (O_2). This is called photolysis. The molecular oxygen (O_2) goes into the atmosphere. The electrons and protons released by the photolysis of water are used up in the production of assimilatory power in the form of NADPH and ATP (reduced nicotinamide adenine dinucleotide phosphate and adenosine triphosphate).

Dark Phase Dark Reaction Blackmann's Reaction:

The assimilatory power, generated in the light phase of photosynthesis is used during this phase where carbon-di-oxide (CO_2) of atmosphere is utilized in the production of carbohydrate. This step was discovered in details by Calvin, Benson and Bassam in the year 1955. It catalyzes assimilation of CO_2 to carbohydrates. These reactions are called dark reactions. It occurs in stroma or matrix of chloroplasts. The reactions do not require light. They are also known as Path of Carbon or CO_2 fixation. Here CO_2 first of all combines with RuBP (Ribulose biphosphate). After a series of reactions, carbohydrate is synthesized and RuBP is regenerated.

1.9 CAM (CRASSULACEAN ACID METABOLISM) PATHWAY:

In succulent members belonging to family Cactaceae and Crassulaceae etc, the stomata remain closed during day time in order to reduce transpiration and the stomata open during night (scotoactive opening).Thus the diurnal acidification and deacidification during the night and day time respectively is called CAM (Crassulacean Acid Metabolism). All reactions of CAM occur in mesophyll cells and it is important for the survival of succulents..

(a) Absorption of CO_2 during night and its storage as organic acid (malic acid) is called acidification.

(b) During daytime malic acid undergoes oxidative decarboxylation and CO_2 is released.

(c) Liberation of CO_2 from an organic acid during daytime is called deacidification.

1.10 NUTRITION IN ANIMALS

As we know that plants are autotrophic organisms and can manufacture their own food. So, plants don't have to look to others for getting their food. They are food producers themselves, while animals depend on other organisms for their food. Thus, animals need an external source of food as hetero trophic. Animals cannot make their own food and depend on handy or readymade food. This handy food comes either from plants or from other animals. Thus, animals obtain their food from plants or other animals (which they eat). We, human beings are also animals, and obtain their foods like wheat, rice, pulses, fruits and vegetables from plants, and the foods like milk, curd, cheese and eggs are obtained from animals. Some people also eat meat, chicken and fish as food. These foods are also obtained from animals. Many other animals obtain their food by eating the flesh of other animals. As the fish, birds, snakes and insects, all obtain their food from other animals. The big fish eats small fish; the birds eat worms and insects; the snake eats frogs and the insects eat dead bodies of animals. The non-green plants (which cannot make their own food by photosynthesis) also obtain their food from other plants and animals. Yeast plant is one such example. Even the plants can eat insects. For example, the pitcher plant and the Venus fly-trap are the two plants which eat insects. All the animals can be divided into three groups on the basis of their food habits (or eating habits) such as (a) Herbivores (b) Carnivores and (c) Omnivores.

(a) Herbivores:

Some animals eat only plants or their products. Those animals which eat only plants are called herbivores. The herbivores may eat grasses, leaves, grains, fruits or the bark of trees. Some of the examples of herbivores are such as Goat, Deer, Cow, Buffalo, Camel, Ass, Sheep, Horse, Ox, Elephant, Grasshopper, Monkey, Squirrel,

Fig. 1.17 Cows eat only palnt food i.e. grass, so it is a herbivore

Rabbit and Hippopotamus (Fig 1.17). Thus, herbivores are plant eaters. Herbivores are also called herbivorous animals.

(b) Carnivores:

Some animals eat only other animals. They do not eat plant food at all. Those animals which eat only other animals as food are called carnivores. Carnivores eat only the meat or flesh of other animals. Some of the examples of carnivores are such as Vulture, Kingfisher, Lizard, Tiger, Frog, Wolf, Snake, Hawk and Lion (Fig 1.18).

Fig. 1.18 Lion eat only meat, so it is a carnivore

(c) *Omnivores*:

Some animals eat both, plants as well as other animals as food. Those animals which eat both, plants and animals are called omnivores. Some of the examples of omnivores are such as man, dog, crow, sparrow, bear, mynah and ant. Man is called an omnivore because he eats the plant food (such as grains, pulses, fruits and vegetable) as well as the meat of animals (such as goat, chicken and fish). Thus, omnivores are plant eaters as well as meat eaters. Omnivores are also called omnivorous animals (Fig 1.19).

Fig. 1.19 Human eat plant food as well as meat, so they are omnivores

Plants use the energy of sun and prepare food by photosynthesis. The plants utilize this food for maintaining their life. These plants and their products are also eaten up by herbivores and omnivores as food. And the carnivores eat herbivores as food. In this way, it is the energy of the sun which provides food for plants, and animals (herbivores, carnivores and omnivores). Thus, an organism either makes its own food from raw materials as green plants do or takes in readymade food as animals do(Fig. 1.20). The process of obtaining food and then using it for obtaining energy, growth and repair of the body, is called as nutrition.

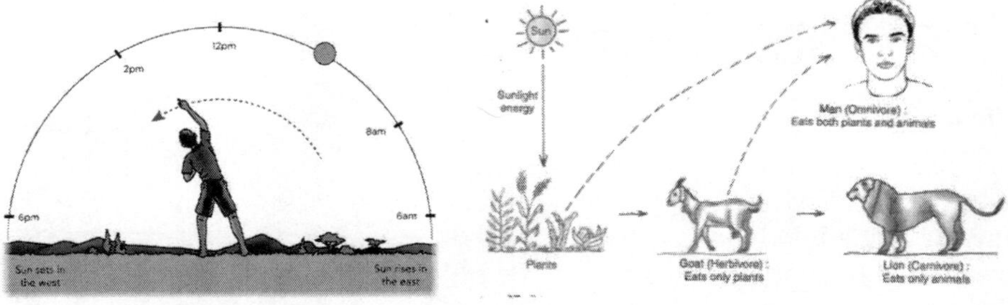

Fig. 1.20 Diagrams show (A) The sun is different positions, and (B) How all living things i.e. plants and animals depends on sun for their food.

1.10.1 Kinds of Nutrition in Animals:

There are five main processes such as Ingestion, Digestion, Absorption, Assimilation and Egestion, which are concerned with the use of food by animals such as:

1. Ingestion:

In order to provide the energy necessary for growth and carry on life's activities, we must 'eat food' or 'take food into the body'. The process of taking food into the body is called ingestion. In most simple terms, ingestion means 'eating of food' by the animal. When we put food into our mouth with hands, we are ingesting (the food).

2. Digestion:

The food of most animals consists of large insoluble molecules which cannot be absorbed by the animal's body. So, before the food can be used by the animal for various functions like getting energy or for growth, it must be broken down into smaller and water soluble molecules which can be absorbed by the body. The process in which the food containing large and insoluble molecules is broken down into smaller and water soluble molecules (which can be absorbed by the body) is called as digestion. In most simple terms, digestion makes the food soluble so that it can be utilized by the animal's body. Most animals use both, physical and chemical methods for digesting (breaking up) the large food molecules. **(i) Physical methods** include chewing and grinding the food in mouth and **(ii) Chemical methods** include the addition of digestive juices (enzymes) to food by the body itself.

3. Absorption:

After digestion, the food molecules become small and soluble. The soluble food molecules can pass through the walls of our intestine and go into blood. The process in which the digested food passes through the intestinal wall into blood stream is called absorption.

4. Assimilation:

Blood carries the absorbed food to all the parts of the body. The food then enters each and every cell of the body where it is used for producing energy and for making materials for the growth and repair of the body. The process in which the absorbed food is taken in by body cells and used energy, growth and repair, is called assimilation.

5. Egestion:

The whole food which we eat is not digested by our body. A part of the food which we eat remains undigested or insoluble which cannot be used by the body. This undigested part of the food is then removed from the body in the form of faeces when we go to toilet. The process in which the undigested food is removed from the body is called egestion.

1.10.2 Nutrition in Simple Animals:

Amoeba and *Paramecium* are two simple animals having body of single cell only, so called as unicellular animals. In these animals, all the processes of nutrition are performed by the single cell.

1.10.3 Nutrition in *Amoeba*:

Amoeba eats tiny (microscopic) plants and animals as foods which float in water in which it lives. The mode of nutrition in Amoeba is holozoic. The process of obtaining food by *Amoeba* is called phagocytosis, which means "cell feeding". The various steps involved in the nutrition of *Amoeba* are ingestion, digestion, absorption, assimilation, and egestion, those are given below (Fig 1.21).

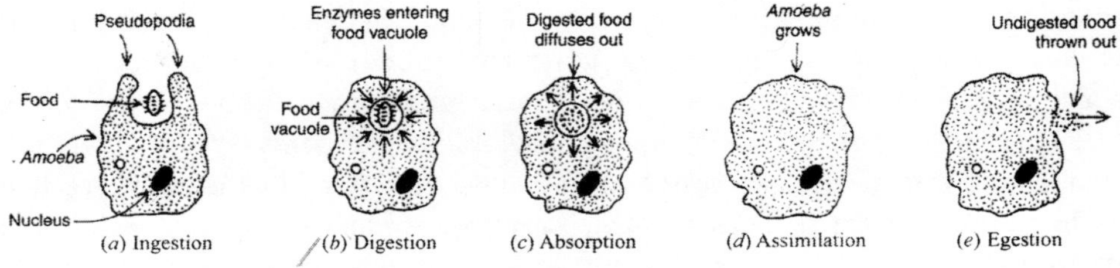

Fig. 1.21 Various stages of nutrition in *Amoeba*

1. Ingestion:

Amoeba has no mouth or a fixed place for the ingestion or intake of food, and ingests food by using its pseudopodia. When a food particle comes near *Amoeba*, then Amoeba ingests this food particle by forming temporary finger-like projections called pseudopodia around it. The food is engulfed with a little surrounding water to form a food vacuole inside the *Amoeba*. This food vacuole can be considered to be a 'temporary stomach' of *Amoeba*.

2. Digestion:

In *Amoeba*, food is digested in the food vacuole by digestive enzymes. The enzymes from surrounding cytoplasm enter into the food vacuole and break down the food into small and soluble molecules by chemical reactions. Thus, digestion in *Amoeba* takes place inside the food vacuole due to which the food dissolves (or food becomes soluble).

3. Absorption:

The digested food present in the food vacuole of *Amoeba* is absorbed directly into the cytoplasm of *Amoeba* cell by diffusion. Since *Amoeba* consists of only one small cell, it does not require blood system to carry the digested food. The digested food just spreads out from the food vacuole into the whole of Amoeba cell. After absorption of food, the food vacuole disappears.

4. Assimilation:

A part of the food absorbed in *Amoeba* cell is used to obtain energy through respiration. The remaining part of absorbed food is used to make the parts of *Amoeba* cell which lead to the growth of Amoeba. Thus, on assimilating food and amoeba grows in size, then can reproduce by dividing into two daughter cells.

5. Egestion:

Amoeba has no fixed place like anus for removing/discarding the undigested part of food. When a considerable amount of undigested food collects inside the *Amoeba* then its cell membrane suddenly ruptures at any place and the undigested food is thrown out of the body of *Amoeba*.

Paramecium is also a tiny unicellular animal which lives in water. *Paramecium* uses its hair like structures called cilia to sweep the food particles from water and put them into its mouth. The *Paramecium* has thin, hair-like cilia all over its body. The cilia move back and forth rapidly in water. When the cilia present around the mouth region of *Paramecium* move back and forth, they sweep the food particles present in water into the mouth of *Paramecium* (Fig 1.22). This is the first step in water into the nutrition of *Paramecium* which is called ingestion. Ingestion is followed by other steps such as digestion, absorption, assimilation and egestion (as explained in the case of *Amoeba*).

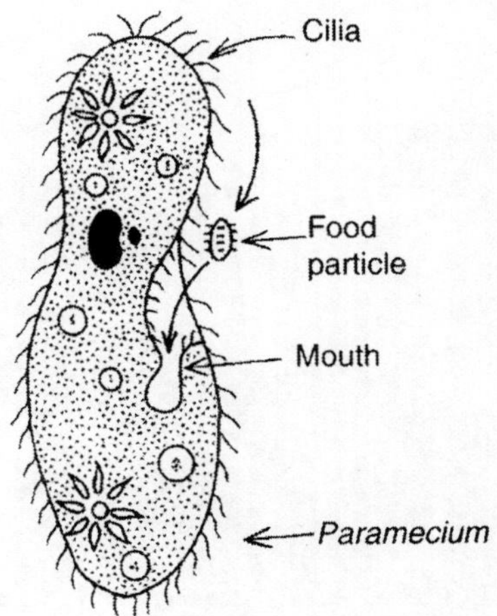

Fig. 1.22 *Paramecium* taking food into its mouth through cilia

1.10.4 Nutrition in Complex Multicellular Animals:

In the complex multicellular animals like human, fish, grasshopper and frog, etc., all the processes involved in nutrition are performed by a combination of digestive organs, called as digestive system. A long tube running from mouth to anus of a human being or other animals in which digestion and absorption of food takes place is called alimentary canal or gut.

1.10.5 Nutrition in Human beings (Human Digestive System):

The nutrition in human beings takes place through human digestive system. The human digestive system consists of the alimentary canal and its associated glands. The various organs of the human digestive system in sequence are such as mouth, oesophagus or food pipe, stomach, small intestine and large intestine. The glands which are associated with the human digestive system are salivary glands, liver and pancreas. The human alimentary canal which runs from mouth to anus is about 9 mtrs long tube. The ducts of various glands open into the alimentary canal and pour the secretions of the digestive juices into the alimentary canal (Fig 1.23)

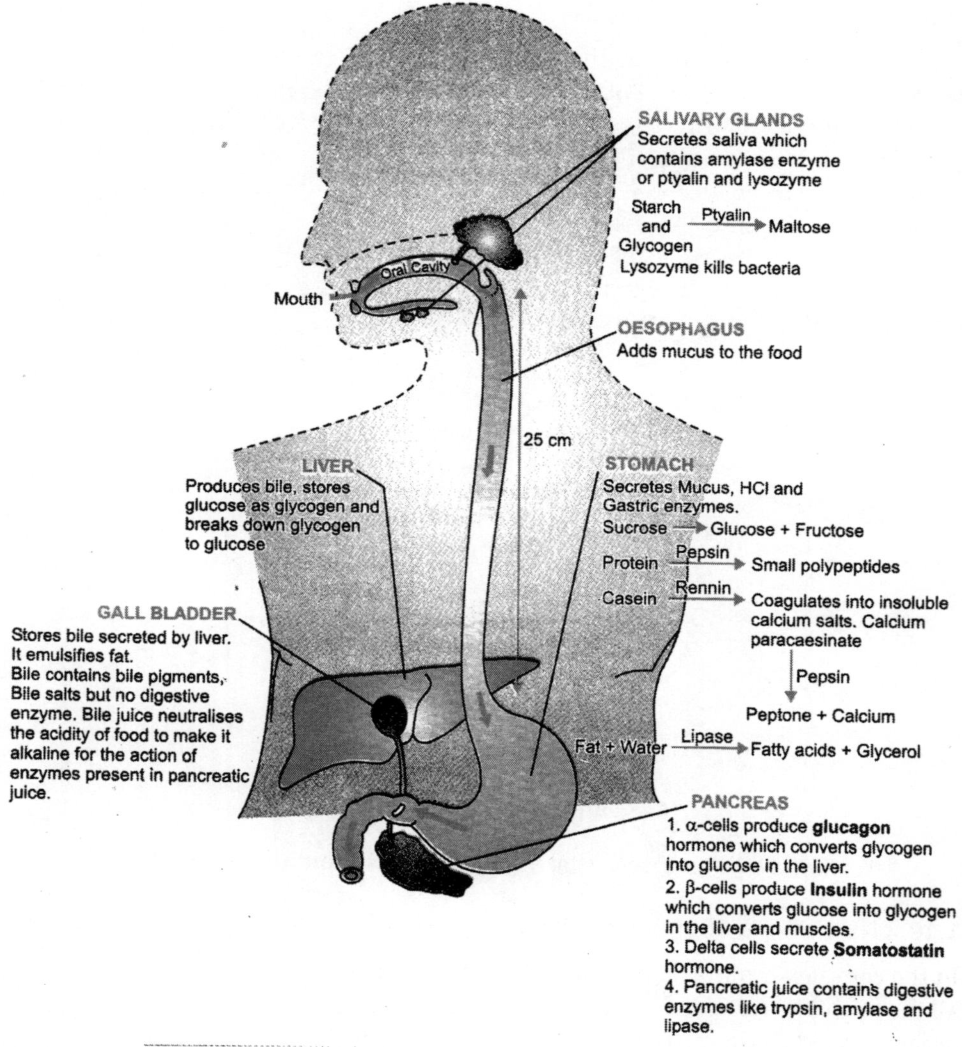

Fig. 1.23A Diagram showing the path of food through Oral cavity, Oesophagus and stomach

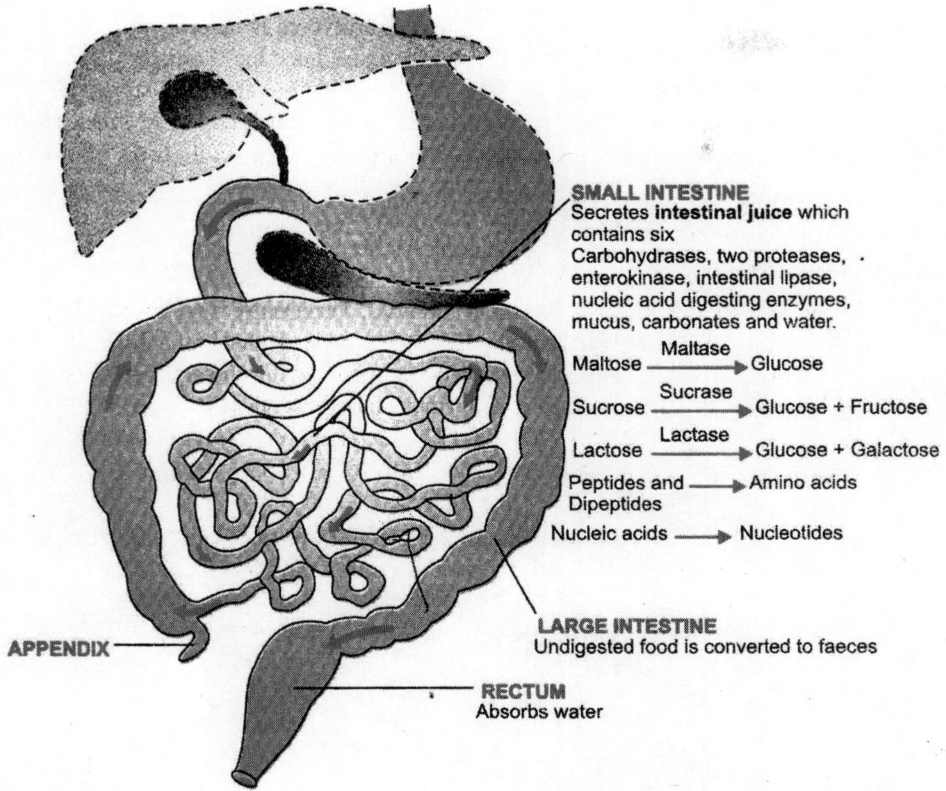

Fig. 1.23B Diagram showing movement through small and large intestine

1. Ingestion:

The human beings have a special organ for the ingestion of food, called as mouth. So, in human beings, food is ingested through the mouth. The food is put into the mouth with the help of hands.

2. Digestion:

In fact, the digestion of food starts as soon as we put food in our mouth. The mouth cavity or buccal cavity contains teeth, tongue and salivary glands. The teeth cut the food into small pieces, chew and grind it. So, the teeth help in physical digestion. The salivary glands in our mouth produce saliva, which wets the food and tongue helps in mixing this saliva with food. The wetted food can be swallowed more easily. The salivary glands help in chemical digestion by secreting enzymes. The human saliva contains an enzyme called salivary amylase which digests the starch present in food into sugar. Thus, the digestion of starch (carbohydrate) begins in the mouth itself. Since the food remains in the mouth only for a short time, so the digestion of food remains incomplete in mouth.

The slightly digested food in the mouth is swallowed by the tongue and goes down the food pipe called oesophagus. It carries food to the stomach. The walls of food pipe have muscles which can contract and expand alternately. When the slightly digested food enters the food pipe, the walls of food pipe start contraction and expansion movements. The contraction and expansion movement of the walls of food pipe is called peristaltic movement. This peristaltic movement of oesophagus pushes the slightly digested food into the stomach. The stomach is a J-shaped organ present on the left side of the abdomen. The food is churned in the stomach for about 3 hrs. During this time, the food breaks down into still smaller pieces and forms a semi-solid paste. The stomach wall contains three tubular glands in it walls, those secrete the gastric juice. The gastric juice contains three substances such as hydrochloric acid, the enzyme pepsin and mucus.

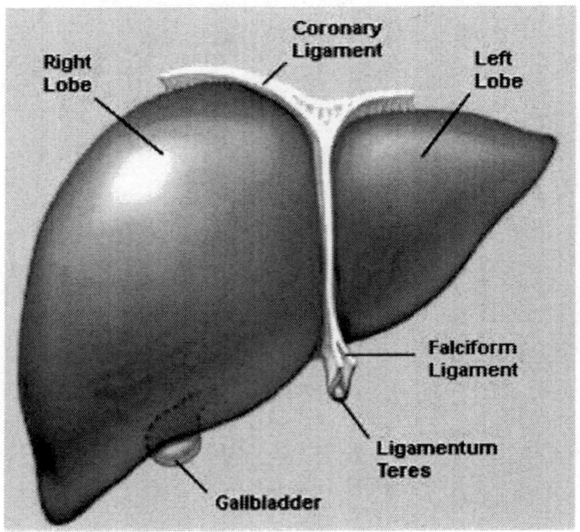

Fig. 1.24 Liver

The gastric juice is acidic in nature due to the presence of hydrochloric acid (HCl). In the acidic medium, the enzyme pepsin the digestion of proteins present in food to form smaller molecules. Thus, the protein digestion begins in the stomach. The protein digesting enzyme pepsin is active only in the presence of an acid. So, the function of hydrochloric acid in the stomach is to make the medium of gastric juice acidic so that the enzyme pepsin can digest the proteins properly. Another function of hydrochloric acid is that it kills the bacteria which may enter the stomach with food. The mucus helps to protect the stomach wall from its own secretions of hydrochloric acid. If mucus is not secreted, hydrochloric acid will cause the erosion of inner lining of stomach leading to the formation of ulcers in the stomach. The partially digested food then goes from the stomach into the small intestine. The exit of food from stomach is regulated by a sphincter muscle which releases it in small amounts into the small intestine. Partially digested food enters into the small intestine from the stomach.

The small intestine is the largest part of the alimentary canal. It is about 6.5 mtrs long in an adult man. Though the small intestine is very long, it is called small intestine because it is very narrow. The small intestine is arranged in the form of a coil in our belly. The length of the small intestine differs in various animals depending on the type of food they eat. For example, meat is a food which is easier to digest. So, the carnivorous animals like tigers which eat meat have a shorter small intestine, whereas cellulose is a carbohydrate food which is digested with difficulty. So, the herbivorous animals like cow which eat grass need a longer intestine to allow the cellulose present in grass to be digested completely. The small intestine receives the secretions of two glands as liver and pancreas.

(a) **Liver** secretes bile. Bile is a greenish yellow, alkaline liquid made in the liver which is normally stored in the gall bladder. It contains no enzyme, and, thus, has no chemical action on food. Its salts, namely, sodium glycocholate and sodium taurocholate, reduce the surface tension of large fat droplets and break them into many small ones. The process is called emulsification, as it produces a fine emulsion of fats in the aqueous intestinal contents. The small fat droplets present larger surface area to lipase. It contains salts which help to emulsify or break the fats (or lipids) present in the food. Thus, bile performs two functions such as (i) bile salts break the fats present in the food into small globules making it easy for the enzymes to act and digest them. (ii) It makes the acidic food coming from the stomach alkaline so that pancreatic enzymes can act on it.

(b) **Pancreas** is a large gland which lies parallel to and beneath the stomach It secreted pancreatic juice which contains digestive enzymes like pancreatic amylase, trypsin and lipase. The enzyme amylase breaks down the starch, the enzyme trypsin digests the proteins and the enzyme lipase breaks down the emulsified fats. Pancreatic lipase hydrolyses the fats in to following three stages such as:

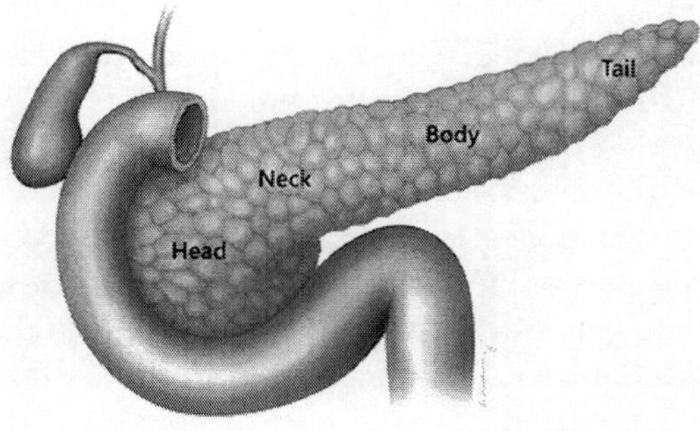

Fig. 1.25 Pancreas

- **Stage 1:** Lipase separates one fatty acid molecule, changing the emulsified triglyceride into a diglyceride.
- **Stage 2:** The diglyceride is broken down into another fatty acid molecule and a monoglyceride.
- **Stage 3:** The monoglyceride is hydrolyzed to another fatty acid molecule and a glycerol molecule.

The walls of small intestine contain glands which secrete intestinal juice. The intestinal juice contains a number of enzymes which complete the digestion of complex carbohydrates into glucose, proteins into amino acids and fats into fatty acids and glycerol. Glucose, amino acids, fatty acids and glycerol are small water soluble molecules. In this way, the process of digestion converts the large and insoluble food molecules into small water soluble molecules. The chemical digestion of food is brought about by biological catalysts called enzymes. Intestinal glands mainly secrete intestinal lipase. This enzyme occurs mainly in the intestinal epithelial cells and only a small amount is released in the intestinal juice.

- The lipase found in the intestinal juice hydrolyses some triglycerides, diglycerides and monoglycerides to fatty acids and glycerol like the pancreatic lipase.
- The intestinal lipase present in the intestinal epithelial cells hydrolyses the absorbed triglycerides and diglycerides to monoglycerides and fatty acids.

The Fatty acids, glycerol and monoglycerides are the end products of fat digestion.

3. Absorption:

Now the molecules of food become so small that they can pass through the walls of the small intestine (which contain blood capillaries) and go into our blood. It is called absorption. The small intestine is the main region for the absorption of digested food. In fact, the small intestine is especially adapted for absorbing the digested food. The inner surface of small intestine has millions of tiny, finger-like projections called villi. The presence of villi gives the inner walls of the small intestine a very large surface area, and the large surface area of small intestine helps in the rapid absorption of digested food. The digested food which is absorbed through the walls of the small intestine goes into our blood.

4. Assimilation:

The blood carries digested and dissolved food to all the parts of the body where it becomes assimilated as part of the cells. This assimilated food is used by the body cells for obtaining energy as well as for growth and repair of the body. The energy is released by the oxidation of assimilated food in the cells during respiration. The digested food which is not used by our body immediately is stored in the liver in the form of a carbohydrate called glycogen. This stored glycogen can be used as a source of energy by the body as and when required.

5. Egestion:

A part of the food which we eat cannot be digested by our body. This undigested food cannot be absorbed in the small intestine. This undigested food passes from the small intestine into a wider tube called large intestine The walls of large intestine absorb most of the water from the undigested food with the help of villi. Due to this, the undigested part of food becomes almost solid. The last part of the large intestine called rectum, which stores this undigested food for some time. At last, when we go to the toilet, then this undigested food is passed out or egested from our body through anus in the form of faeces or stool. The act of expelling the faeces is called egestion or defection. The exit of faeces is controlled by the anal sphincter.

1.11 DENTAL CARIES:

The hard outer covering of a tooth is called enamel, which is harder than even bones in our body. The part of tooth below enamel is called dentine. Dentine is similar to bone. Inside the dentine is pulp cavity. The pulp cavity contains nerves and blood vessels. The formation of small cavities or holes in the teeth due to the action of acid-forming bacteria and improper dental care is called dental caries (Fig 1.26A). At the time of taking sugary food, the bacteria in our mouth act on sugar to produce acids. These acids first dissolve the calcium salts from the tooth enamel and then from dentine forming small cavities or holes in the tooth over a period of time. The formation of cavities reduces the distance between the outside of the tooth and the pulp cavity which contains nerves and blood vessels. The acids produced by bacteria irritate the nerve endings inside the tooth and cause toothache. If the cavities caused by dental decay are not cleaned and filled by a dentist, the bacteria will get into the pulp cavity of tooth causing inflammation and infection leading to severe pain. If the teeth are not cleaned regularly, they become covered with a sticky, yellowish layer of food particles and bacteria cells called dental plaque. Since plaque covers the teeth forming a layer over them, the alkaline saliva cannot reach the tooth surface to neutralize the acid formed by bacteria and hence tooth decay sets in. Brushing the teeth regularly, after eating food, removes the plaque before bacteria produces acids and prevent dental caries or tooth decay.

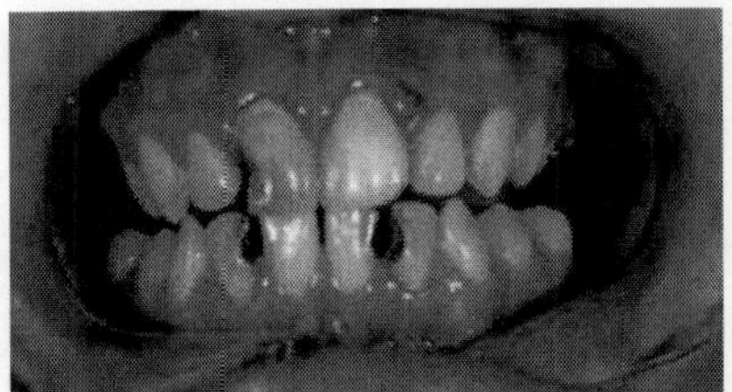

Fig. 1.26A Teeth Caries

A normal adult mouth has 32 teeth (except for wisdom teeth vestigial organ) has erupted by about age 13.

1. Incisors (8 total) – The middle most 4 teeth on the upper and lower jaw.

2. Canines (4 total) – The pointed teeth just outside the incisors teeth.
3. Premolars (8 total) – Teeth between the canines and molars.
4. Molars (8 total) – Flat teeth in the rear of mouth, best at grinding food.
5. Wisdom teeth – On third molars (4 total) these teeth erupt at around age 18, but are often surgically removed to prevalent displacement of other teeth.

1.11.1 Structure of Tooth:

Teeth are embedded in the jaws. Man has a fixed upper jaw and a movable lower jaw. Each tooth consists of three parts such as crown, neck and root (Fig 1.26B).

(i) **Crown:** It is the exposed portion of tooth above the gums (gingival). The gingival is a specialized region of the oral mucosa surrounding the neck of the teeth. Crown is covered with the hardest substance called enamel that protects the crown.

(ii) **Neck:** it is a narrow portion at the gum line.

(iii) **Root:** It is embedded in the jaw bone and holds the tooth securely in place.

Beneath the enamel is present dentine (secreted by odontoblast) which is made up of a hard substance similar to bone. It is also tough but not as tough as enamel and can decay. It has numerous canaliculi that pass radially from the pulp cavity towards the enamel. Dentine is regenerable living tissues that form the bulk of the tooth. There is a pulp cavity inside the dentine. It is a jelly like substance and carries the nerves fibres, blood vessels and sensory cells. The root is fixed in the alveolus of the jaw bone by periodontal membrane and cementum. There are two main types of cells (i) dentine forming odontoblasts and (ii) enamel forming ameloblasts.

Cement: It is a bone like hard substance around the root that holds a tooth in its socket and periodontal membrane covers the cement. These are live cells.

Salivary Glands: There are three pairs of salivary glands: parotid, sublingual and submaxillary or submandibular.

(i) **Parotid Glands:** These are the largest salivary glands. They lie on sides of the face, just below and in front of the ears. The parotid ducts, also called Stenson's ducts, open into the vestibule opposite the upper second molar teeth. Viral infection of the parotid glands, causing swelling and pain, is the disease called mumps.

(ii) **Sublingual Glands:** These lie under the front part of the tongue. The sublingual ducts, also termed ducts of Rivinus, also open under the tongue.

(iii) **Submaxillary Glands:** These lie at the angles of the lower jaw. The submaxillary ducts, also known as Wharton's ducts, open under the tongue.

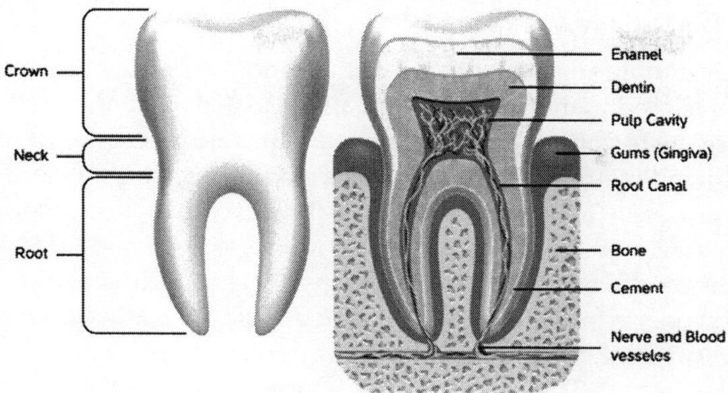

Fig. 1.26B Tooth Anatomy

1.12 RESPIRATION:

As we studied that the digested food is assimilated into the body of the living organisms. The assimilated food is used mainly for two purposes:

- Used as a fuel to get energy for various life processes, and
- Used as a material for the growth and repair of the body.

We will now describe how energy is released from the food which is absorbed and assimilated in the cells of the body. The food is the fuel for energy production in cells. Most living things need oxygen (of air) to obtain energy from food. This oxygen reacts with the food molecules like glucose present in the body cells and burns them slowly to release energy. The energy thus released is stored in ATP molecules in the cells. The body can use this stored energy whenever it wants to do so. The process of releasing energy from food is called respiration. When oxygen burns the food in the cells of the body to release energy, then carbon-di-oxide and water are produced as waste products which are to be eliminated from the body. The process of respiration involves taking in oxygen into the cells, using it for releasing energy by burning food, and then eliminating the waste products (carbon-di-oxide and water) from the body. The process of respiration can be written in the form of a word equation as follows:

Food + Oxygen → Carbon-di-oxide + Water + Energy

The process of respiration which releases energy takes place inside the cells of the body, called also cellular respiration. The process of cellular respiration is common to all the living organisms. It provides energy to the cells. There are two by-products of cellular respiration such as carbon dioxide and water. Out of these only carbon dioxide is considered the real waste product of respiration because its accumulation in the body is harmful to the organism. Water produced during respiration is not harmful to the body. It is rather beneficial for the body.

1.12.1 Breathing and Respiration:

Respiration is a more complex process. Respiration includes breathing as well as the oxidation of food in the cells of the organism to release energy. The mechanism by which organisms obtain oxygen from the air and release carbon dioxide is called breathing. Breathing is a physical process whereas respiration also includes biochemical process of oxidation of food. The process of breathing involves the lungs of the organism whereas the process of respiration also involves the mitochondria in the cells where food is oxidized to release energy (Fig 1.27). Respiration is actually a biochemical process which occurs in stages and required many enzymes. The main purpose of respiration is the release of energy from the oxidation of simple food molecules like glucose. The energy released during respiration is used for carrying out the biological functions which are necessary for the maintenance of life and survival of an organism. Respiration is just opposite of photosynthesis. Because photosynthesis makes food like glucose by using carbon dioxide, water sunlight energy, and releasing oxygen; whereas respiration breaks food like glucose by using oxygen, and releasing carbon dioxide, water and energy.

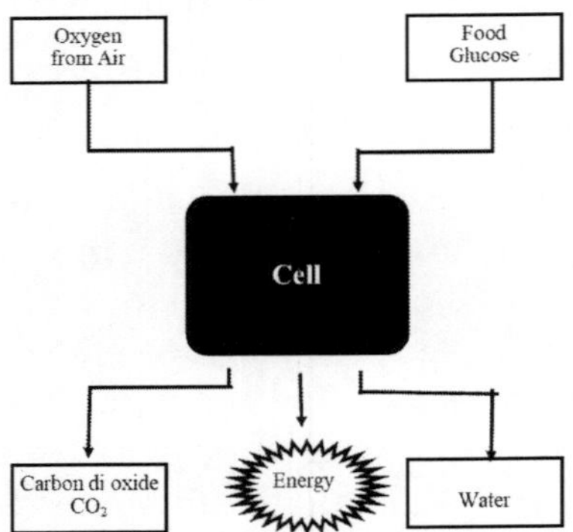

Fig. 1.27 Concept of Respiration

1.12.2 Storage of Released Energy during Respiration in the form of ATP:

All the energy released during respiration is not used immediately by an organism (plant or animal). The energy produced during respiration is stored in the form of ATP molecules in the cells of the body and used by the organism as and when required. ADP is a substance called Adenosine Di-Phosphate. The molecules of ADP are present in a cell. ADP has low energy content. ATP (Adenosine Tri-Phosphate) is the chief energy carrier molecule in all type of organisms. ATP is required for the anabolic (generative) process of an organism by which organic macromolecules required for the growth are synthesized. ATP is formed by the catabolic (degradative) process in which macromolecules are broken and energy is generated. The structure of ATP was first detected by Lohman in 1930 and confirmed by Alexander Todd et al., in 1948. ATP consists of an adenosine (adenine plus ribose sugar) and three phosphate groups (Fig 1.28). These three phosphate groups are bound by high energy bonds to adenosine unit and whenever one or two of these phosphate groups are removed from the ATP, large amount of energy is released.

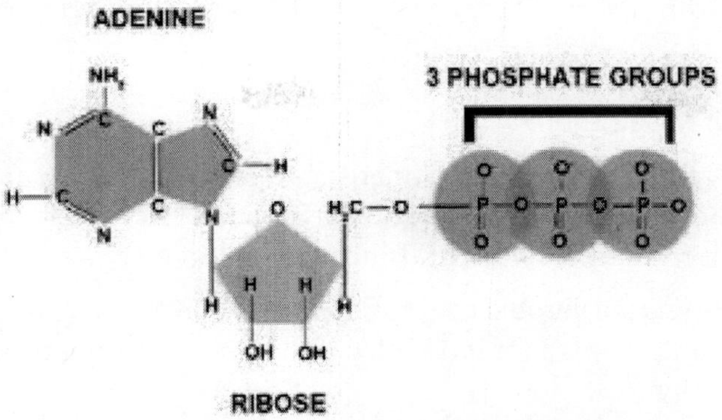

Fig. 1.28 Structure of ATP Molecule

ATP (Adenosine triphosphate) ADP (Adenosine diphosphate) +Pi + Energy

ADP $\rightarrow$ AMP (Adenosine monophosphate) + Pi + Energy

ATP has high energy content. Inorganic phosphate is a substance which contains a phosphate group made up of phosphorus and oxygen. Inorganic phosphates are also present in a cell. Inorganic phosphate can be represented by writing just 'Phosphate'. The inorganic phosphate can also be represented by the symbol Pi (where P stands for phosphate and i for inorganic). ADP contains two phosphate groups whereas ATP contains three phosphate groups in its molecule.

(i) The energy released during respiration is used to make ATP molecules from ADP and inorganic phosphate. ADP combines with inorganic phosphate by absorbing the energy released during respiration to form ATP molecules. Thus, energy is stored in the cells in the form of ATP.

ADP + Phosphate + Energy ATP

(Low energy) (From respiration) (High energy)

(ii) When the cell needs energy, then ATP can be broken down using water to release energy.

ATP $\rightarrow$ ADP + Phosphate + Energy

(For use in cells)

The energy equivalent to 30.5 kJ/mole is released in this process. The energy released by ATP is used to carry out all the endothermic reactions taking place in the cells. ADP can be converted to ATP by absorbing energy produced during respiration, and ATP can be converted back to ADP releasing energy to be used by the cells, again and again. This ensures a continuous supply of energy to the organism. Like battery it can provide electrical energy for different purposes such as lighting, heating, running radio and computer, etc., in the same way, the energy stored in ATP is used by the body cells for various purposes like contraction of muscles, conduction of nerve impulses, synthesis of proteins and many other activities related to the functioning of cells. In fact, ATP is known as the energy currency of cells.

1.12.3 Types of Respiration:

In most of the cases, the organisms (plants and animals) carry out respiration by using oxygen (called **aerobic respiration**). There are, however, some organisms which carry out respiration without using oxygen (called **anaerobic respiration**).

Before we describe aerobic respiration and anaerobic respiration, we should keep the following points in mind which will help us in understanding the two types of respiration.

(a) Glucose ($C_6H_{12}O_6$) is a six carbon atom compound, which is oxidized in the cells of organisms during respiration.

(b) The oxidation of glucose to pyruvic acid (or pyruvate) is called glycolysis. It occurs in the cytoplasm of a cell and not in mitochondria. The oxidation of glucose to pyruvic acid does not require oxygen. One molecule of glucose on glycolysis produces two molecules of pyruvic acid (or pyruvate).

(c) Pyruvic acid is a three carbon atom compound. It is also called pyruvate. It is a ketonic carboxylic acid.

(d) The fate of pyruvate formed during respiration depends on whether oxygen is present in the cells or not. If oxygen is present in the cells, then pyruvate is completely oxidized to carbon dioxide and water, and a lot of energy is produced in the form of ATP. If, however, oxygen is not present in the cell (that is, the absence of oxygen), pyruvate is converted to ethanol and carbon-di-oxide of lactic acid depending on whether such a process is taking place in a plant cell or an animal cell. Much less energy is released in this case.

(e) Lactic acid is also a three carbon atom compound. It is also called lactate. It is a hydroxyl carboxylic acid.

1.12.4 Types of Respiration:

So far we have studied that respiration takes place in the presence of oxygen (of air). Respiration can, however, also take place in the absence of oxygen (of air) through it is very rare. This means that oxidation of food to obtain energy can occur in the presence of oxygen as well as in the absence of oxygen. Thus, there are two types of respiration such as aerobic respiration and anaerobic respiration.

1.12.4.1. Aerobic Respiration:

The respiration which used oxygen is called aerobic respiration. It is called aerobic respiration because it used air which contains oxygen ('aerobic' means 'with air'). In aerobic respiration, the glucose food is completely broken down into carbon dioxide and water by oxidation. Aerobic respiration produces a considerable amount of energy for use by the organism which gets stored in the ATP molecules.

The breaking down of glucose (food) during aerobic respiration (which is carried out by most of the organisms) can be represented as follows :

$$\underset{\text{(1 Molecule)}}{\text{Glucose}} \xrightarrow[\text{In Cytoplasm}]{\text{Glycolysis}} \underset{\text{(2 Molecules)}}{\text{Pyruvate Pyruvicacid}} \xrightarrow[\text{in Mitochondria}]{\text{Oxygen (Kreb'sCycle)}} 6CO_2 + 6H_2O + 38 \text{ ATP Energy}$$

Please note that during aerobic respiration (shown above), I molecule of glucose (food) produces 38 energy-rich ATP molecules. All the organisms which obtain energy by aerobic respiration cannot live without oxygen (of air). This is because if there is no oxygen, they cannot get energy from the food which they eat. Mitochondria are the sites of aerobic respiration in the cells (Fig 1.29). Thus, the breakdown of pyruvate to give carbon dioxide, water and energy takes place in mitochondria.

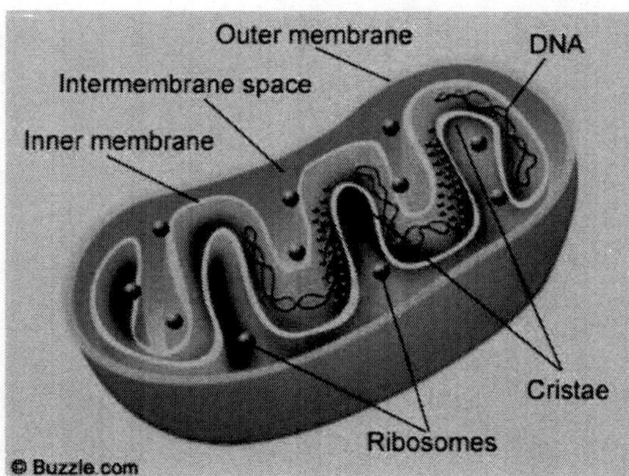

Fig. 1.29 Structure of Mitochondria

The energy released during aerobic respiration is used by the organism. Most of the living organisms carry out aerobic respiration (by using oxygen of air). For example, humans (man), dogs, cats, lions, elephants, cow, buffaloes, goat, deer, birds, lizards, snakes, earthworms, frogs, fish, and insects aerobic respiration by using oxygen of air (to obtain energy)

1.12.4.2. Anaerobic Respiration:

The respiration which takes place without oxygen is called anaerobic respiration. It is called anaerobic respiration because it takes place without air which contains oxygen (anaerobic means 'without air'). The microscopic organisms like yeast and some bacteria obtain energy by anaerobic respiration (which is called fermentation. In anaerobic respiration, the micro-organisms like yeast break down glucose (food) into ethanol and carbon dioxide, and release energy. This energy is then used by the micro-organisms. Anaerobic respiration produces much less energy which gets stored in the ATP molecules. The breaking down of glucose (food) during anaerobic respiration carried out by yeast (plants) can be represented as follows:

$$\underset{\text{(1 molecule)}}{\text{Glucose}} \xrightarrow[\text{(in cytoplasm)}]{\text{Glycolysis}} \begin{array}{c}\text{Pyruvate}\\ \text{(Pyruvic acid}\\ \text{(2 molecule)}\end{array} \xrightarrow[\text{(Yeast Fermentation)}]{\text{In absence of } O_2} \underset{\text{Ethanol}}{2C_2H_5OH} + 2CO_2 + \underset{\text{Energy}}{\text{2 ATP}}$$

Please note that during anaerobic respiration (shown above), 1 molecule of glucose (food produces only 2 energy-rich ATP molecules. A few organisms such as yeast plants and certain bacteria (called anaerobic bacteria) can obtain energy from food in the absence of oxygen by the process of anaerobic respiration. Please note that all the organisms which obtain energy by anaerobic respiration can live without oxygen (of air). For example, the single-celled, non-green plant called 'yeast' can live without oxygen because it obtains energy by the process of anaerobic respiration. From this discussion we conclude that all the cells do not use oxygen to produce energy. Energy can be produced in cells even without oxygen. Please note that the whole process of anaerobic respiration takes place in the cytoplasm of cells.

We can carry out the fermentation of sugar by using the anaerobic respiration of yeast as follows:

Take some sugar solution (or fruit juice) in a test-tube and add a little of yeast to it. Close the mouth of the test-tube with a cork and allow it to stand for some time. Now, open the cork and smell. A characteristic smell of ethanol (ethyl alcohol) is obtained from the test-tube. A gas is also evolved during this process. When this gas is passed through lime-water, the lime-water turns milky showing that it is carbon dioxide gas. This experiment tells us that the products of fermentation of sugar brought about by yeast are ethanol and carbon dioxide.

We (the human beings) obtain energy by aerobic respiration. But anaerobic respiration can sometimes take place in our muscles (or the muscles of other animals). For example, anaerobic respiration takes place in our muscles during vigorous physical exercise when oxygen gets used up faster in the muscle cells than can be supplied by the blood. When anaerobic respiration takes place in human muscles (or animal muscles), then glucose (food) is converted into lactic acid with the release of a small amount of energy. The breaking down of glucose (food) during anaerobic respiration in muscles can be represented as follows:

$$\underset{\text{(1 molecule)}}{\text{Glucose}} \xrightarrow[\text{(in cytoplasm)}]{\text{Glycolysis}} \begin{array}{c}\text{Pyruvate}\\ \text{(Pyruvic acid}\\ \text{(2 moleculces)}\end{array} \xrightarrow[\text{(muscle tissue)}]{\text{In absence of oxygen}} \text{2 Lactic acid} + \underset{\text{Energy}}{\text{2 ATP}}$$

The sudden buildup of lactic acid in our muscles during vigorous physical activity can cause muscular 'cramps'. (The painful contractions of muscles are called cramps). During heavy physical exercise (or any other heavy physical activity), most of the energy in our muscles in produced by aerobic respiration. Anaerobic respiration in muscles provides only some extra energy which is needed under the conditions of heavy physical activity (like running very fast or running for a long time) the people running a long distance race. The anaerobic respiration by muscles brings about partial breakdown of glucose (food) to form lactic acid. This lactic acid accumulates in the muscles. The accumulation of lactic acid in the muscles causes muscle cramps. Thus, muscle cramps occur due to the accumulation of lactic acid in muscles when the muscles respire anaerobically (without oxygen) while doing hard physical exercise. We can get relief from cramps in muscles caused by heavy exercise by taking a hot water bath or a massage. Hot water bath (or massage) improves the circulation of blood in the muscles. Due to improved blood flow, the supply of oxygen to the muscles increases. This oxygen breaks down lactic acid accumulated in muscles into carbon dioxide and water, and hence gives us relief from cramps. The anaerobic respiration does not take place only in the muscles of human beings, it also takes place in the muscles of other animals such as lion, tiger, cheetah, deer and many other animals when then run very fast and require much more energy than normal. This means that even the animals like lion, tiger, cheetah and deer, etc., can get leg cramps due to the accumulation of lactic acid in leg muscles if they run very fast for a considerable time. Please not that:

(a) The anaerobic respiration in plants (like yeast) produces ethanol and carbon dioxide as end products.

(b) The anaerobic respiration in animal muscle tissue produces lactic acid as the end product.

The similarity between aerobic respiration and anaerobic respiration is that in both the cases, energy is produced by the breakdown of food like glucose. The main difference between aerobic and anaerobic respiration are given here.

Table 1.4 Differences between Aerobic and Anaerobic Respiration

S/No.	Aerobic Respiration	Anaerobic Respiration
1.	Aerobic respiration takes place in the presence of oxygen.	Anaerobic respiration takes place in the absence of oxygen.
2.	Complete breakdown of food occurs in aerobic respiration.	Partial breakdown of food occurs in anaerobic respiration.
3.	The end products in aerobic respiration are carbon dioxide and water.	The end products in anaerobic respiration may by ethanol and carbon dioxide (as in yeast plants), or lactic acid (as in animal muscles).
4.	Aerobic respiration produces a considerable amount of energy.	Much less energy is produced in anaerobic respiration.

1.13 PHOSPHORYLATION:

The trapping of chemical energy released by oxidative reactions of the cell in the form of ATP, is called Phosphorylation. It is of different types such as:

1.13.1 Photophosphorylation:

It occurs in the presence of light by photosynthetic organisms (mainly in the photoautotroph and photo heterotrophs). In the photosynthetic cells, green pigment (chlorophyll) present in the chloroplast or bacterio-chloroplast present in the cell membranes of thylakoids, trap energy from solar radiations which become activated, releasing electrons. These electrons then pass through a series of carriers. Thus, the energy released is trapped in the form of phosphate bonds in ATP. Photophosphorylation is of two types such as:

(i) Cyclic Photophosphorylation:

In this type of photophosphorylation, the electrons are released from chlorophyll molecules (due to excitement by light energy), and return back to the same chlorophyll molecules by the same route. Such pathway is cyclic in nature as the electron pass. So, it is known as cyclic photophosphorylation and energy is released which is trapped in the form of ATP.

(ii) Non - Cyclic Photophosphorylation:

In this type of photophosphorylation, the electrons are released from chlorophyll molecules and don't return back to the same chlorophyll molecules, but instead are received by nicotinamide adenine dinucleotide phosphate (NADP). From NADP electrons enter the electron transport chain, where by oxidative phosphorylation occurs and ATP is formed. So, it is known as non-cyclic photophosphorylation. The electrons released from chlorophyll molecules are replenished in the chlorophyll molecules by photolysis of water.

$$2H_2O \rightarrow 4H^+ + 4e^- + O_2$$

Thus we can see that oxygen is released in this type of phosphorylation, and this type of phosphorylation is not found among anaerobic microorganisms because oxygen is toxic for anaerobes.

1.13.2 Substrate Level Phosphorylation:

In substrate level phosphorylation, ATP is generated by the transfer of high energy phosphate bond from any other metabolic compound of ADP. For example:-

GTP (Guanosine triphosphate) + ADP $\rightarrow$ ATP + GDP (Guanosine biphosphate)

1.13.3 Oxidative Phosphorylation:

In this type of phosphorylation, the electrons are collected by certain electron carriers like NAD, NADP and FAD from various sources are passed into electron transport chain. Finally electron reaches to oxygen or some other inorganic molecules like Fe and nitrate etc, which act as final electron acceptors. The transfer of electron from one carrier to another carrier released energy, which is used to generate ATP from ADP. Oxidative phosphorylation occurs in plasma membrane of prokaryotes and in the inner membrane of mitochondria of eukaryotes. One molecule of NADPH generates three molecule of ATP, when it enters the electron transport chain for oxidative phosphorylation. However, one molecule of FADH generates only two molecules of ATP. It is due to fact that FADH enters the electron transport chain (ETC) later than NADPH.

1.14 ANAEROBIC RESPIRATION (FERMENTATION):

It has been seen that pyruvic acid (formed by glycolysis) is converted either into ethyl alcohol (alcoholic fermentation) or lactate (lactic acid fermentation or muscle glycolysis) in the absence of oxygen (anaerobic condition) (Fig 1.29 and 1.30). Several unifying themes should be kept in mind when microbial fermentations are examined as:

- NADH is oxidized to NAD+
- Electron acceptor is often either pyruvic acid or pyruvic derivative. In fermentation, the substrate is partially oxidized, ATP is formed by substrate level phosphorylation only, and oxygen is not needed.

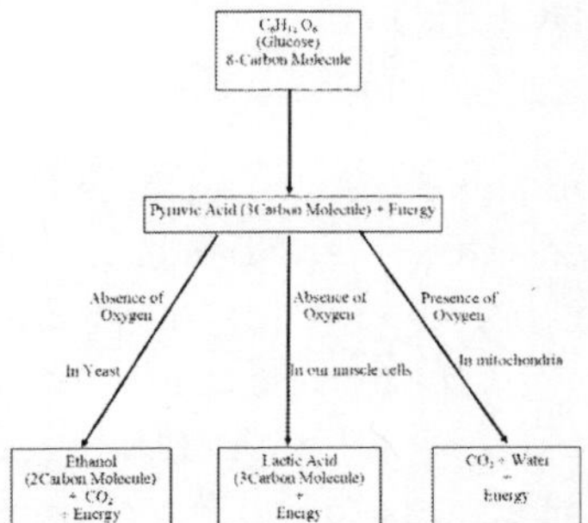

Fig. 1.30 Breakdown of glucose by various ways

1.14.1 Alcoholic Fermentation:

Many fungi and some bacteria, algae and protozoa ferment sugars to ethanol and carbon di oxide by alcoholic fermentation process. The pyruvate is first decarboxylated to acetaldehyde in the presence of the enzyme pyruvate decarboxylated. Thiamine pyro phosphate (TTP) is required as co- factor in the reaction. Acetaldehyde is then reduced to ethyl alcohols by the enzyme alcohol dehydrogenase and coenzyme NADH (produced in glycolysis) is oxidized in this reaction.

$$2CH_3CO.CHOO \xrightarrow[\text{TPP}]{\text{Pyruvate decarboxylase}} 2CH_3CHO + 2CO_2$$

(Pyruvate (Acetaldehyde)

Alcohol dehydrogenase)

$$2CH_3CHO + 2NADH^+ \xrightarrow[\text{Ethyl alcohol}]{\text{Alcohol dehydroenase}} H^+2CH_3CH_2OH + 2NADH_2$$

(Acetaldehyde)

The overall equation for man aerobic respiration involving alcoholic fermentation is as follows:

$$C_6H_{12}O_6 \rightarrow 2CH_3CH_2OH + 2CO_2$$

(Glucose) (Alcohol)

1.14.2 Lactic Acid Fermentation:

Here, the pyruvate is converted into lactate by the enzyme lactate dehydrogenase. Coenzyme NADH (produced in glycolysis) is oxidized.

Lactate-dehydrogenase

$$2CH_3CO.COOH + 2NADH_2 \rightarrow 2CH_3CHOHCOOH + 2NAD^+$$

(Pyruvate) (Lactate)

The overall equation for anaerobic respiration involving lactic acid fermentation is as follows:

$$C_6H_{12}O_6 \rightarrow 2CH_3CHOH.COOH$$

(Glucose) (Lactate)

It can be separated into two groups such as (i) homolactic fermentation and (ii) heterolactic fermentation. Many bacteria especially members of family Enterobacteriaceae can metabolite pyruvate in to formic acid and other products in a process sometimes called the formic acid fermentation. Formic acid may be converted into H_2 and CO_2 by formic dehydrogenase. There are two types of formic acid fermentation results in the excretion of ethanol and a complex mixture of acids, particularly acetic, lactic, succinic and formic acid. The second type is butanediol form (*Enterobacter, Serratia, Erwinia* and *Bacillus sps*). Pyruvate is converted acetoin, while which is then reduced to 2,3 butanediol with NADH. A large amount of ethanol also produced together with small amount of acids found in a mixed acid fermentation.

1.15 AEROBIC RESPIRATION:

This type of respiration is completed in the presence of oxygen. It is completed in to following manner:

1.15.1 Glycolysis (Breakdown of Glucose):

Glycolysis (from the greek word gkykys, meaninig 'sweet' and lysis meaning splitting) is the process of the catabolism of glucose either in the presence of oxygen such as in the liver (aerobic glycolysis) or in the lack of O_2 such as in skeletal muscle (anaerobic glycolysis). The process of glycolysis is completed in the cytoplasm and this cycle is discovered by **Embeden Meyernof** and **Parnas,** so it is also known as **EMP Scheme** (Fig 1.31). It is completed in to two phases such as:

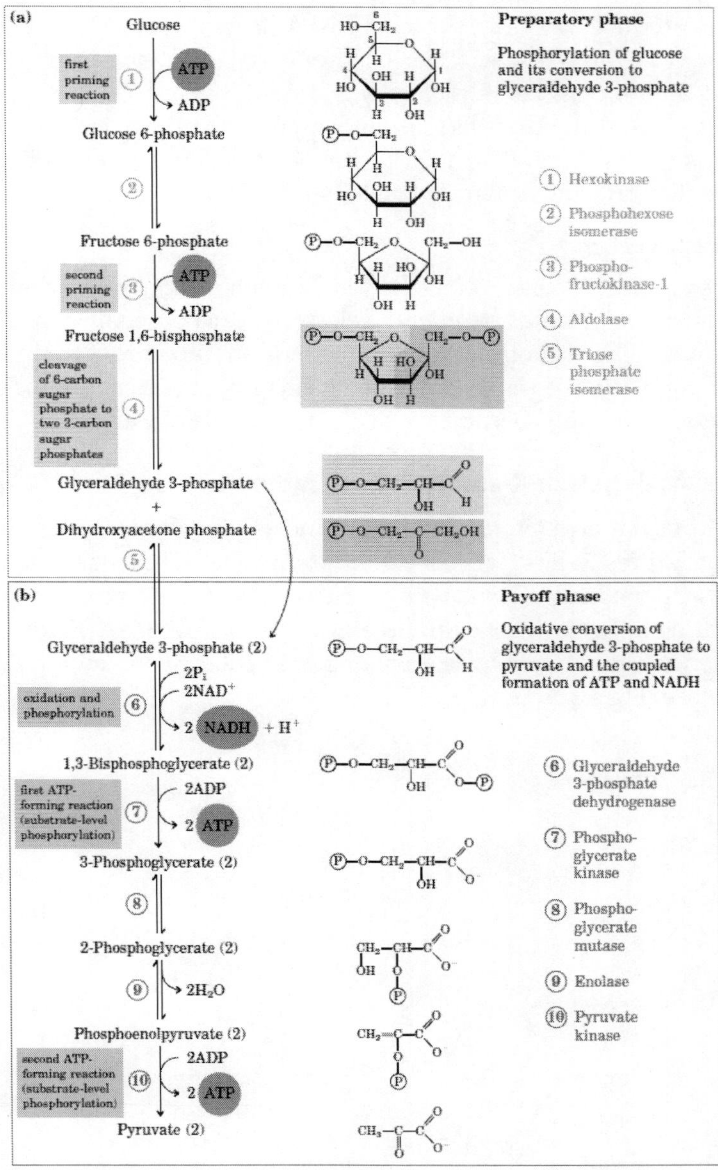

Fig. 1.31 Glycolysis (Break down of Glucose) or EMP Scheme

(i) Preparatory Phase

(ii) Pay off Phase

Energy Production during Glycolysis:

Total number of ATP produced during glycolysis varies, depending upon the conditions of the reactions and concentrations of the substrates.

Under Anaerobic Condition:

Since each of the energy-yielding steps, i.e. the conversion of 1, 3-bisphosphoglycerate to pyruvate, produces one ATP, a total of two ATP are produced for each molecule of the triose oxidized, i.e. 4 ATP per hexose. As two ATP are used, in the initial reactions of the process, i.e. in the conversion of glucose to glucose-6-phosphate and fructose-6-phoshpate to fructose-1, 6-disphoshpate, the net yield per molecule of glucose is only two ATP. In the erythrocytes, when glycolysis occurs via 2, 3-diphophoglycerate, the net energy yield is zero.

Under Aerobic Condition:

In this process, since NADH produced during the conversion of 3-phosphoglycerate to 1, 3-bisphosphoglycerate can enter the electron transport chain and release three ATP (during the process of oxidative phosphorylation), additionally 6 ATP are produced per molecule of glucose. The total yield under these conditions is therefore 10 ATP. The net yield thus becomes 8 ATP per molecule of glucose.

1.15.2 Citric Acid Cycle or Krebs Cycle or Tricarboxylic Acid Cycle (TCA)

Pyruvate or pyruvic acid (3C) is the end product of glycolysis. It does not enter the citric acid cycle directly. Firstly, this 3-carbon atom molecule changed into a 2-carbon atom acetic acid molecule. One carbon is released as CO_2 (decarboxylation). Acetic acid on entering the mitochondria unites with coenzyme A (CoA) to form acetyl-Co A. This reaction is catalyzed by several enzymes collectively known as pyruvate dehydrogenase (Fig 1.32).

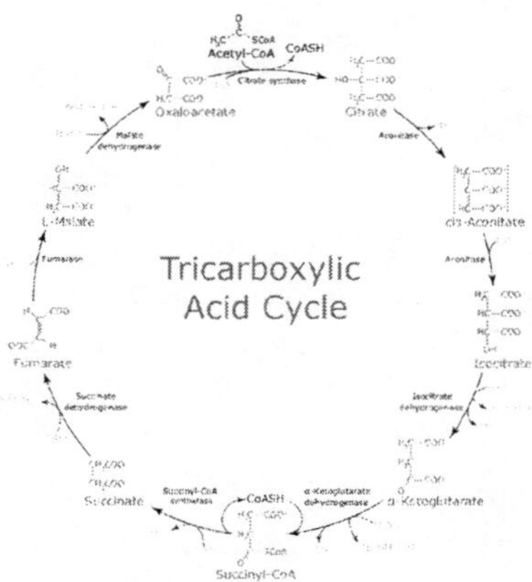

Fig. 1.32 Krebs Cycle or Tricarboxylic Acid Cycle

Pyruvate Dehydrogenase Complex (PDC) is a complex of three enzymes that converts pyruvate into acetye–COA by a process called pyruvate decarberay pation. Acetye COA may then beused in the litric acid oycle to carryout cellular respiration, and this complex links the glycolysis metabolic pathway to the citric acid cycle. Pyruvate decarboxylation is also knows the 'pyruvate dehydrogenase reaction' because it also involves the oxidation of pyruvate.

1.15.3 Electron Transport System or Chain:

It is now a well-established fact that the electron carries in the respiratory chain function in a specific sequence. The following evidences support the statement:

(a) Firstly, as expected their standard redox potentials are more positive going toward oxygen, since electrons tend to flow from electronegative to electropositive systems, causing decrease in free energy.

(b) Secondly, each member of the chain is specific for a given electron donor and acceptor. For instance, NADH can transfer electrons to NADH dehydrogenase, but cannot transfer them directly to cytochrome c.

(c) Lastly, four structured complexes of functionally related electron carriers have been isolated from mitochondria membrane.

1.16 RESPIRATION IN PLANTS:

Plants also need energy like animals. The plants get this energy by the process of respiration. Thus, the respiration in plants also involves the exchange of oxygen and carbon dioxide. The respiration in plants differs from that in animals in following three respects:

(a) All the parts of a plant like root, stem and leaves perform respiration individually, whereas an animal performs respiration as a single unit.

(b) A little transport of respiratory gases is found from one part of the plant to the other in plants, whereas, respiratory gases are usually transported over long distances inside an animal during respiration.

(c) The respiration in plants occurs at a slow rate, whereas in animals it occurs at a much faster rate.

Plants get Oxygen by Diffusion:

Plants have a branching shape, so they have quite a large surface area in comparison to their volume. Therefore, diffusion alone can supply all the cells of the plants with as much oxygen as they need for respiration. Diffusion occurs in the roots, stems and leaves of plants.

1.16.1. Respiration in Roots:

Air is present in between the particles of soil. The roots of a plant take the oxygen required for respiration from the air present in between the soil particles by the process of diffusion. The extensions of the epidermal cells of a root are called root hair, which are in contact with the air in the soil. Oxygen diffuses into root hairs and reaches all the other cells of the root for respiration. Carbon-di-oxide gas produced in the cells of the root during respiration moves out through the same root hairs by the process of diffusion. Thus, the respiration in roots occurs by the diffusion of respiratory gases through the root hairs. If their roots remain waterlogged for a considerable time, the land plants die. This is because too much water expels all the air from in between the soil particles (Fig 1.33). Due to this, oxygen is not available to the roots for aerobic respiration. Under these conditions, the roots will respire anaerobically, producing alcohol. This may kill the plant. In order to understand the respiration in stems of plants we should remember that the soft stems of small, herbaceous plants have stomata in them whereas the hard and woody stems of large plants and trees have lenticels in them. Lenticel is a small area of bark in a woody stem where the cells are loosely packed allowing the gaseous exchange to take place between the air and the living cells of the stem.

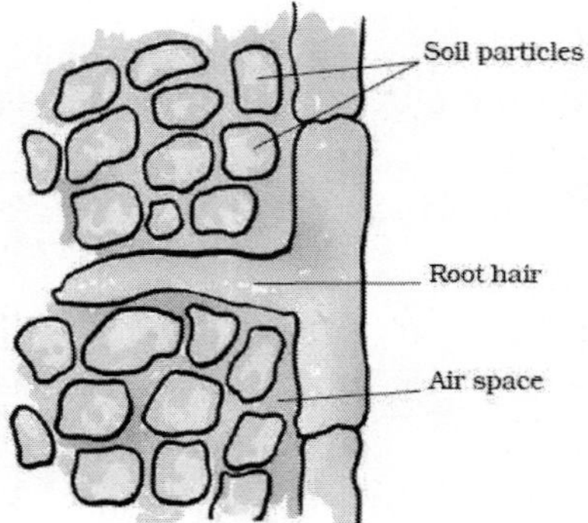

Fig. 1.33 Roots absorb oxygen from air present between soil particles through root hairs

1.16.2. Respiration in Stems:

The stems of herbaceous plants have stomata. So, the exchange of respiratory gases in the stems of herbaceous plants takes place through stomata. The oxygen from air diffuses into the stem of herbaceous plant through stomata and reaches all the cells for respiration. The CO_2 gas produced during respiration diffuses out into the air through the same stomata. The hard and woody stems of big plants or trees do not have stomata. In woody stems, the bark (outer covering of stem) has lenticels for gaseous exchange. The oxygen from air diffuses into the stem of a woody plant through lenticels and reaches all the inner cells of the stem for respiration (Fig 1.34). The carbon dioxide gas produces in the cells of the stem during respiration diffuses out into the air through the same lenticels.

Fig. 1.34 Woody stems of plants or trees having lenticles for respiratory gases exchange.

1.16.3. Respiration in Leaves:

The leaves of a plant have tiny pores called stomata. The exchange of respiratory gases in the leaves takes place by the process of diffusion through stomata. Oxygen from air diffuses into a leaf through stomata and reaches all the cells where it is used in respiration. The carbon dioxide produced during respiration diffuses out from the leaf into the air through the same stomata (Fig 1.35). It should be noted that respiration in leaves occurs during the day time as well as at night. On the other hand, photosynthesis occurs only during the day time (no photosynthesis occurs at night). Due to this, the net gaseous exchange in the leaves of a plant is as follows:

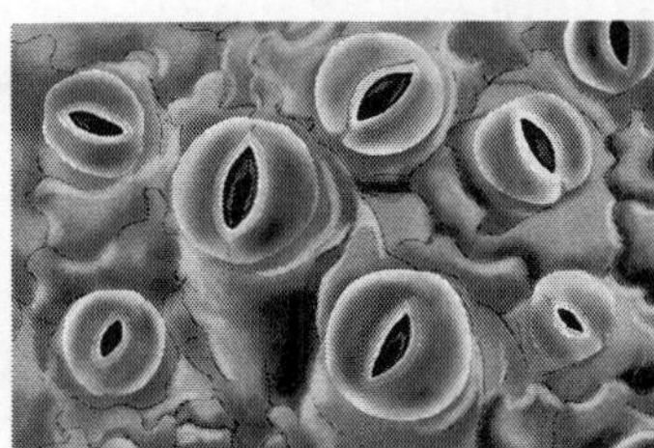

Fig. 1.35 Exchange of Respiratory gases in leaves through stomata

(i) **During the day time,** when photosynthesis occurs, oxygen is produced. The leaves use some of this oxygen for respiration and the rest of oxygen diffuse out into air. Again, during day time, carbon-di-oxide produced by respiration is all used up in photosynthesis by leaves. Even more carbon-di-oxide is taken in from air. Thus, the net gas exchange in leaves during day time is: O_2 diffuses out; CO_2 diffuses in.

(ii) **During the night time,** when no photosynthesis occurs and hence no oxygen is produced, oxygen from air diffuses into leaves to carry out respiration, and the carbon-di-oxide produced by respiration diffuses out into air. So, the net gas exchange in leaves at night is: O_2 diffuses in; CO_2 diffuses out.

1.17 RESPIRATION IN ANIMALS:

Different animals have different modes of respiration as:

(a) By Cell Membrane: In simple unicellular animals like Amoeba, respiration takes place by the simple diffusion of gases through the cell membrane.

(b) **By Skin:** The animals like earthworms which live in the soil use their skin to absorb oxygen from air and remove carbon dioxide. So, the respiratory organ in the earthworm is the skin.

(c) **By Gills:** The aquatic animals like fish, prawns and mussels have gills as the respiratory organs which extract oxygen dissolved in water and take away carbon dioxide from the body.

(d) **By Spiracles:** In the insects like grasshopper, cockroach, housefly and a mosquito, the tiny holes called spiracles on their body and the air tubes called tracheae are the respiratory organs.

(e) **By Lungs:** The respiratory organs of the land animals such as man (humans), birds, lizard, dog, frogs etc. have the lungs. (Frogs, however, breathe both by lungs and skin).

Once the air (oxygen) enters the skin or lungs, blood absorbs the oxygen and transports it to various tissues of the animal. Blood also picks up the carbon-di-oxide from the tissues and brings it back to the skin or lungs for throwing it out into the air. All the respiratory organs whether skin, gills, trachea or lungs have following three common features such as:

- Have a large surface area to get enough oxygen.
- Have thin walls for easy diffusion and exchange of respiratory gases.
- Have a rich blood supply for transporting respiratory gases (only in the tracheal system of respiration, air reaches the cells directly) in the case of skin, gills and lungs.

The animals which live in water (known as aquatic animals) use the oxygen dissolved in water to carry out respiration. Since the amount of dissolved oxygen in water is low as compared to the amount of oxygen in the air, therefore, the rate of breathing in aquatic animals in much faster than in terrestrial or land animals. A faster rate of breathing provides more oxygen to the aquatic animal. The terrestrial animals use the oxygen of air or atmosphere for breathing and respiration. Thus, a terrestrial animal has an advantage over an aquatic animal in regard to obtaining oxygen for respiration that it is surrounded by an oxygen-rich atmosphere from where it can take any amount of oxygen.

1.17.1 Respiration in *Amoeba*:

Amoeba is a single-celled animal, and lives in water. It depends on simple diffusion of gases for breathing. The diffusion of gases takes place through the thin cell membrane of *Amoeba*. The oxygen from water diffuses into the body of *Amoeba* through its cell membrane since the *Amoeba* is very small in size, so the oxygen spreads quickly into the whole body of *Amoeba* (Fig 1.36). This oxygen is used for respiration (energy release) inside the *Amoeba* cell. The process of respiration produces carbon-di-oxide gas continuously. This carbon-di-oxide gas diffuses out through the membrane of *Amoeba* into the surrounding water. Thus, the breathing or respiratory surface of *Amoeba* is its cell surface membrane.

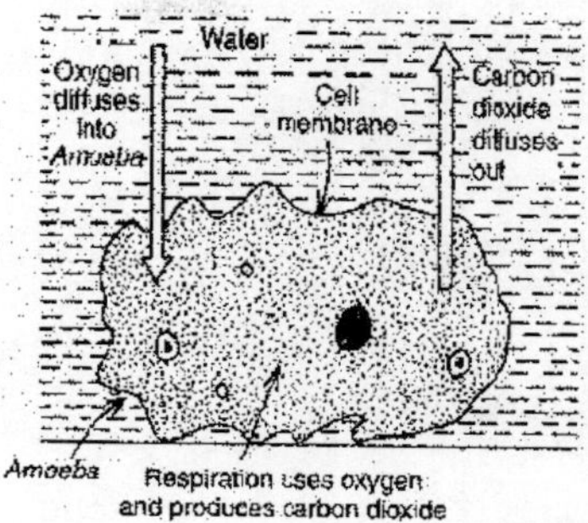

Fig. 1.36 Process of Gaseous exchange or breathing in *Amoeba*

In addition to *Amoeba*, other simple animals like Paramecium and Planaria also breathe through their cell membranes and depend on the simple diffusion of gases for breathing and respiration. As we have already discussed that oxygen can be introduced quickly into the whole body of *Amoeba* by the process of diffusion due to the smallness of *Amoeba* cell. But this is not so in larger animals like earthworms, grasshopper, fish and man (humans), etc. in large animals, the volume of body is so big that oxygen cannot diffuse into all the cells of the body quickly. This is because in these cases the oxygen has to travel a very large distance to reach each and every cell of the body. So, in large animals, there is a blood circulatory system to carry oxygen to all the parts of the body quickly and remove carbon-di-oxide. The blood contains respiratory pigments which take up oxygen from air and carry it to the body cells.

1.17.2 Respiration in Fish:

The fish has special organs of breathing called gills; those are found on both the sides of its head. These are not visible from outside due to covering from gill cover. The fish breathes by taking in water through its mouth and sending it over the gills. When water passes over the gills, the gills extract dissolved oxygen from this water. The water then goes out through the gill slits (hidden under the gill cover). Thus, the dissolved oxygen is extracted from water by the fish when it flows over the gills (Fig 1.37). The extracted oxygen is absorbed by the blood and carried to all the parts of the fish. The carbon-di-oxide produced by respiration is brought back by the blood into the gills for expelling into the surrounding water. The fish has no lungs like human beings; the gaseous exchange in fish takes place in the gills. So, the respiratory surface of a fish is the surface of its gills. It is a common observation that when fish is taken out from water it dies soon (even though there is a lot of oxygen in the air around it). This is because a fish does not have lungs to utilize the oxygen of air for breathing and respiration. The fish has gills which can extract only dissolved oxygen from water and provide it to fish. Gills cannot take in the oxygen from air on land. Since fish does not get oxygen for breathing when taken out of water, it dies. In addition to fish, many other aquatic animals like prawns and mussels also have special organs called 'gills' for breathing and respiration.

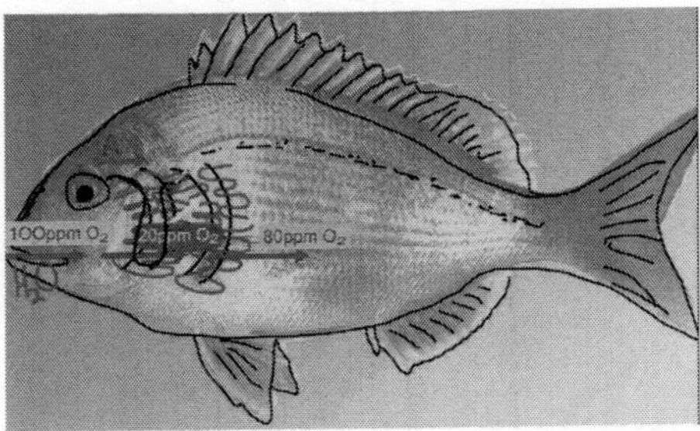

Fig. 1.37 Fish in water showing gills which for extraction of dissolved oxygen from water

1.17.3 Respiration in Earthworm:

The earthworm exchanges the gases through its skin. This means that the respiratory surface of an earthworm is its skin. The skin of an earthworm is quite thin and moist, and has a good blood supply. So, the earthworm absorbs the oxygen needed for respiration through its moist skin (Fig 1.38). The oxygen is then transported to all the cells of the earthworm by its blood where it is used in respiration. The carbon dioxide produced during respiration is carried back by the blood. This carbon dioxide is expelled from the body of the earthworm through its skin which is thin and moist. Just like earthworms, the leeches also absorb the oxygen needed for respiration through their skin, and carbon dioxide produced inside the leeches during respiration also goes out through the skin.

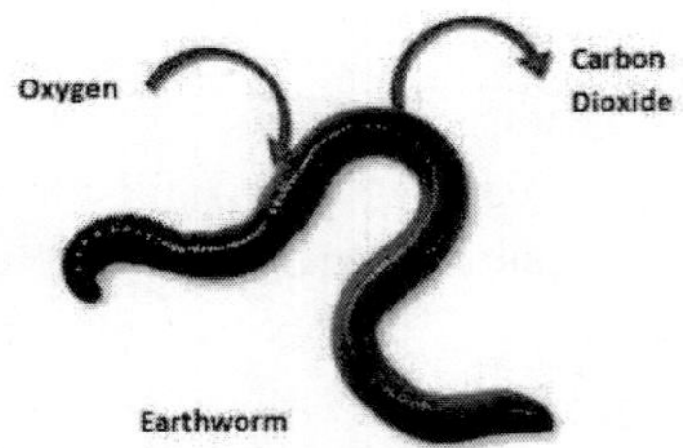

Fig. 1.38 Gaseous exchange in Earthworm

1.17.4 Respiration in Humans:

Like other land animals, human beings are air breathers. The human beings need oxygen to stay alive and get this oxygen by breathing in air. The oxygen helps to break down the food absorbed in the body to release energy. This energy is used for maintaining our life. The process by which energy is released from food in our body is called respiration. Carbon-di-oxide and water are the two by-products of respiration. The process of respiration takes place inside the cells of our body. It involves our respiratory system. The function of respiratory system is to breathe in oxygen for respiration (producing energy from food), and to breathe out carbon dioxide produced by respiration. The breathing organs of human beings are lungs. It is in the lungs that the gases are exchanged between the blood and air (Fig 1.39).

The gases exchanged between blood and air is oxygen and carbon dioxide.

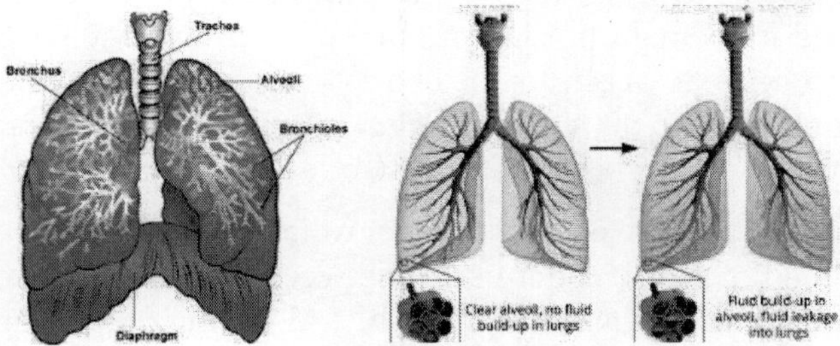

Fig. 1.39 Human Lungs

Breathing is the process by which air rich in oxygen is taken inside the body of an organism and air rich in carbon dioxide is expelled from the body with the help of breathing organs. The breathing in human beings takes place though the organs called lungs. The taking in of air rich in oxygen into the body during breathing is called inhalation and giving out (or expelling) the air rich in carbon dioxide is called exhalation. Both, inhalation and exhalation take place regularly during breathing. A breath means 'one inhalation plus one exhalation'. Thus, we 'breathe in' air to supply oxygen to the cells of our body for the breakdown of food to release energy, and we 'breathe out' to remove waste product carbon dioxide from our body which is produced during the breakdown of food in the cells.

1.17.5 Mechanism of Breathing:

The process of breathing takes place in our lungs. Lungs are connected to our nostrils (holes in the nose) through nasal passage (or nasal cavity) and windpipe. When we inhale air, it enters our nostrils, passes through nasal passage and windpipe, and reaches our lungs. Our two lungs hang in an airtight space in our body called chest cavity. Around the sides of the chest cavity is the rib cage with sheets of muscles between the ribs. The rib cage encloses the lungs in it. At the bottom of the chest cavity is a curved sheet of muscle called diaphragm (Fig 1.40). Diaphragm forms the floor of chest cavity. Breathing involves the movements of the rib cage the diaphragm. This happens as follows:

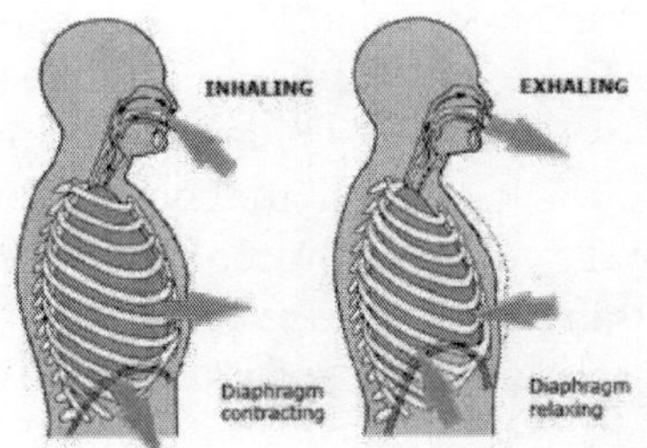

Fig. 1.40 Mechanism of Breathing

(a) **Breathing in:** When we breathe in (or inhale), then two things happen at the same time as (i) muscles between the ribs contract causing the rib cage to move upward and outward, and (ii) the diaphragm contracts and moves downward. The upward and outward movement of rib cage, as well as the downward movement of diaphragm, both increase the space in the chest cavity and make it larger. As the chest cavity becomes larger, air is sucked in from outside into the lungs. The lungs get filled up with air and expand (Fig 1.40).

(b) **Breathing out:** When we breathe out (or exhale), even then two things happen at the same time as (i) the muscles between the ribs relax causing the rib cage to move downward and inward, and (ii) the diaphragm relaxes and moves upward. The downward and inward movement of rib cage, as well as the upward movement of diaphragm, both decrease the space in out chest cavity and make it smaller. As the chest cavity becomes smaller, air is pushed out from the lungs(Fig 1.40).

1.18 RESPIRATORY SYSTEM IN HUMANS (MAN):

In human beings, many organs take part in the process of respiration (Fig 1.41). The main organs of human respiratory system are as Nose, Nasal passage or cavity, Trachea, Bronchi, Lungs and Diaphragm. The human respiratory system begins from the nose. Our nose has two holes in it which are called nostrilsgure. There is a passage in the nose behind the nostrils which is called nasal passage or cavity. The air for respiration is drawn into our body through the nostrils present in the nose. This air then goes into nasal passage. The nasal passage is separated from the mouth cavity known as buccal or oral cavity by a hard, bony palate so that we can breathe in air even when we are eating food (and the mouth cavity is filled with food). The nasal passage is lined with fine hair and mucus. Mucus is secreted by the glands inside the nasal passage. When air passes through the nasal passage, the dust particles and other impurities present in it are trapped by nasal hair and mucus so that clean air goes into the lungs. The part of throat between the mouth and wind pipe is called pharynx. From the nasal passage, air enters into pharynx and then goes into the wind pipe (or trachea). The trachea is a tube which is commonly known as wind pipe. The air coming from the nostrils during breathing passes through trachea. Trachea does not collapse even when there is no air in it because it is supported by rings of soft bones called cartilage. The upper end of trachea has a voice box called larynx. The trachea runs down the neck and divides into two smaller tubes 'bronchi' at its lower end. (singular bronchus). The two bronchi are connected to the two lungs. The lungs lie in the chest cavity or thoracic cavity which is separated from abdominal cavity by a muscular partition called diaphragm. The lungs are covered by two thin membranes called pleura. The lungs are enclosed in a 'rib cage' made of bones called 'ribs'. Each bronchus divides in the lungs to form a large number of still smaller tubes called 'bronchioles'. The smallest bronchioles have tiny air-sacs at their ends. The pouch-like air-sacs at the ends of the smallest bronchioles are called 'alveoli' (singular alveolus). The walls of alveoli are very thin and they are surrounded by very thin blood capillaries. It is in the alveoli that oxygen is taken into the body and carbon dioxide is eliminated.

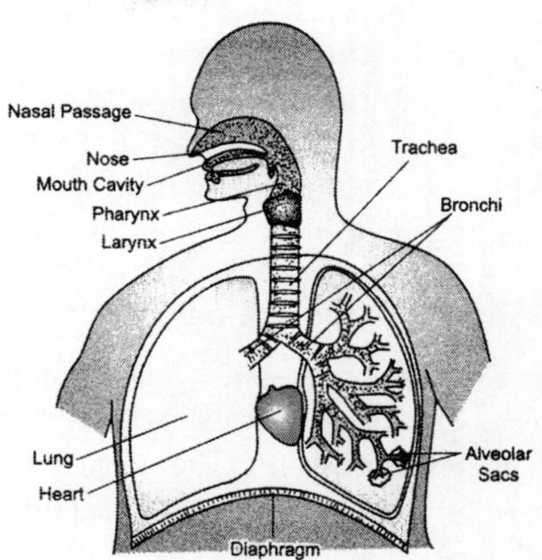

Fig. 1.41 The Human Respiratory System

Human lungs have been designed to maximize the exchange of gases as follows:

There are millions of alveoli in the lungs. The presence of millions of alveoli in the lungs provides a very large area for the exchange of gases and the availability of large surface area maximizes the exchange of gases. For example, if all alveoli from the two human lungs are unfolded, they would give an area of about 80 mtr^2 (which is nearly the size of a tennis court!). The diaphragm is a sheet of muscle below the lungs. It helps in 'breathing in' and 'breathing out'. The muscles of chest also help in breathing in and breathing out. When we breathe in air, the diaphragm and muscles attached to the ribs contract due to which our chest cavity expands. This expansion movement of the chest increases the volume inside the chest cavity. Due to increase in volume, the air pressure decreases inside the chest cavity and air from outside (being at higher pressure) rushes into the lungs through the nostrils, trachea and bronchi. In this way, during the process of 'breathing in' the air sacs or alveoli of the lungs get filled with air containing oxygen. The alveoli are surrounded by very chin blood vessels called capillaries carrying blood in them. So, the oxygen of air diffuses out from the alveoli walls into the blood. The oxygen is carried by blood to all the parts of the body by a red pigment called haemoglobin present in blood. As the blood passes through the tissues of the body, the oxygen present in it diffuses into the cells (due to its higher concentration in the blood). This oxygen combines with the digested food (glucose) present in the cells to release energy. Carbon dioxide gas is produced as a waste product during respiration in the cells of the body tissue. This carbon dioxide diffuses into the blood (due to its higher concentration in body tissues). Blood carries the carbon dioxide back to the lungs where it diffuses into the alveoli. When we breathe out air, the diaphragm and the muscles attached to the ribs relax due to which our chest cavity contracts and becomes smaller. This contraction movement of the chest pushes out carbon dioxide from the alveoli of the lungs into the trachea, nostrils and then out of the body into air. In this way the process of gaseous exchange is completed in the human respiratory system. During the breathing cycle, when air is taken in (or inhaled) and let out (or exhaled), the lungs always contain a certain residual volume of air so that there is sufficient time 'for the oxygen to be absorbed' into the blood and 'for the carbon dioxide to be released' from the blood. Another point to be noted is that carbon dioxide is more soluble in water (than oxygen), so it is mostly transported in the dissolved form in our blood.

1.18.1 Experiment to show that Carbon-di-oxide is produced during Respiration:

We know that carbon dioxide gas turns lime-water milky. The fact that carbon-di-oxide is produced during respiration can be shown by demonstrating the effect of inhaled air and exhaled air on lime-water. The apparatus consists of two boiling tubes A and B fitted with two-holed corks. The boiling tubes A and B are connected through a special type of glass tube C. The left arm of glass tube C is short which goes in the boiling tube A. The right arm of glass tube C is long and dips in lime-water in boiling tube B. The boiling tube A has another bent glass tube D whose longer side dips in lime-water contained in it. The boiling tub B has also another short, bent tube E in it which does not dip in lime-water (Fig 1.42).

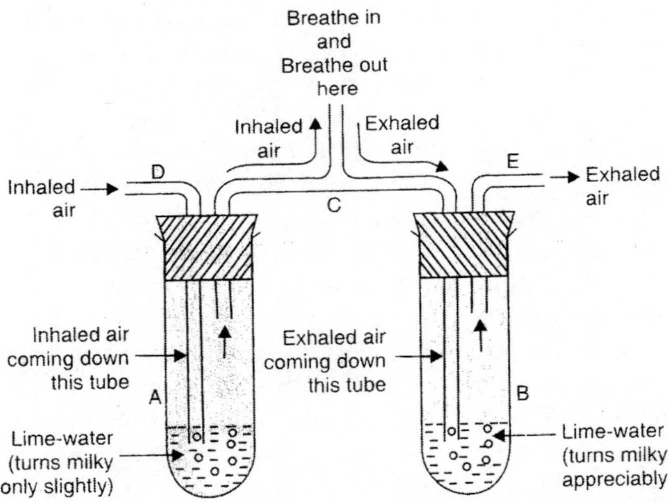

Fig. 1.42 Diagram shows Test inhaled air and exhaled air for Carbon dioxide

To perform the experiment, we put the top end of the tube C in mouth and 'breathe in' and 'breathe out' gently. When we breathe in, then the inhaled air (fresh air) enters the glass tube D and passes through the lime-water in boiling tube A. When we breathe out, then the exhaled air (coming from our lungs) passes through the lime-water in boiling tube B. We continue to breather in and breathe out for about 5 minutes. We will find that the lime water in boiling tube A (in which inhaled air is passed) turns milky only slightly but the lime-water in boiling tube B (in which exhaled air is passed) turns milky appreciably. It shows that less carbon dioxide is present in inhaled air but much more carbon dioxide is present in exhaled air. We conclude from this observation that carbon dioxide is produced during respiration (which comes out in exhaled air).

The air which we 'inhale' is a mixture of gases and the air which we 'exhale' is also a mixture of gases. The only difference in the inhaled air and exhaled air is that they contain different proportions of oxygen, carbon dioxide and water vapour. The proportion of nitrogen gas in the inhaled air and exhaled air remains the same, 78%, because it is neither used up in respiration nor produced during respiration. The proportions of oxygen, carbon dioxide and water vapour in 'inhaled air' and 'exhaled air' are given in below table:

Components	Inhaled Air	Exhaled Air
Oxygen	21%	16.4%
Carbon di oxide	0.04%	4.4%
Vapors	A little	A lot

These figures show that the air which we inhale contains a greater proportion (21%) of oxygen. Now, some of the oxygen of inhaled air is used up in breaking down glucose food during respiration, so the exhaled air which comes out after the process of respiration contains a lower proportion (16.4%) of oxygen. The air which we inhale contains a lower proportion (0.04%) of carbon dioxide. Now, during respiration, when oxygen breaks down glucose food, then a lot of carbon dioxide is produced, so the exhaled air which comes out after respiration contains a much higher proportion (4.4%) of carbon-di-oxide. Again, the air which we inhale contains only a little of water vapour. Now, when glucose food is broken down by oxygen during respiration, then water is also produced (along-with carbon-di-oxide), so, the exhaled air contains a lot more water vapour than inhaled air.

1.18.2 Rate of Breathing:

The process of breathing pumps in oxygen into our body (and removed carbon dioxide). Breathing occurs involuntarily (on its own) but the rate of breathing is controlled by the respiratory system of brain. The average breathing rate in an adult man at rest is about 15 to 18 times per minute. This breathing rate increases with increased physical activity. For example, if we do some physical exercise (like sit-up exercise), then our breathing rate goes up considerably. This is because when we do some physical exercise, then our body needs more energy, and to produce more energy through respiration, our body requires more oxygen gas. Rapid breathing supplies more oxygen to body cells for producing more energy required for doing physical exercise. Thus, we breathe faster after exercise so as to produce more energy to compensate the loss of energy suffered by our body in doing exercise. We all breathe through nose. We can, however, not breathe inside water when we are diving. This is because water does not have free air or oxygen for us to breathe (and we do not have gills like the fish to utilize oxygen dissolved in water). So, the deep sea divers carry oxygen gas cylinders with them for breathing when they go under the sea. We have just studied that oxygen required for breathing and respiration (release of energy) is carried by haemoglobin present in our blood. The normal range of haemoglobin in the blood of a healthy adult person is from 12 to 18 gms per deciliter (g/dL) of blood. The deficiency of haemoglobin in the blood of a person reduces the oxygen-carrying capacity of blood resulting in breathing problems, tiredness and lack of energy. The person looks pale and loses weight. Many times we have heard of carbon monoxide poisoning.

Carbon monoxide (CO) gas is formed whenever a fuel burns in an insufficient supply of air. For example, if coal or charcoal is burned in a closed space (like a room with closed doors and windows), then a lot of carbon monoxide is formed. Carbon monoxide is also produced when petrol burns in a car engine. Now, we know that haemoglobin present in our blood carries oxygen to all the parts of our body. Haemoglobin has more affinity (or attraction) for carbon monoxide than oxygen. So, if carbon monoxide gas is inhaled by a person, then this carbon monoxide binds very strongly with haemoglobin in the blood and prevents it from carrying oxygen to the brain and other parts of the body. Due to lack of oxygen, the person cannot breathe properly. If carbon monoxide is inhaled for a long time, then the person becomes unconscious and can even die due to oxygen starvation. The persons having breathing problems (or respiratory problems) are given oxygen masks to facilitate breathing. In serious cases, the patient is put on a machine called 'ventilator' in which a tube is inserted directly into the trachea (or wind pipe) of the patient to help him in breathing comfortably.

1.19 TRANSPORT IN PLANTS:

Transport system in plants is less elaborate than in animals including human beings. Plants are less active, so their cells do not need to be supplied with materials so quickly. All the cells of a plant can get oxygen for respiration and carbon dioxide for photosynthesis directly from the air by diffusion due to the branching shape of a plant. Therefore, the only substances which are to be supplied to a plant through a transport system are water and minerals (Fig 1.43).

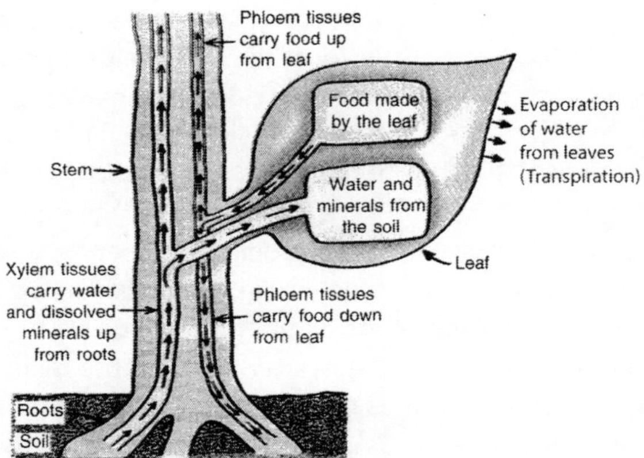

Fig. 1.43 Transport in Plants

1.19.1 Types of Conducting Systems in Plants:

The plants have two types of conducting tissues called xylem and phloem .

- Xylem which carries water and minerals in plants, and
- Phloem which carries the food materials which the plant makes (Phloem also carries the hormones made by the plants in their root and shoot tips).

1.19.2 Transport of Water and Minerals:

Plants required water for making food by photosynthesis. Plants also need mineral salts for various purposes such as making proteins, etc. Water and minerals are absorbed from the soil by the roots of the plant and transported to the various parts of the plant like stem, leaves and flowers. In the leaves, water is used in making food by photosynthesis. The water and minerals dissolved in it move from the roots of the plant to its leaves through the two kinds of elements of the xylem tissue called xylem vessels and tracheids. Xylem vessels and tracheids are both non-living conducting tissues, which have thick walls.

1. Xylem Vessels:

The xylem vessel is a non-living, long tube which runs like a drainpipe through the plant. It is made of many hollow, dead cells (called vessel elements), joined end to end. The end walls of the cells have broken down so a long, open tube is formed. Xylem vessels run from the roots of the plant right up through the stem and reach the leaves. The xylem vessels branch into every leaf of the plant Xylem vessels do not contain the cytoplasm or nuclei. The walls of xylem vessels are made of cellulose and lignin. Lignin is a very hard and strong substance, so xylem vessels also provide strength to the stems and help to keep the plant upright. Wood is made almost entirely of lignified xylem vessels. Xylem vessels have pits in their thick cell walls. Pits are not open pores. Pits are the thin areas of the cell wall where no lignin has been deposited. Pits have unthickened cellulose cell wall. In flowering plants, either only xylem vessels transport water or both xylem vessels and tracheids transport water (Fig 1.44A).

2. Tracheids:

Tracheids are long, thin, spindle shaped cells with pits in their thick cells walls. Water flows from one tracheid to another through pits. Tracheids are dead cells with lignified walls but they do not have open ends, so they do not form vessels. They are elongated cells with tapering ends. Even though their ends are not completely open, tracheids have pits in their walls, so water can pass from one tracheid to another through these pits. Although all the plants have tracheids, they are the only water conducting tissue in non-flowering plants (Fig 1.44B).

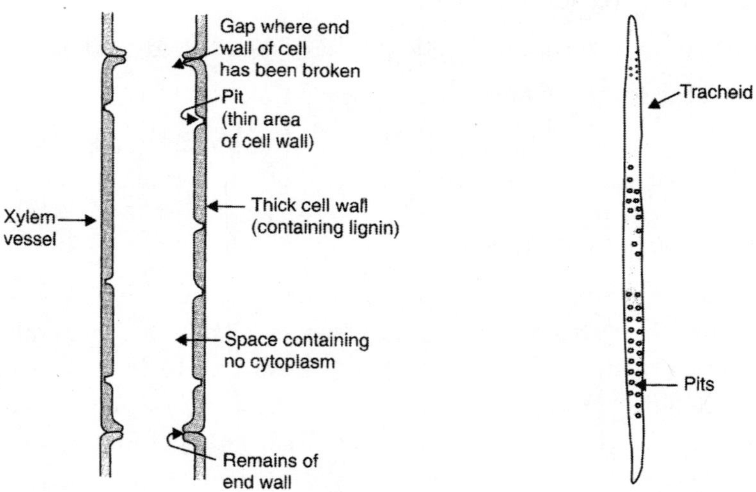

Fig. 1.44 A Xylem vessel **Fig. 1.44 B Tracheid**

1.19.3 Mechanism of Transport of Water and Minerals in a Plant:

The outer layer of cells in the root is called epidermis. Epidermis is only one cell thick. The layer of cells around the vascular tissues (xylem and phloem) in the root is called endodermis. The part of root between the epidermis and endodermis is called root cortex. And the xylem tissue present in roots is called root xylem. Please note that in a root, the root hair are at its outer edge but the root xylem vessels (which carry water to the other parts of plant) are at the centre of the root, and in-between the root hair and root xylem are epidermis, root cortex and endodermis. So, before water absorbed by root hair from the soil reaches the root xylem, it has to pass through the epidermis, root cortex and endodermis. It is also to be noted that the minerals needed by the plants are taken up by the plants in inorganic form such as nitrates and phosphates. These minerals are present in the soil. The minerals present in soil dissolve in water to form an aqueous solution. So, when water is transported by the root of the plant to its leaves, then the minerals dissolved in it also transported along with it.

The plants take in water (containing dissolved minerals) from the soil through their roots. This water containing minerals called xylem sap is carried by the xylem vessels to all the parts of the plant. The roots of a plant have hair called root hairs. The function of root hairs is to absorb water and minerals from the soil. The root hairs are directly in contact with the film of water in-between the soil particles. Water and dissolved minerals get into the root hairs by the process of diffusion. The water and minerals absorbed by the root hair from the soil pass from cell to cell by osmosis through the epidermis, root cortex; endodermis and reach the root xylem. The xylem vessels of the root of the plant are connected to the xylem vessels of its stem. So, the water enters from the root xylem vessels into stem xylem vessels. The xylem vessels of the stem branch into the leaves of the plants. So, the water and minerals carried by the xylem vessels in the stem reach the leaves through the branched xylem vessels which enter from the petiole (stalk of the leaf) into each and every part of the leaf. In this way, the water and minerals from the soil reach through the root and stem to the leaves of the plant (Fig 1.45). Only about 1 to 2% of the water absorbed by the plant is used up by the plant in photosynthesis and other metabolic activities. The rest of water is lost as water vapour to the air through transpiration.

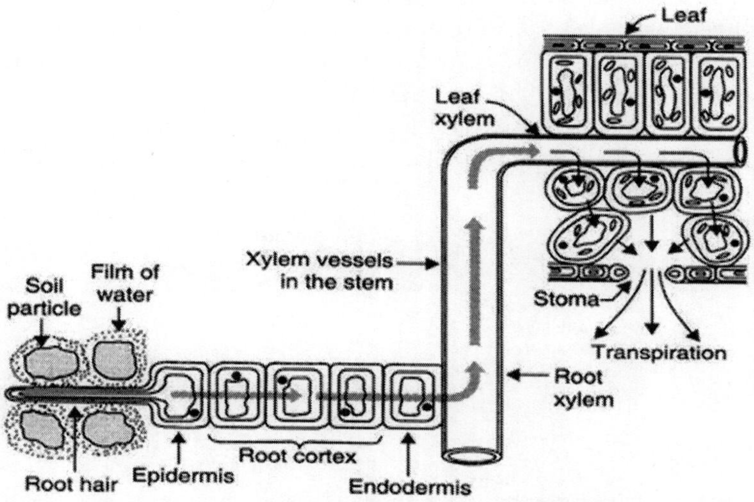

Fig. 1.45 Mechanism of Transport of Water and Minerals in a Plant

1.19.4 Water is sucked up by the Xylem Vessels:

Water moves up into xylem vessels in the same way that a cold drink moves up a straw when we suck at the upper end of the straw. Now, when we suck a straw, we are reducing the pressure at the top of the straw. The cold drink at the bottom of the straw is at a higher pressure (which is atmospheric pressure), so the cold drink flows up the straw into our mouth. The same thing happens with the water in the xylem vessels. The pressure at the top of the xylem vessels (in the leaves) is lowered whereas the pressure at the bottom of the xylem vessels remains high. Due to this water flows up the xylem a vessel into the leaves.

1.19.5 Transport of Food and Other Substances:

The food made by leaves is in the form of simple sugar (glucose)by the process of photosynthesis. Other types of substances called plant hormones are made in the tips of roots and shoots. Now, every part of the plant needs food. So, food made in the leaves of a plant has to be transported to all the parts of the plant like branches, stem and roots, etc. The food manufactured by the leaves of a plant is transported to its all other parts through a kind of tubes called phloem (which are present in all the parts of a plant). The transport of food from the leaves to other parts of the plant is called translocation. Thus, phloem translocates the food (or sugar) made in the leaves. The movement of food materials (and other substances like hormones) through phloem depends on the action of living cells called sieve tubes.

Phloem Sieve Tubes:

Like xylem vessels, phloem is made of many cells joined end to end to form long tubes. However, the end walls of the cells which form phloem are not completely broken down. The end walls of cells in the phloem from sieve plates, which have small holes in them. These holes in the sieve plates allow the food to pass along the phloem tubes. The cells of phloem are called sieve tubes or elements. Sieve tubes which form phloem are living cells which contain cytoplasm but no nucleus. The sieve tube cells do not have lignin in their walls. Each sieve tube cell has a companion cell next to it. The companion cell has a nucleus and many other organelles. Companion cells supply the sieve tubes with some of their requirements (Fig 1.46).

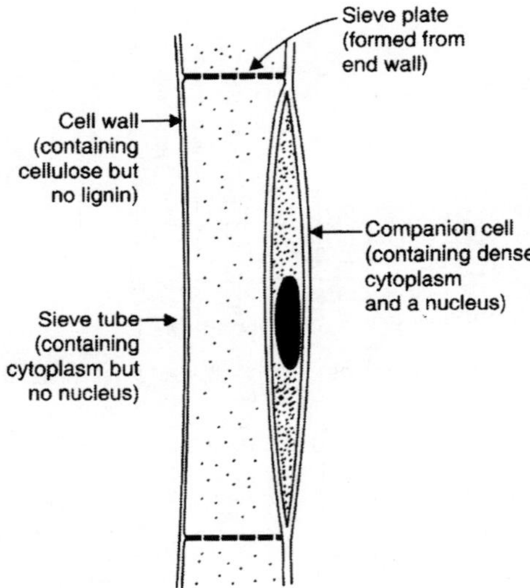

Fig. 1.46 Structure of Phloem containing Sieve tubes

The food is made in the mesophyll or photosynthetic cells of a leaf. The food like sugar made by the mesophyll cells of a leaf enters into the sieve tubes of the phloem. Interconnected phloem tubes are present in all the parts of the plant. So, once the food like sugar enters the phloem tubes in the leaves, it is transported to all other parts of the plant by the network of phloem tubes present in all the parts of the plant like stem and roots. The translocation (transport of food from leaves to other parts of the plant) is necessary because every part of the plant needs food for obtaining energy, for building its parts and maintaining its life. Please note that when food is transported in a plant through a network of phloem tubes, then other substances made by the plant like hormones are also carried by the phloem tubes from one part of the plant to its other parts.

We have already studied that the movement of water and dissolved salts is xylem is always upwards from soil to leaves and it is caused by the suction of water at the top because of low pressure created by transpiration from leaves. The movement of food in phloem can be, however, upwards or downwards depending on the needs of the plant.

1.19.6 Mechanism of Transport of Food in a Plant:

The movement of food in phloem or translocation takes place by utilizing energy. This happens as follows: The food made in leaves is loaded into the sieve tubes of phloem tissue by using energy from ATP. Water now enters into sieve tubes containing sugar by the process of osmosis due to which the pressure in the phloem tissue rises. This high pressure produced in the phloem tissue moves the food to all the parts of the plant having less pressure in their tissues. This allows the phloem to transport food according to the needs of the plant. For example, in spring, even the sugar stored in the root or stem tissue of a plant would be transported through phloem to the buds which need energy to grow.

Transport system in plants is less elaborate than in animals including human beings. Plants are less active, so their cells do not need to be supplied with materials so quickly. The plants have two types of conducting tissues called xylem and phloem, those help in the transport of water and mineral (by xylem) and food material (by phloem). The movement of food in phloem or translocation takes place by utilizing energy.

Discussion:

How is the pressure at the top of the xylem vessels reduced? The pressure at the top of xylem vessels in a plant is reduced due to transpiration.

The evaporation of water from the leaves of a plant is called transpiration. The leaves of a plant have tiny pores on their surface which are called stomata. A lot of water from the leaves keeps on evaporating into the air through the stomata. This loss of water as water vapour from the leaves of a plant is called transpiration. Since the cells of the leaf are losing water by transpiration, so water from the xylem vessels in the leaf will travel to the cells by osmosis to make up this loss of water. Thus, water is constantly being taken away from the top of the xylem vessels in the leaves to supply it to the cells in the leaves. This reduces the effective pressure at the top of the xylem vessels, so that water flows up into them (from the soil). Thus, the continues evaporation of water (or transpiration) from the cells of a leaf creates a kind of suction which pulls up water through the xylem vessels. In this way, the process of transpiration helps in the upward movement of water (and dissolved minerals) from the roots to the leaves through the stem.

1.20 EXCRETION:

All the organisms (plants and animals) are made up of cells. These cells work all the time for sustaining the life of the organism. Most of the work of the cells is in the form of biochemical reactions which they carry out all the time. The biochemical reactions taking place in the cells of an organism may produce toxic wastes or poisonous wastes in the body. The accumulation of toxic wastes in the body harms an organism. So, for an organism to lead a normal life, the toxic wastes being produced in its body must be removed continuously. The process of removal of toxic wastes from the body of an organism is called excretion. It takes place in plants as well as in animals.

1.20.1 Excretion in Plants:

Like animals, plants also produce a number of waste products during their life processes. As compared to animals, the plants produce waste products very slowly and in very small amounts. The plants have no special organs for waste removal like the animals. The plants remove their waste products by different methods. Some of the important methods are given here.

The main waste products produced by plants are carbon dioxide, water vapour and oxygen. Carbon dioxide and water vapour are produced as wastes during respiration by plants whereas oxygen is produced as a waste during photosynthesis. The gaseous wastes of respiration and photosynthesis in plants (carbon dioxide, water vapour and oxygen) are removed through the stomata (in leaves) and lenticels (in stems) and released to the air. The plants excrete carbon dioxide produced as a waste during respiration only at night time. This is because the carbon dioxide produced during respiration in day time is all used up by the plant itself in photosynthesis. The plants excrete oxygen as a waste only during the day time (because oxygen is produced by photosynthesis only during the day time when the sunlight is there). Water vapour produced as a waste by respiration is, however, excreted by plants all the time (day as well as night). This waste water is got rid of by transpiration.

The plants also store some of the waste products in their body parts. For example, some of the waste products collected in the leaves, bark and fruits of the plants or trees. The plants get rid of these wastes by shedding of leaves, peeling of bark and dropping of fruits. So, when the dead leaves, bark and ripe fruits fall off from a tree, then the waste products contained in them are got rid of (Fig 1.47A). Some of the plant wastes get stored in the fruits of the plant in the form of solid bodies called raphides. These wastes are removed when the fruits get detached from the plant. For example, the fruit called 'yam' (zamikand) had needle-shaped raphides on its surface. The plants secrete their wastes in the form of gum and resins from their stems and branches. The plants also excrete some waste substances into the soil around them (Fig 1.47B). From this discussion, it is concluded that the various methods used by the plants to get rid of their waste products are the follows:

(A) Plants show fall of old leaves in autumn

(B) Secretion of waste product in the form of gum from the plant

Fig. 1.47 Excretion in Plants

(a) The plants discard wastes by secreting them in the form of gums and resins.

(b) The plants discard gaseous waste products through stomata in leaves and lenticels in stem.

(c) The plants discard stored solid and liquid wastes by the shedding of leaves, peeling of bark and dropping of fruits.

(d) The plants also excrete some waste substances into the soil around them.

1.20.2 Excretion in Animals:

Different animals have different organs or arrangements for the excretion, which depend on the constitution of the animal. For example:

- **In earthworms:** The tubular structures called nephridia are the excretory organs. In addition to nephridia, the moist skin of earthworm also acts as an excretory organ.
- **In Amoeba and other single celled animals:** The waste material carbon dioxide is removed by diffusion through the cell membrane, but nitrogenous wastes like ammonia and excess water are removed by the contractile vacuole.
- **In human beings:** the microscopic thin tubules form nephron, which functions as excretory unit. About 1 million nephrons taken together from the excretory organ of human beings called kidney.

The process of removal of toxic wastes from the body of an organism is called excretion. It takes place in plants as well as in animals. The main waste products produced by plants are carbon dioxide, water vapour and oxygen. The major wastes produced by the human body are the carbon dioxide and urea. Different animals have different organs or arrangements for the excretion, which depend on the constitution of the animal. For example: nephridia in earthworm, cell membrane in amoeba, nephron in human beings

1.20.3 Removal of Waste Products in Humans:

All the time (even when we are asleep), our body produces waste substances. The major wastes produced by the human body are the carbon dioxide and urea. Carbon dioxide is produced as a waste by the oxidation of food during the process of respiration. Urea is produced as a waste by the decomposition of unused proteins in the liver. Our body must get rid of these waste materials because their accumulation in the body is poisonous and harms us. Waste removal is called excretion. The human body has different organs for the removal of wastes from the body. These are our lungs and kidneys. Our lungs excrete carbon dioxide, and kidneys excrete urea. The kidneys are the main excretory organs of the human body. So, the main excretory system in human beings involves the kidneys.

1.20.4 Excretion of Carbon dioxide by Lungs:

The lungs remove respiratory waste carbon dioxide. Carbon dioxide is produced as a waste product in the body by the oxidation of food during respiration. This carbon dioxide enters from the body tissues into the blood stream by diffusion. Blood carries this carbon dioxide to the lungs. When we breathe out, then the lungs excrete carbon dioxide which goes into the air through nostrils. Thus, our lungs act as the excretory organs for removing the waste product carbon dioxide from the body.

1.20.5 Main Excretory System of Human body:

The excretory system of human beings collects the liquid wastes of the body and helps it get rid of them. **The excretory system of human consists of the following main organs such as (i) Two kidneys (ii) Two ureters (iii) Bladder and (iv) Urethra.** The kidneys are bean shaped organs towards the back of our body just above the waist. Every person has two kidneys. The blood in our body is constantly passing through our kidneys. The renal artery or kidney artery brings in the dirty blood (containing waste substances) into the kidneys. The function of kidneys is to remove the poisonous substance urea, other waste salts and excess water from the blood and excrete them in the form of yellowish liquid called urine. Thus, kidneys clean our blood by filtering it to remove unwanted substances present in it. The cleaned blood is carried away from the kidneys by the renal vein or kidney vein. The ureters or excretory tubes, one from each kidney, open into urinary bladder. Ureters are the tubes which carry urine from the kidneys to the bladder (Fig 1.48). Urine is stored in the bladder. The bladder is a bag which stores urine till the time we go to the toilet. The urethra is a tube. The urine collected in the bladder is passed out from the body through the urethra. Each kidney is made up of a large number of excretory units called nephrons (Fig 1.49). The nephron has a cup-shaped bag at its upper end which is called Bowman's capsule. The lower end of Bowman's capsule is tube-shaped and it is called a tubule. The Bowman's capsule and the tubule taken together make a nephron. One end of the tubule is connected to the Bowman's capsule and its other end is connected to a urine-collecting duct of the kidney. The Bowman's capsule contains a bundle of blood capillaries which is called glomerulus (plural glomeruli). One end of the glomerulus is attached to the renal artery which brings the dirty blood containing the urea waste into it. The other end of glomerulus comes out of Bowman's capsule as a blood capillary, surrounds the tubule of nephron and finally joins a renal vein (putting urea-free clean blood into it).

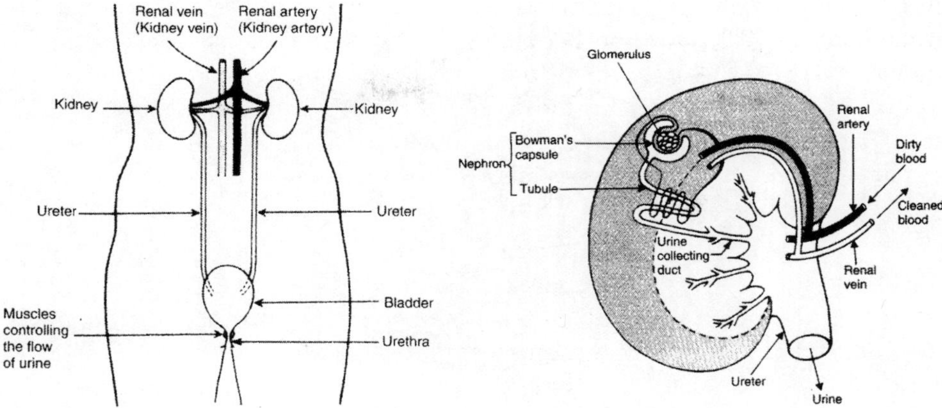

Fig. 1.48 The Human Excretory System **Fig. 1.49 Structure of Kidney**

The function of glomerulus is to filter the blood passing through it. Only the small molecules of substances present in blood like glucose, amino acids, salts, urea and water, etc., pass through the glomerulus and collect as filtrate in the Bowman's capsule. The large molecules like proteins and blood cells cannot pass out through the glomerulus capillaries and hence remain behind in the blood. The function of tubule of nephron is to allow the selective reabsorption of the useful substances like glucose, amino acids, salts and water into the blood capillaries (which surround it). But the waste material like urea remains behind in the tubule. It does not get reabsorbed into blood capillaries.

1.20.6 Working of Excretory System of Humans:

The dirty blood containing waste like urea (brought by renal artery) enters the glomerulus.The glomerulus filters this blood. During filtration, the substances like glucose, amino acids, salts, water and urea, etc., present in the blood pass into Bowman's capsule and then enter the tubule of nephron (Fig 1.50). When these filtrate containing useful substances as well as the waste substances passes through the tubule, then the useful substances like all glucose, all amino acids, most salts, and most water, etc. are reabsorbed into the blood through blood capillaries surrounding the tubule. Only the waste substances urea, some unwanted salts and excess water remain behind in the tubule. The liquid left behind in the tubule of nephron is urine. The nephron carries this urine into the collecting duct of the kidney from where it is carried to ureter. From the ureter, urine passes into urinary bladder. Urine is stored in the bladder for some time and ultimately passed out of the body through urethra. Human urine contains water, some salts and nitrogenous substances, most of which is urea (and some uric acid).

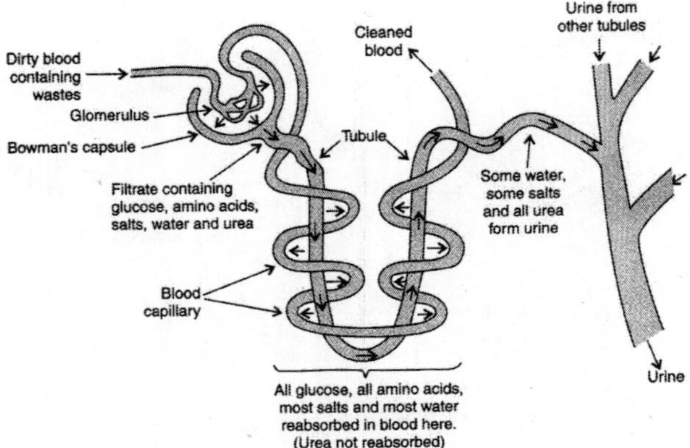

Fig. 1.50 Working of Human Excretory System

1.20.7 Kidney or Renal Failure and the Survival Technology:

Sometimes, a person's kidneys may stop working or failed. An infection in the kidneys, an injury to kidneys, very high blood pressure, very high blood sugar or restricted blood flow to the kidneys, can damage the kidneys due to which they stop working. Complete failure of the kidneys allows the urea and other waste products to build up in the blood. Even the amount of water in the body is not regulated. This will cause death if the patient is not given immediate treatment.

The best long term solution for kidney failure is the kidney transplant. The damaged kidney is removed and a matching kidney donated by a healthy person is transplanted in its place by a surgical operation. If a kidney transplant is not possible due to some reasons, then the patient with kidney failure is treated periodically on a kidney machine by a procedure called dialysis. This is because a kidney machine can do the work of damaged kidneys. The kidney machine is sometimes called artificial kidney. An artificial kidney is a device to remove nitrogenous waste products from the blood through dialysis.

Dialysis:

The blood of a person having kidney failure can be cleaned regularly by using a kidney machine or dialysis machine. The procedure used for cleaning the blood of a person by separating the waste substance (urea) from it is called dialysis.

Principle of Dialysis:

The blood from an artery in the patient's arm is made to flow into the dialyzer of a dialysis machine made of long tubes of selectively permeable membrane like cellulose which are coiled in a tank containing dialyzing solution. The dialyzing solution contains water, glucose and salts in similar concentrations to those in normal blood. As the patient's blood passes through the dialyzing solution, most of the wastes like urea present in it pass through the selectively permeable cellulose tubes into the dialyzing solution. The clean blood is pumped back into a vein of the patient's arm.

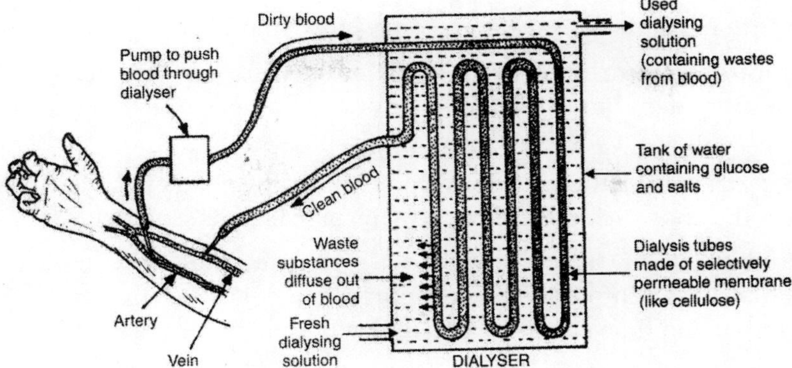

Fig. 1.51 Principle of Kidney Dialysis Machine

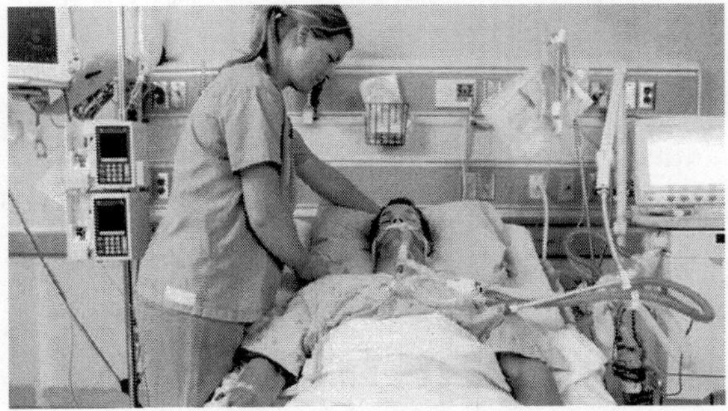

Fig. 1.52 Patient for Kidney Dialysis

REMEMBER SOCKETS

- Animals can move their body parts and can move from one place to another i.e. a bird moves when it flies in the sky and a fish moves when it swims in water.
- Non-living things (which are not alive) cannot move by themselves.
- All living organisms perform some basic functions or processes in order to remain alive.
- The basic functions performed by living organisms to maintain their life on this earth are called life processes.
- Nutrition is a process by which the organisms obtain and utilize the nutrients (food).
- Nutrients are inorganic and organic substances which the organisms obtain from their surroundings in order to synthesize their body constituents and use them as a source of energy.

- Nutrients are the source of energy which comes from outside the body of the individual organism.
- Nutrition in animals is heterotrophic, mostly holozoic.
- Saprotrophic nutrition is that nutrition in which an organism obtains its food from decaying organic matter of dead plants and animals and rotten bread, etc.
- The parasitic nutrition is that nutrition in which an organism derives its food from the body of another living organism (called its host) without killing it.
- The holozoic nutrition in which an organism takes the complex organic food materials into its body by the process of ingestion, the ingested food is digested and then absorbed into the body cells of the organism.
- The holozoic animals may be herbivorous, carnivorous or omnivorous. Human beings are omnivorous.
- Amoeba engulfs food by the process of phagocytosis using pseudopodia.
- Human beings have an elaborated digestive system consisting of the (i) alimentary canal and (ii) associated digestive glands.
- Human alimentary canal comprises of (a) mouth (b) buccal cavity (c) pharynx (d) oesophagus (e) stomach (f) small intestine (g) large intestine and (h) anus.
- Digestive glands include gastric glands, salivary glands, liver, pancreas and intestinal glands.
- Buccal cavity contains tongue and teeth.
- Saliva contains starch digesting enzyme ptyalin and mucin.
- Stomach receives gastric juice from numerous microscopic gastric glands.
- Gastric juice contains (a) HCl, which disinfects the food and activates protein degradation (b) Pepsin and rennin break proteins into proteoses and peptones (c) Gastric converts some fats into monoglycerides and fatty acids.
- Small intestine is a long, greatly coiled tube. Small intestine receives bile and pancreatic juice by bile-pancreatic duct.
- Bile contains bile salt and bile pigments.
- Bile salts emulsify fats to facilitate the action of lipase.
- Pancreatic juice has three inactive proteases, pancreatic amylase and pancreatic lipase.
- Larger part of small intestine releases intestinal juice which completes the digestion of food.
- The digested food is absorbed in small intestine and goes into the blood.
- Blood stream carries digested food into the cells where it is utilized. The utilization of food is called assimilation.

- The small intestine opens into large intestine where water is absorbed from the digested food.
- Finally the faecal matter of rectum is egested out through anus.
- Respiration is a biochemical process in which the food (respiratory substrate) is oxidized to release energy in the form of ATP.
- The first step in respiration is breathing, a kind of ventilation, which involves the intake of oxygen and release of carbon dioxide.
- The biochemical part of respiration occurs inside the cells. It is called cellular respiration in which glucose is oxidized to release energy and CO_2.
- ATP was first deduced by Lohman in 1930 and confirmed by Alexander Todd *et al.*, in 1948.
- ATP consists of an adenosine (adenine plus ribose sugar) and triphosphate groups.
- The trapping of chemical energy released by oxidative reactions of the cell in the form of ATP is called Phosphorylation.
- Glycolysis is the process of the catabolism of glucose either in the presence of oxygen such as in the liver (aerobic glycolysis) or in the lack of O_2 such as in skeletal muscle (anaerobic glycolysis).
- The process of glycolysis is completed in the cytoplasm
- Glycolysis was discovered by Embeden Meyernof and Parnas, so it is also known as EMP pathway.
- Glycolysis is completed in two phases such as Preparatory Phase and Pay off Phase
- Pyruvate or pyruvic acid (3C) is the end product of glycolysis.
- Pyruvic acid does not enter the citric acid cycle directly. Firstly, this 3-carbon atom molecule changed into a 2-carbon atom acetic acid molecule. One carbon is released as CO_2 (decarboxylation). Acetic acid on entering the mitochondria unites with coenzyme A (CoA) to form acetyl-CoA. This reaction is catalyzed by several enzymes collectively known as pyruvate dehydrogenase.
- Respiration in presence of oxygen is called aerobic respiration.
- The respiratory substrates of aerobic respiration are completely broken down to release carbon dioxide and water.
- Respiration in absence of oxygen is called anaerobic respiration.
- The respiratory substrates of anaerobic respiration are incompletely broken down to release ethyl alcohol or lactic acid.
- Animals have specific organs for gaseous exchange such as skin, trachea, book lung, gills and lungs.
- Gaseous exchange in fishes takes place through gills.

- Earthworm uses skin
- Frogs use both skin and lungs for gaseous exchange.
- Insects have an elaborate tracheal system for gaseous exchange.
- Human respiratory system consists of two major components (a) respiratory tract and (b) respiratory organs.
- Respiratory tract includes external nostrils, nasal cavities, pharynx and trachea.
- Respiratory organs are a pair of lungs.
- Air passing inward through the nasal chambers is warmed, moistened, disinfects and cleaned of dust particles to keep the lungs healthy.
- The voice box (larynx) is an enlarged upper part of trachea.
- Each lung is enclosed in two membranes (the pleura).
- Lungs bronchi are divided into bronchioles.
- Each bronchiole divides into alveolar ducts which enter the alveolar sacs.
- Breathing involves inhalation and exhalation.
- Exchange of environmental oxygen with carbon dioxide of the blood in the respiratory organs (lungs) is called external respiration.
- Internal respiration involves uptake of oxygen by tissue cells from blood; oxidation of food in cells; production of carbon dioxide, water and energy; diffusion of carbon dioxide from cells into blood; and conservation of energy in ATP.
- Energy is needed to perform life activities.
- All living organisms need nutrients to build up most of their own body molecules and to get energy for doing work.
- Depending upon the modes of obtaining nutrients, the organisms are autotrophic or heterotrophic.
- The autotrophic nutrition occurs in all green plants and some bacteria.
- In heterotrophic nutrition, the organisms get organic molecules by taking plants or animals (living or dead) or their products.
- Heterotrophic mode of nutrition may be saprophytic, parasitic or holozoic.
- Green plants synthesize organic food (carbohydrates) from inorganic raw materials (CO_2 and Water) by the process of photosynthesis.
- The actual site of photosynthesis is green coloured plastids, called chloroplasts.
- The plants use the energy in sunlight to prepare food from carbon-di-oxide and water in the presence of chlorophyll. The carbon dioxide gas produces in the cells of the stem during respiration diffuses out into the air through the same lenticels.

- The exchange of respiratory gases in the leaves takes place by the process of diffusion through stomata.
- The fish has special organs of breathing called gills; those are found on both the sides of its head. These are not visible from outside due to covering from gill cover.
- The earthworm exchanges the gases through its skin.
- The normal range of haemoglobin in the blood of a healthy adult person is from 12 to 18 grams per deciliter (g/dL) of blood.
- The plants have two types of conducting tissues called xylem and phloem
- Chlorophyll is present in the green coloured bodies called 'chloroplasts' inside the plant cells.
- Carbon dioxide and water are the raw materials for photosynthesis.
- Photosynthetic process occurs in two steps such as (a) light reaction and (b) dark reaction.
- Light reaction occurs in the grana thylakoids of chloroplasts, while dark reaction occurs in the stroma.
- In light reaction, the green pigments – chlorophylls absorb light, become excited, emit electrons which travel through electron transport chain and water is photooxidised to evolve molecular O_2.
- In dark reaction, the carbon dioxide is reduced to carbohydrate by the Calvin-Benson Cycle.
- The important factors which affect the process of photosynthesis are light, temperature, carbon dioxide and water.
- The rate of photosynthesis increases with increase in temperature upto 40°C.
- Above 40°C temperature, there is a decrease in the photosynthesis. Similarly, low temperature also inhibits photosynthesis.
- The pigment chlorophyll present in the chloroplasts absorbs visible light and after absorption creates such condition that water breaks into protons (H+), electron (e-) and molecular oxygen (O_2). This is called photolysis.
- Dark reactions are also known as Path of Carbon or CO_2 fixation.
- The raw materials for photosynthesis are: (i) Carbon-di-oxide and (ii) Water.
- The rate of photosynthetic process is affected by several external factors (light, temperature, carbon-di-oxide and water) as well as internal factors (chlorophyll, protoplasm and accumulation of photosynthetic products).
- Light is the most essential and important factors to control the rate of photosynthesis as light supplies the energy necessary for this process. Light intensity, quality (wave length) and duration affect the process of photosynthesis.

- Excretion is a biological process by which an organism gets rid of excess of toxic waste products of metabolism.
- Waste products are nitrogenous materials (such as ammonia, urea and uric acid), carbon dioxide, pigments, inorganic salts, excess of water, etc.
- Osmoregulation is a process that maintains the amount of water and proper ionic balance in the body.
- The major excretory products of plants are excess of oxygen, carbon dioxide, water, excess of salts, latex, gums, resins, calcium oxalate and many other toxic substance.
- Amoeba has contractile vacuole, which serves the function of excretion and osmoregulation.
- Amoeba is a single-celled animal, and lives in water. It depends on simple diffusion of gases for breathing.
- The excretory organs of earthworm are nephridia.
- Excretory system in human beings consists of a pair of kidneys, a pair of ureters, a urinary bladder and a urethra.
- There are two kidneys in human beings. Each kidney contains about a million nephrons. A nephron is the structural and functional unit of excretion.
- The impure blood entering into the glomerulus is filtered out in the Bowman's capsule. The glomerular filtrate passes through a long tube where reabsorption of useful materials takes place.
- The fluid and dissolved waste substances excreted by the kidneys is called urine.
- The failure of both kidneys immediately needs medical intervention. In case of kidney failure, a matching kidney from a healthy person may be transplanted.
- An artificial kidney machine may be employed in order to get rid of metabolic wastes from the blood.
- The artificial kidney functions on the same principle as the normal kidney and the procedure is called dialysis.

VERY SHORT ANSWERS TYPE QUESTIONS:

1. Define the following terms (i) Life process (ii) Respiration (iii) Transport (iv) Excretion (v) Response of stimuli
2. Give any two differences between autotrophic and heterotrophic nutrition.
3. Give the full form of CAM.
4. Give the types of heterotrophic nutrition.
5. Give any two differences chlorophyll a and chlorophyll b.

6. Define the term Photosynthesis.
7. Define the term Photolysis.
8. Define the term Herbivores and give one example.
9. Define the term Carnivores and give one example.
10. Define the term omnivores and give one example.
11. Give the example of physical method of digestion.
12. What is enamel?
13. Define the term respiration.
14. Define the term breathing.
15. Give the full form of ATP.
16. Give the full form of ADP.
17. What is the end product of aerobic respiration?
18. What is the end product of anaerobic respiration?
19. Define the term fermentation.
20. Define the term phosphorylation.
21. Define the term photophosphorylation.
22. What is the end product of glycolysis
23. What is known as power house of energy.
24. Full form of ETC.
25. Name the cellular organelles where the following processes occur:
    (a) Photosynthesis
    (b) Cellular aerobic respiration
26. Which war material is primarily responsible for the release of molecular O_2 in photosynthesis?
27. Which substance is absorbed in large intestine?
28. What is the function of bile?
29. Which food product is preferred as primary substrate for respiration?
30. What are the end products of anaerobic respiration in yeast?
31. What are the end products of aerobic respiration?
32. What is the total area for gaseous exchange in lungs of normal adult man?
33. What is the vital capacity of lungs?
34. Which organelle serves the function of excretion and osmoregulation in *Amoeba*?
35. Which substance is absorbed in large intestine?

36. What is the function of bile?
37. Which food product is preferred as primary substrate for respiration?
38. What are the end products of anaerobic respiration in yeast?
39. What are the end products of aerobic respiration?
40. Why do the herbivores have longer whereas carnivores have a shorter small intestine?
41. Name the gas transporting protein molecule present in the blood. Why this molecule is employed to transport oxygen, but not CO_2?
42. Write one function of each of the following components of the transport system in human being?
43. Define the term transportation.
44. What is the function of xylem?
45. What is the function of phloem?

SHORT ANSWERS TYPE QUESTIONS:

1. Define the term nutrition. What are the different modes of nutrition?
2. Why energy is needed for life processes.
3. Give the properties of chlorophyll.
4. Mention the major events during photosynthesis.
5. How light affecting the process of photosynthesis.
6. Give the role of water in the process of photosynthesis.
7. Define the two stages of photosynthesis.
8. How will you prove that sunlight/chlorophyll/CO_2 is necessary for photosynthesis?
9. Give the raw materials used for photosynthesis.
10. Write short notes on (i) Site of chloroplast (ii) Chloroplast Pigments (iii) CAM Plants (iv) Assimilatory power (v) Thylakoids(vi) Hill Reactions (vii) Blackmans Reactions
11. Give the differences between (i) Living and Non-living things (ii) Saprotrophic and Parasitic Nutrition (iii) Parasitic and Holozoic Nutrition (iv) Light and Dark reactions
12. Define the term nutrition and give the name of all five main processes which are concerned with the use of food by animals.
13. How pancreatic lipase hydrolyses the fats.
14. Give one example of each (i) Autotroph (ii) Herbivore (iii) Carnivore (iv) Omnivore

15. Give the types of salivary gland with their roles.
16. Give the difference between digestion and egestion.
17. Give the role of HCl in digestion.
18. Give the role of gastric Juice.
19. Give the difference between respiration and breathing.
20. Give the difference between aerobic and anaerobic Respiration.
21. Give the difference between cyclic and non-cyclic Photophosphorylation.
22. Give the difference between homolactic and heterolactic fermentation.
23. Give the modes of respiration in animals.
24. Give the mechanism of breathing.
25. Give the experiment to show that Carbon-di-oxide is produced during Respiration.
26. Write short notes on (i) Phagocytosis (ii) Liver (iii) Pancreas (iv) Dental Caries
27. Write short notes on (i) Pyruvic Acid (ii) Glycolysis (iii) Krebs Cycle (viii) ETC
28. Give the difference between xylem and phloem.
29. Give the difference between xylem vessels and tracheids.
30. Write short notes on (i) Phloem Sieve tubes (ii)Mesophyll cells

LONG ANSWERS TYPE QUESTIONS:

1. Define the term life process and give the various aspects of life forms.
2. Explain the reason for demanding the energy for nutrition.
3. Define the term nutrition and give the types of nutrition.
4. Explains the types of heterotrophic nutrition.
5. Explain the process of photosynthesis.
6. Describe any two conditions necessary for photosynthesis with the suitable examples.
7. Explain the factors affecting the process of photosynthesis.
8. Explain the mechanism of nutrition in *Amoeba*.
9. Explain the mechanism of nutrition in human beings.
10. Explain the structure of tooth with suitable diagram.
11. Define the term phosphorylation and explain the all types of Phosphorylation.
12. Explain the all steps of aerobic respiration.

13. Explain the process of respiration in roots and stems.
14. Human lungs have been designed to maximize the exchange of gases.
15. Describe the process of digestion of food in humans.
16. Explain the process of nutrition in *Amoeba*.
17. Describe the alimentary canal of man.
18. Describe the process of digestion of food in humans.
19. Define nutrition. What are the different modes of nutrition?
20. What happens to the rate of breathing during vigorous exercise and why?
21. Explain double circulation of blood in human heart.
22. Why do the herbivores have longer whereas carnivores have a shorter small intestine?
23. Name the gas transporting protein molecule present in the blood. Why this molecule is employed to transport oxygen, but not CO_2?
24. Explain the process of nutrition in *Amoeba*.
25. Describe the alimentary canal of man.
26. Explain the process of breathing in man.
27. Explain the importance of soil for plant growth.
28. Draw the diagram of alimentary canal of man and label the following parts. Mouth, Oesophagus, Stomach, Intestine.
29. How do carbohydrates, proteins and fats get digested in human beings?
30. Explain the three pathways of breakdown in living organisms.
31. Describe the process of digestion of food in humans.
32. Explain the process of nutrition in plants.
33. Describe the process of digestion of food in humans.
34. Explain the mechanism of transport of water in a plant.
35. Explain the mechanism of transport of food in a plant.

MULTIPLE CHOICE QUESTIONS:

1. The process of taking food into the body is called as

   (a) Ingestion (b) Digestion (c) Absorption (d) Egestion

2. The process in which the food containing large and insoluble molecules is broken down into smaller and water soluble molecules called as

   (a) Ingestion (b) Digestion (c) Absorption (d) Egestion

3. The process in which the digested food passes through the intestinal wall into blood stream is called as
   (a) Ingestion (b) Digestion (c) Absorption (d) Egestion
4. The process in which the undigested food is removed from the body is called as
   (a) Ingestion (b) Digestion (c) Absorption (d) Egestion
5. The process in which the absorbed food is taken in by body cells and used energy, growth and repair is called as
   (a) Assimilation (b) Digestion
   (c) Absorption (d) Egestion
6. In which nutrition an organism obtains its food from decaying organic matter of dead plants and animals and rotten bread, etc.
   (a) Parasitic (b) Saprophytic (c) Holozoic (d) None of these
7. In which nutrition, an organism derives its food from the body of another living organism (called its host) without killing it.
   (a) Parasitic (b) Saprophytic (c) Holozoic (d) None of these
8. In which nutrition, an organism takes the complex organic food materials into its body by the process of ingestion, the ingested food is digested and then absorbed into the body cells of the organism.
   (a) Parasitic (b) Saprophytic (c) Holozoic (d) None of these
9. Which organism engulfs food by the process of phagocytosis using pseudopodia?
   (a) Earthworm (b) *Amoeba* (c) *Bacillus* (d) *Mycorrhiza*
10. The process of obtaining food by Amoeba is called ass
    (a) Assimilation (b)Phagocytosis (c) Absorption (d) Egestion
11. Alimentary canal has diameter of about
    (a) 3 mtr (b) 6.5 mtr (c) 9.5 mtr (d) 10.5 mtr
12. Movement of food through the Oesophagus is due to
    (a) Peristalsis (b) Lubrication by Saliva
    (c) Gravitational (d) All of these
13. The human saliva contains an enzyme which digests the starch present in food into sugar.
    (a) Pepsin (b) Renin
    (c) Salivary amylase (d) Carboxylase
14. The gastric juice contains
    (a) Hydrochloric acid (b) Enzyme Pepsin
    (c) Enzyme Mucus (d) All of these

15. The gastric juice is in nature
    (a) Acidic (b) Basic (c) Neutral (d) Alkaline
16. Gastric digestion takes place efficiently in
    (a) Alkaline medium (b) Neutral medium
    (c) Acidic medium (d) Highly alkaline medium
17. Where is bile produced?
    (a) Liver (b) Spleen (c) Blood (d) Gall bladder
18. The small intestine receives the secretions of glands
    (a) Liver (b) Pancreas (c) Both a and b (d) None of these
19. Which of the following is not a salivary gland?
    (a) Parotid (b) Submaxillary (c) Sublingual (d) Lacrymal
20. The hard outer covering of a tooth is called as
    (a) Cement (b) Enamel (c) Both a and b (d) None of these
21. The assimilated food is used mainly for
    (a) Used as a fuel to get energy for various life processes
    (b) Used as a material for the growth and repair of the body
    (c) Both a and b
    (d) None of these
22. Dentine is the
    (a) Hardest part of the tooth
    (b) Softest part of the tooth
    (c) Part of the tooth between enamel and pulp
    (d) None of these
23. Respiration is the process in which
    (a) Energy is released and stored in the form of ATP
    (b) Energy is not released at all
    (c) Energy is stored in the form of ADP
    (d) Energy is used up
24. How many molecules at ATP are gained during anaerobic respiration of one molecule of glucose?
    (a) 2 (b) 36 (c) 4 (d) 38
25. How may ATP molecules are produced during oxidation of 1 glucose molecule in aerobic respiration.
    (a) 30 ATP (b) 32 ATP (c) 38 ATP (d) 36 ATP

26. Glucose gets converted into pyruvate in
    (a) Yeast (b) Cytoplasm (c) Mitochondria (d) Muscle cells
27. ATP was first deduced by
    (a) Krebs in 1940 (b) Lohman in 1930
    (c) Myerhoffin 1940 (d) Alexander Todd in 1930
28. The oxidation of glucose to pyruvic acid is called as
    (a) Krebs Cycle (b) Glycolysis (c) ETC (d) C_3 Path way
29. In anaerobic respiration pyruvic acid in muscles form
    (a) Glucose (b) Lactic acid (c) Alcohol (d) None of these
30. Respiration in *Amoeba* is carried out by
    (a) Cell Membrane (b) Skin
    (c) Gills (d) Lungs
31. Respiration in earthworm is carried out by
    (a) Cell Membrane (b) Skin
    (c) Gills (d) Lungs
32. Respiration in Fish is carried out by
    (a) Cell Membrane (b) Skin
    (c) Gills (d) Lungs
33. Respiration in Animals is carried out by
    (a) Cell Membrane (b) Skin
    (c) Gills (d) Lungs
34. The average breathing rate in an adult man at rest is about
    (a) 12 to 18 times per minute (b) 15 to 18 times per minute
    (c) 18 to 21 times per minute (d) None of these
35. Which is the correct sequence of parts in human alimentary canal?
    (a) Mouth → oesophagus → stomach → small intestine → large intestine
    (b) Mouth → stomach → oesophagus → large intestine → small intestine
    (c) Mouth → oesophagus → stomach → large intestine → small intestine
    (d) Mouth → stomach → small intestine → oesophagus → large intestine
36. Choose the function of the pancreatic juice from the following
    (a) Trypsin digests emulsified fats and lipase proteins
    (b) Trypsin digests proteins and lipase carbohydrates
    (c) Trypsin digests proteins and lipase emulsified fats
    (d) Trypsin and lipase digest fats

37. Based on nutrition all the organisms can be categorized into
    (a) Autotrophic (b) Heterotrophic
    (c) Both a and b (d) None of these
38. Type of Heterotrophic Nutrition is/are
    (a) Saprotrophic Nutrition (b) Parasitic Nutrition
    (c) Holozoic Nutrition (d) All of these
39. A raw material for photosynthesis is/are:
    (a) Carbon-di-oxide (b) Water
    (c) Both a and b (d) None of these
40. The green plants need ________ for photosynthesis
    (a) Chlorophyll (b) CO_2 and water (c) Light (d) All of these
41. Which of the following equations most accurately sums up the photosynthesis?
    (a) $CO_2 + H_2O \xrightarrow{\text{Light}} CH_2O + O_2$
    (b) $C_6H_{12}O_6 + 6O_2 \rightarrow 6CO_2 + 6H_2O$
    (c) $6CO_2 + 6H_2O \rightarrow C_6H_{12}O_6 + 6O_2$
    (d) $6CO_2 + 12H_2O \xrightarrow[C_6H_{12}O_6 + 6O_2 + 6H_2O]{\text{Light Chlorophyll}} C_6H_{12}O_6 + 6O_2$
42. Chlorophyll activates on exposure to
    (a) Light (b) CO_2 (c) Water (d) Soil
43. The process in which water splits during photosynthesis is called as
    (a) Plasmolysis (b) Hydrolysis (c) Photolysis (d) None of these
44. Chlorophyll is present
    (a) On the surface of chloroplasts
    (b) In the grana of chloroplasts
    (c) Dispersed throughout the chloroplasts
    (d) In the stroma
45. Molecular formula of Chlorophyll a is
    (a) $C_{55}H_{72}O_5N_4Mg$ (b) $C_{56}H_{70}O_6N_5Mg$
    (c) C_2H_5OOH (d) $C_6H_{12}O_6$
46. Molecular formula of Chlorophyll b is
    (a) $C_{55}H_{72}O_5N_4Mg$ (b) $C_{56}H_{70}O_6N_5Mg$
    (c) C_2H_5OOH (d) $C_6H_{12}O_6$

47. Which gas is produced as a by-product during the preparation of food by photosynthesis?

    (a) CO_2 (b) O_2 (c) H_2S (d) None of these

48. The food prepared by the green leaves of a plant is in the form of a simple sugar called as

    (a) Protein (b) Glucose (c) Starch (d) Cellulose

49. Chlorophyll pigments are present in the thylakoids.

    (a) Mitochondrion (b) Cytoplasm
    (c) Lenticels (d) Thylakoids

49. Which factors of light the affect the process of photosynthesis.

    (a) Light intensity (b) Light Quality
    (c) Light duration (d) All of these

50. Light Phase takes place in which region of chloroplast

    (a) Grana (b) Stroma (c) Both a and b (d) None of these

51. Dark Phase takes place in which region of chloroplast

    (a) Grana (b) Stroma (c) Both a and b (d) None of these

52. The length of human man alimentary canal which runs from mouth to anus is about

    (a) 3mtrs (b) 5mtrs (c) 7 mtrs (d) 9 mtrs

53. The small intestine is about

    (a) 5.5 mtrs (b) 6.5 mtrs (c) 7.5 mtrs (d) 8.5 mtrs

54. The inner surface of small intestine has millions of tiny, finger-like projections called as

    (a) Grana (b) Stroma (c) Villi (d) Vein

55. The exit of faeces is controlled by the anal sphincter

    (a) Intestine (b) Stomach (c) Duodenum (d) Analsphincter

56. Each tooth consists of

    (a) Crown (b) Neck (c) Root (d) All of these

57. In aerobic respiration, one molecule of glucose (food) produces how many energy-rich ATP molecules

    (a) 30 (b) 32 (c) 36 (d) 38

58. In anaerobic respiration, one molecule of glucose (food) produces how many energy-rich ATP molecules

    (a) 2 (b) 3 (c) 4 (d) 5

59. The sudden buildup of lactic acid in our muscles during vigorous physical activity can cause

(a) Muscular 'cramps (b) Anemia
(c) Fever (d) None of these

60. The oxygen is carried by blood to all the parts of the body by

(a) Muscular cramps (b) Cough
(c) Haemoglobin (d) None of these

61. The normal range of haemoglobin in the blood of a healthy adult person is ...grams per deciliter (g/dL) of blood.

(a) 8 to 10 (b) 10 to 12 (c) 12 to 15 (d) 12 to 18 grams

62. Opening and closing of stomata is due to

(a) Changes of turgor pressure in guard cells
(b) Genetic constitution
(c) Pressure of gases inside the leaves
(d) Effect of hormones

63. Which one of the following is connected with transport of water in plants?

(a) Xylem (b) Cambium (c) Phloem (d) Epidermis

64. Translocation of solutes primarily takes place through

(a) Xylem (b) Pith (c) Phloem (d) Cortex

65. Maintenance of proper amount of water and ionic balance in the body is called as

(a) Excretion (b) Dehydration (c) Osmoregulation (d)None of these

ANSWER KEY

1. (a)	2. (b)	3. (c)	4. (d)	5. (a)	6. (b)	7. (a)	8. (c)
9. (b)	10. (b)	11. (b)	12. (a)	13. (c)	14. (d)	15. (a)	16. (c)
17. (a)	18. (c)	19. (d)	20. (b)	21. (c)	22. (a)	23. (a)	24. (a)
25. (c)	26. (b)	27. (b)	28. (b)	29. (b)	30. (a)	31. (b)	32. (c)
33. (d)	34. (a)	35. (a)	36. (c)	37. (b)	38. (d)	39. (c)	40. (d)
41. (c)	42. (a)	43. (c)	44. (c)	45. (a)	46. (b)	47. (b)	48. (b)
49. (d)	50. (a)	51. (b)	52. (d)	53. (b)	54. (c)	55. (d)	56. (d)
57. (d)	58. (a)	59. (a)	60. (c)	61. (d)	62. (a)	63. (a)	64. (c)
65. (c)							

❖❖❖

2

CHAPTER

CONTROL AND CO-ORDINATION IN ANIMALS AND PLANTS

Tropic movements in plants; Introduction of plant hormones; Control and co-ordination in animals: Nervous system; Voluntary, involuntary and reflex action; Chemical co-ordination: animal hormones.

2.1 CONTROL AND COORDINATION:

All the living organisms (plants and animals) respond and react to changes in the environment around them. The changes in the environment to which the organisms respond and react are called stimuli (singular-stimulus). The living organisms show response to stimuli such as light, heat, cold, sound, smell, taste, touch, pressure, pain, water and force of gravity etc. Both, plants and animals respond to various stimuli around them. But the method of reacting to stimuli is not similar in plants and animals. They react to stimuli in different ways. The response of organisms to a stimulus is usually in the form of some movement of their body part. For example, if a man touches a very hot utensil accidently, the quickly pulls his hand away from the hot utensil. Here, heat is the stimulus and the man reacts by moving his hand away from the hot utensil. Similarly, when the sun is bright, we close our eyes. In this case, light is the stimulus and response is closing the eyes. For example, plants bend towards light but animals do not bend towards light. The animal *Amoeba* reacts to the presence of food by moving towards the food particle. Similarly, *Amoebae* tend to aggregate (collect together) in moderately warm water which is their reaction to the stimulus called heat. *Amoeba* and other protozoa react to the mechanical obstacles by avoiding them. We find that the *Amoeba* (which is an animal) can react to different stimuli in different ways.

The animals can react to stimuli in many different ways because they have a nervous system and an endocrine system involving hormones. The plants, however, react to stimuli in a very limited way. This is because the plants do not have a nervous system like the animals have. The plants use only the hormones for producing reaction to external stimuli. Sunflower always faces the sun. Thus, sunlight is the stimulus and sunflower reacts by bending (or moving) towards the sun (Fig 2.1A). In the case of hunger of animal, hunger is the stimulus and we react by eating food. It is concluded here that the reaction to stimuli is a characteristic property of the living organisms (Fig 2.1B). Another word which is also used in place of 'reaction' is 'response'. So, we can also say that the response to stimuli is a characteristic property of the living organisms. From all these examples, it is concluded that when a stimulus acts on our body, then we react or respond in a manner which is in the best interest of our body. The reaction which we give to the stimulus involves many organs of our body (Fig 2.1C). It is, therefore, necessary that all the concerned organs should work with one another in a systematic manner so as to produce the required reaction. In other words, the various organs should cooperate with one another to provide proper reaction to the stimulus. The working together of the various organs of an organism in a systematic manner so as to produce a proper response to the stimulus is called coordination.

(A) Sunflower always faces the sun

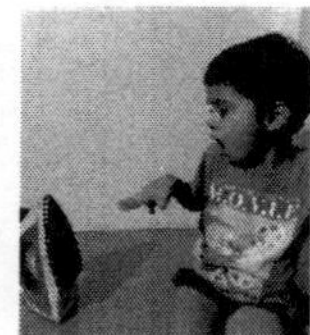

(B) We pull back our hand when touched with not object

(C) Animals respond towards food

Fig. 2.1 Examples of various types of stimuli

2.2 CONTROL AND COORDINATION IN PLANTS:

The plants do not have a nervous system and sense organs such as eyes, ears, or nose, skin etc., like the animals, but they can still sense things. The plants can sense the presence of stimuli like light, gravity chemicals, water, and touch, etc., and respond to them by the action of hormones in them. These stimuli like are called as environmental changes. The hormones in plants coordinate their behavior by affecting the growth of a plant. The effect on plants growth can result in the movement of a part of the plant like shoot or root etc.

2.2.1 Plant Movements:

Plants generally show movements at a very slow rate. The movements in plants are not as apparent as in case of animals. The higher plants are fixed to the substratum by means of roots. They cannot move from one place to another. They, therefore, show movements of their organs only. The plants are fixed at a place with their roots in the ground, so they cannot move from one place to another. That is, plants do not show locomotion. However, movements of the individual parts of organs of a plant like shoot, root, leaves, etc. are possible when they are subjected to some external stimuli like light, force of gravity, chemical substances, water, and touch, etc. These movements of the plant part are usually caused by an unequal growth in the two regions by the action of plant hormones, under the influence of the stimulus.

Plants are fixed at a place with their roots in the ground, so they cannot move from one place to another, while moment of the individual parts of organs of a plant (like shoot, root, leaves, etc.) are possible when they are subjected to some external stimuli like light, force of gravity, chemical substances, water, and touch, etc. These movements of the plant can be autonomic (spontaneous) or paratonic (induced) movements.

2.2.2 Movements of Curvature:

These are changes in orientation of some plant parts in relation to others caused by internal (intrinsic) or external (extrinsic) stimuli. Curvature movements can be growth movements or turgor movements. These can be autonomic (spontaneous) or paratonic (induced) movements.

2.2.3 Classification of Induced Plant Movements:

Induced or paratonic plant movements made in response to external stimuli fall into two main categories such as tropisms and nastic movements.

- Nastic movements
- Tropic movements

Nastic movements:

These are non-directional induced variation movements that occur due to turgor are not directional movements. Here, the part of the plant does not respond towards or away from the stimulus. Nastic movements are further classified such as:

(i) **Seismonastic Movements:** Such movements occur in response to touch (shock). These movements are very quick and are best seen in touch-me-not plant (*Mimosa pudica*), also called 'Lajwanti' or 'Chhui-mui' or 'Sensitive plant'. If we touch the leaves of the chhui-mui plant with our finger, we find that all its leaves immediately fold up and droop. After sometime, the leaves regain their original status. Here, no growth is involved. Instead, plant cells change shape by changing the amount of water in them (turgor changes), resulting in folding up and drooping of leaves (Fig. 2.2). As the 'touch-me-out' plant responds to touch stimulus, this phenomenon is also commonly called thigmonasty.

Before touch **After touch**

Fig. 2.2 *Mimosa Pudica* plant (Chhui-mui)

(ii) Nyctinastic Movements: The movements involving the diurnal variation in the position of flowers and leaves of many plants in day and night are called nyctinastic or sleep movements. Nyctinastic movements include photonastic movements and thermonastic movements, those are given below as:

(a) Photonastic Movements: The non-directional movement of a plant part (usually petals of flowers) in response to light is called Photonasty (Photonastic Movement). A dandelion flower opens up in the morning in bright light but closes in the evening when the light fades and it gets dark. The opening and closing of petals of dandelion flowers in response to the intensity of light is an example of nastic movement in which the stimulus is light. The moonflower behaves exactly opposite to that of dandelion flowers in respect of response to light. The petals of moonflower close during the day when there is bright light but open up at night when it is dark and there is no light. This is also an example of photo nasty (Fig. 2.3A). Please note that the opening and closing of flowers in response to light are growth movements. Petals open when their inner surfaces grow more than their outer surfaces. On the other hand, petals close when their outer surfaces grow more than their inner surfaces.

Moon flower opens during the night

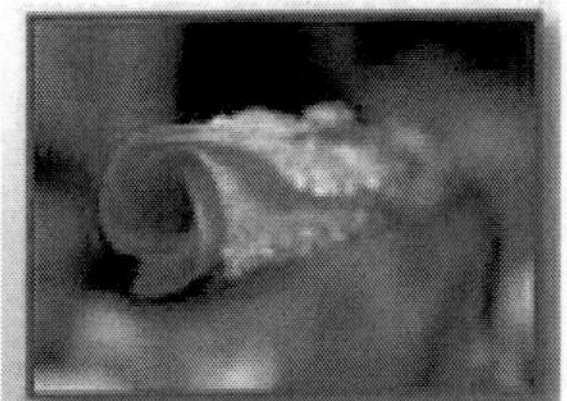

Moon flower closes during the day

Fig. 2.3A Photonastic Movement in Moon flower

(b) Thermonastic Movements: If the diurnal variations in the position of plant parts (e.g., flowers and leaves of plants) are caused by the change in temperature of the surroundings, such non-directional movements are called thermonastic movements. *Tulip* and *Crocus* which open during high temperature and close during low temperature (Fig. 2.3B).

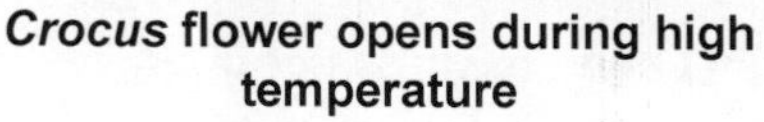

Crocus flower opens during high temperature

Crocus flower closes during low temperature

Fig. 2.3B Thermonastic movements in *Crocus* flower

> Nastic movements are non-directional induced variation movements that occur due to turgor are not directional movements. Nastic movements are further classified into Seismonastic and Nyctinastic Movements

2.2.3.2 Tropisms (Tropic Movements):

Tropic movements are induced growth movements of curvature that occur due to differential growth. These are directional movements are orientations that occur in response to external stimuli such as light, force of gravity, chemicals, water etc. when some stimulus is received by the plant, specific part of the plant shows response in the form of growth movement due to unequal growth in its two sides by the action of plant growth hormones (phytohormones). Directional movements or orientations of specific part of plant revealing growth in response to external stimuli are called tropisms or tropic movements. Tropic movements are very slow. The movement of the plant part can be seen either towards the stimulus or away from the stimulus.

- If the movement of the plant part is away from the stimulus, it is termed as negative tropism.
- If the movement of the plant part is towards the stimulus, it is termed as positive tropism.

Types of Tropism:

Light, chemical substances water are and force of gravity four common stimuli in the environment to which the plants respond. The responses of plants to these stimuli are termed as (a) Phototropism (b) Chemotropism (c) Hydrotropism and (d) Geotropism, respectively.

(a) Phototropism (Light): It is directional movement or orientation of the plant part in response to light stimulus. If the plant part moves towards light, it is called positive phototropism. Alternatively, if the plant part moves away from light, it is called negative phototropism (Fig. 2.4). **Examples:** (i) Stem or shoot of a growing plant moves towards light and thus shows positive phototropism (ii) Roots of a plant move away from light and thus show negative phototropism.

Fig. 2.4 Phototropism (Plants bend towards light)

(b) **Chemotropism (Chemical):** It is the directional movement or orientation of the plant part in response to chemical stimulus. If the plant part moves towards the chemical stimulus, it is called positive chemotropism. On the other hand, if the plant part moves away from the chemical stimulus, it is called negative chemotropism. **Example:** During the process of fertilization, growth of pollen tube towards the ovule in the ovary is an example of positive chemotropism (Fig. 2.5).

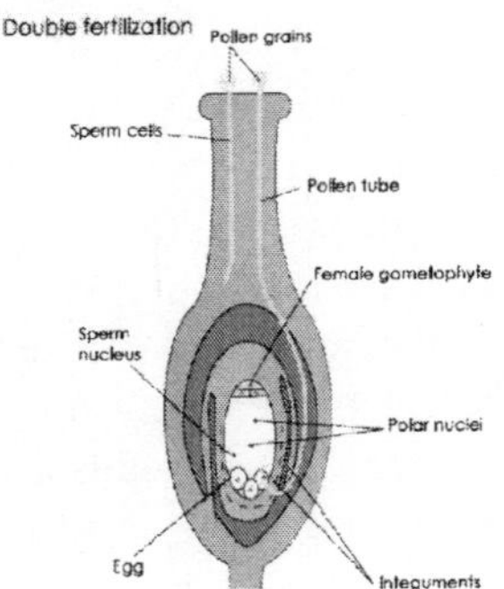

Fig. 2.5 Chemotropism (Fertilization in plants is brought about by chemotropism)

(c) **Hydrotropism (Water):** It is the directional movement or orientation of the plant part in response to water stimulus. If the plant part moves towards the water stimulus, it is called positive hydrotropism. On the other hand, if the plant part moves away from the water stimulus, it is called negative hydrotropism (Fig. 2.6). **Example:** Bending of roots of the plant towards water signifies positive hydrotropism.

Fig. 2.6 Hydrotropism (Bending of roots)

(d) **Geotropism (Gravity):** It is directional movement or orientation of the plant part in response to gravity. If the plant part moves in the direction of gravity, it is called positive geotropism. Alternatively, if the plant part moves against the direction of gravity, it is termed as negative geotropism (Fig. 2.7). **Examples:** (i) Roots of a plant move downward in the soil in the direction of gravity and thus show positive geotropism (ii) The stem or shoot of a plant shows movement against the direction of gravity and thus shows negative geotropism.

Tropic movements are induced growth movements of curvature that occur due to differential growth. Light, chemical substances, water and force of gravity are four conman stimuli in the environment to which the plant respond. The responses of plants to these stimuli are termed as (a) Phototropism (b) Chemotropism (c) Hydrotropism and (d) Geotropism.

Table 2.1 Differences between Tropic and Nastic Movements

Sr. No.	Tropic Movements	Nastic Movements
1.	These are directional movements or orientation of the part of the plant in response to specific stimulus. The movement can be towards the stimulus or away from the stimulus.	These are not directional movements, i.e. these movements are neither towards the stimulus nor away from the stimulus.
2.	These are plant movements of curvature caused by unilateral growth, i.e. one side of an organ grows faster than the other causing curvature.	These movements are caused due to variations in environmental conditions, i.e., these are variation movements. These are also movements of curvature but these occur due to turgor changes.
3.	Bending of stem towards light (phototropism), and downward growth of roots in response to gravity (geotropism) are some examples.	Folding and drooping of leaves of Mimosa plant ('touch-me-out' plant) in response to touch is the example.
4.	The stimulus is generally unilateral	The stimulus is generally diffused

2.2.4 Demonstration of Responses of Plant Parts of various Stimuli:

2.2.4.1. Response of Plants to Light (Phototropism):

Plants manufacture their own food in the presence of sunlight, so known as autotrophs. Therefore, they respond to light by growing towards it. Plants also turn their leaves towards the sun to ensure that the latter get maximum sunlight. This phenomenon can easily be depicted by performing simple experiment (Fig. 2.7).

- Take two potted plants.
- Place one plant in the open so that it receives the sunlight coming from above Place the other in a room near the window in such a way that it receives sunlight from one side only, i.e. through the window.
- After some days, observe both the plants.

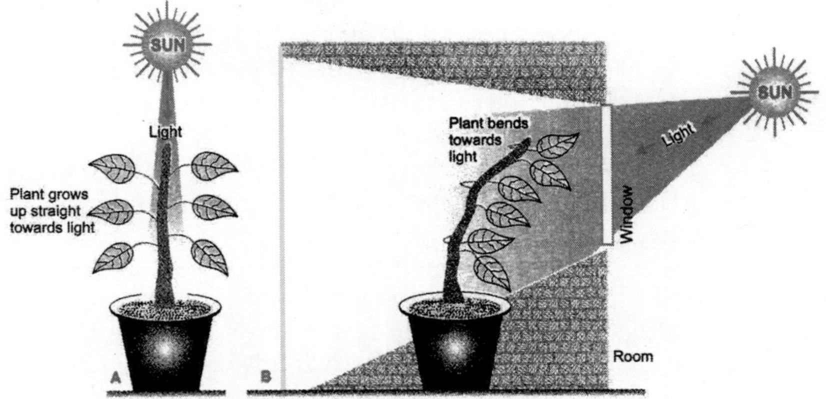

Fig. 2.7 Phototropism : Experiment to show response of a plant to light

Observation:

The first plant (which was kept in the open) has grown up straight towards light. However, the second plant (which was kept in the room and receiving light from one side) has grown by bending towards the light. Thus, the stem of the plant responds to light by showing growth movement towards light (positive phototropism).

Role of Auxin Hormone in Phototropic Movement involving Growth:

In the above experiment, the plants revealed growth by showing movement towards light. This growth movement of the plant part (stem) is caused by the action of auxin (Fig. 2.8). Auxin hormone is synthesized by the meristematic tissue at the tip of the stem (or tip of the shoot). Auxin diffuses uniformly down the stem in plant kept in the open and receiving sunlight auxin moved from the tip of stem and concentrated more on the side not receiving light (shady side, i.e., the side of stem away from light). Due to presence of more auxins the shady side of stem grew faster than the side of stem receiving light. As a result, the stem bent towards the direction of light.

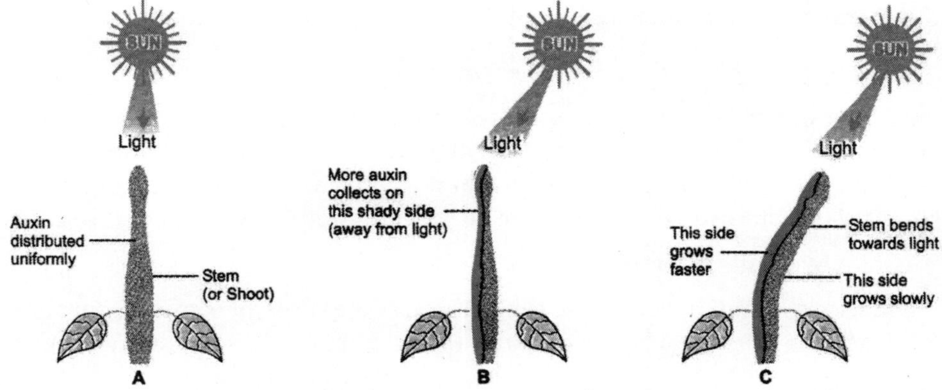

Fig. 2.8 Phototropism : Experiment to show affect of Auxin Hormone in response to light

2.2.4.2. Directional Movement due to growth in response to touch as shown by Pea Plant

Pea plant has weak stem. The latter needs support to climb up. The pea plant develops tendrils which help it to climb up other plants or fences or some other support. These tendrils are sensitive to touch (Fig. 2.9). As the tendrils come in contact with any support, the part of the tendril away from the support grows faster than the part in contact with the support. As a result, the tendril circles around the support and clings to it. This movement of tendril around the support is caused by the action of auxin. More auxin is diffused from its site of synthesis (i.e., from its tip) towards the side of the tendril away from the support. As a result, the side having more auxin grows faster than the one in contact with the support. This causes the tendril to circle around the support.

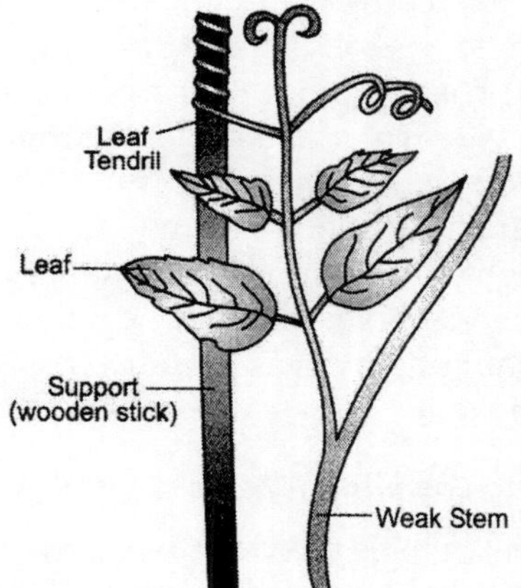

Fig. 2.9 Response of plant part (Tendril) to touch stimulus

2.2.4.3. Response of Plants to Gravity (Geotropism):

Gravity refers to the force with which the earth pulls all the things towards it. It acts in the downward direction. The plant part responds to gravity by either moving towards or away from earth. This phenomenon can easily be depicted by performing a simple experiment (Fig. 2.10).

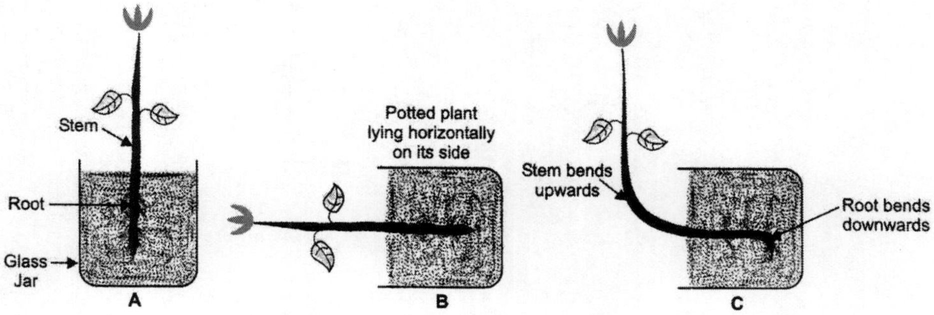

Fig. 2.10 Geotropism : Experiment to show response of a plant to gravity

- Take a potted plant growing in a transparent glass jar.

- Keep the potted plant growing in a transparent glass jar straight on the ground.
- Observe that its stem is growing upward and its roots are growing downward.
- Now, place this potted plant horizontally on its side on the ground and observe the position of the growing stem and the root.
- Keep the plant in this position for a few days.
- After few days, observe the plant.

Observation:

The potted plant is kept horizontally on its side on the ground; the growing stem and the growing root are parallel to the ground. After few days, you find that the stem of this plant has bended upwards away from earth (negative geotropism) and the root of the plant has bended downward towards earth, i.e., positive geotropism. This experiment confirms response of plants to gravity (geotropism). Main functions of the roots of plants are (i) These provide of anchorage (fix the plant to substratum or soil), and (ii) These provide water and minerals from substratum or soil. Roots of the plant therefore show positive geotropism, i.e., they grow downward in response to gravity. The main function of stem (or shoot) of the plant is to expose leaves to maximum sunlight. Therefore, stem or shoot of the plant always grows up, away from the pull of gravity, i.e., it shows negative geotropism. The response of different parts of the plant of gravity is actually caused by the action of various plant hormones including auxin.

2.2.4.4. Response of Plants to Chemicals (Chemotropism):

The ripe stigma of the carpel (female reproductive part) of a flower secretes a sugary substance (chemical) into the style towards the ovary. They pollen grains (which fall on the stigma of the carpel) respond to this sugary substance (chemical stimulus) by growing respective pollen tubes in the downward direction in the style of the carpel. Generally, one pollen grain is successful in elongating its pollen tube to reach the ovule in the ovary of the flower (Fig. 2.11). This happens to ensure fertilization in plants. This response of pollen grain (i.e. growth of pollen tube) to sugary substance (chemical stimulus) secreted by the stigma of the carpel of a flower is an example of chemotropism.

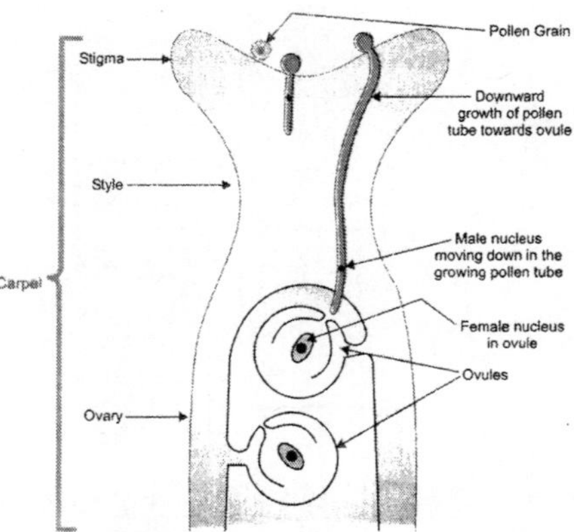

Fig. 2.11 Chemotropism: Response of a pollen grain to chemical, secreted by stigma

2.2.4.5. Response of Roots of Plants to Water (Hydrotropism):

The roots of plants move towards water and, therefore, depict positive hydrotropism. Normally, roots of plants grow downwards. However, in order to reach water in the soil, they can grow sideways or even upwards. Such type of response of roots of a plant towards water can be demonstrated by performing a simple experiment (Fig. 2.12).

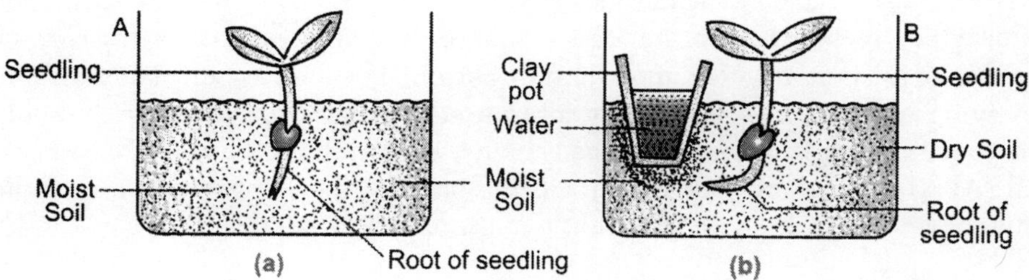

Fig. 2.12 Hydrotropism: Experiment to show the response of root of plant to water

- Take two galls troughs A and B fill them with soil up to two-third level. Plant a tiny seedling in each trough/trench/holder.
- Now, place a small clay pot in the soil trough B. Water the soil daily in trough A uniformly. However, in trough B, put water daily in clay pot only.
- After few days, carefully dig up the seedlings in both the glass without damaging their roots. You will notice that in trough A, the root of seedling is straight. However, in trough B, the root of seedling is bent towards the direction of clay pot containing water.

Observation:

In trough A, soil is watered uniformly daily. Therefore, the root of seedlings gets water equally from all sides. Hence, it grows straight downwards. On the other hand, in trough B, the root of seedling only gets water that oozes out of clay pot buried in soil. Therefore, in trough B, the root of seedling grows by bending towards the direction of clay pot. Thus, it reveals the response of root of seedling (plant part) towards water (positive hydrotropism).

Sr. No.	Movement of leaves of sensitive plant	Movement of a shoot towards light
1.	It is a nastic movement which does not depend on the direction of stimulus applied	It is a tropic movement which depends on the direction of stimulus applied.
2.	The stimulus is 'tough'.	The stimulus is 'light'.
3.	It is caused by the sudden loss of water from the swellings at the base of leaves.	It is caused by the unequal growth on the two sides of the shoot.
4.	It is not a growth movement.	It is a growth movement.

2.3 PLANT HORMONES OR PHYTOHORMONES (PHYTO= PLANT):

The control and coordination in plants is done by plant hormones (or phytohormones). The plant hormones coordinate the activities of the plant by controlling one or the other aspect of the growth of the plant. So, the plant hormones are also known as plant growth substances. The growth of a plant can be divided into three stages such as cell division, cell enlargement and cell differentiation (or cell specialization), and these stages have particular locations in a plant. Apart from these, promotion of dormancy, breaking of dormancy, stomata control, falling of leaves, fruit growth, ripening of fruits and ageing in plants are controlled by the various plant hormones Following are five major types of phytohormones, which are involved in the control and coordination in plants such as Auxins, Gibberellins, Cytokinins, Abscisic acid (ABA) and Ethane (Ethylene). On the basis of their effect, plant hormones (phytohormones) can be divided into two groups such as:

(a) Growth Promoters.

These stimulate the plant growth, e.g. Auxins, Gibberellins, Cytokinins and Ethylene.

(b) Growth Inhibitors:

These inhibit or retard the plant growth e.g. Abscisic acid (ABA).

The synthesis and action of various plant hormones are influenced by external stimuli. Some prominent actions of these plant hormones are mentioned below:

- Apical dominance.
- Dormancy and breaking of dormancy.
- Flowering of plants.
- Growth and ripening of fruits.
- Growth of root, stem and leaves
- Movements in plants (tropic and nastic movements).
- Opening and closing of stomata (stomatal movements) in leaves.
- Parthenocarpy.

2.3.1 Auxins:

Auxins are the plant hormones which promote cell enlargement and cell differentiation in plants. Auxins also promote fruit growth. Auxin hormone controls a plant's response to light and gravity. In other words, auxin hormone is responsible for the phototropic and geotropic responses of plants. Auxin is made by cells at the tip of stems and roots. Auxin moves away from light and towards gravity. Auxin has opposite effect on the growth of stem and roots. Auxin speeds up growth in stem but it slows down growth in roots. Synthetic auxins are applied in agriculture and horticulture. The major functions of auxins are cell enlargement, root formation, movement, apical dominance, fruit growth etc.

2.3.2 Gibberellins:

These are plant hormones which promote cell enlargement and cell differentiation in the presence of auxins. Gibberellins help in breaking the dormancy in seeds and buds. They also promote growth in fruits. Gibberellin hormone is involved mainly in shoot extensions. Gibberellin stimulates elongation of shoots of various plants. The main functions of gibberellins are growth, fruit yield, overcoming dormancy, and replace the long and cold day conditions for flowering

2.3.3 Cytokinins:

These are the plant hormones which promote cell division to plants. Cytokinins also help in breaking the dormancy of seeds and buds. They delay the ageing in leaves. Cytokinins promote the opening of stomata. They also promote fruit growth. The main function of cytokinins is cell division, differentiation, phloem transport, apical dominance etc.

2.3.4 Abscisic acid:

It is a plant hormone which functions mainly as a growth inhibitor. Abscisic acid promotes the dormancy in seeds and buds (this is the opposite of breaking of dormancy). It also promotes the closing of stomata. Abscisic acid promotes the wilting and falling of leaves (which is called abscission). It also causes the detachment of flowers and fruits from the plants. The main functions of abscisic acid are controlling growth, dormancy, wilting, senescence and abscission and flowering in short day plants.

2.3.5 Ethylene (Ethane):

It is a gaseous hormone which promotes transverse growth but inhibits longitudinal growth and several functions. It is mostly formed during ripening of climacteric fruits e.g. apple, banana. Its main function is in fruit ripening, dormancy of rhizomes, bulbs, induces abscission of fruits and leaves.

Plants do not have a nervous system and sense organs like eyes, ears, or nose, skin etc., like the animals, but they can still sense things. The plants can sense the presence of stimuli like light, gravity chemicals, water, and touch, etc., and respond to them.by the action of hormones in them. Plant hormones or phytohormones regulate various functions in plants such as Breaking the dormancy of seeds, Growth of root, stem and leaves, Stomatal movement, Flowering of plants, Ripening of fruits, and Phototropism, geotropism, chemotropism, hydrotropism, thigmotropism and nastic movements. Plant hormones such as Auxins, Gibberellins, Cytokinins, Abscisic acid (ABA) and Ethane (Ethylene) are involved in the control and coordination in plants.

Table 2.2: Types of Phytohormones with their functions

Sr. No.	Plant hormones	Functions
1.	Auxins (Naturally occurring auxin is Indole 3-Acetic Acid)	• These promote cell enlargement and cell differentiation in plants. • These promote stem and fruit growth. • These regulate important plant growth movements, i.e. tropisms. • These induce parthenocarpy (i.e., the formation of seedless fruits without fertilization in number of plants.)
2.	Gibberellins (Gibberellic Acid)	• These promote cell enlargement and cell differentiation in plants in the presence of auxins. • These also promote growth in stems and fruits. • Rosette plants (i.e., plants that show profuse leaf developments but reduced internode growth) show bolting and flowering when treated with gibberellins. • These also induce parthenocarpy in many plants.
3.	Cytokinins	• These promote cell division in plants. • These play vital role in the morphogenesis in plants. • These help in breaking the dormancy of seeds and buds. • These delay the ageing in leaves. • These promote the opening of stomata. • These also promote fruit growth.
4.	Ethylene	• It promotes growth and ripening of fruits. • It helps in breaking the dormancy in buds and seeds. • It stimulates the formation of separation layer (abscission zone) in leaves, flowers and fruits. • It promotes yellowing and senescence of leaves.
5.	Abscisic Acid (ABA)	• It promotes the dormancy in seeds and buds and thus inhibits growth. • It also promotes the closing of stomata and thus affects wilting of leaves. • It also promotes the falling of leaves (abscission) and senescence in leaves.

2.4 DORMANCY AND QUIESCENCE:

Dormancy refers to a resting, inactive condition of the seed when it fails to germinate even though the environmental conditions usually considered favorable for active growth are present. A resting or inactive condition, in which metabolism is most minimum called as the dormancy. A seed must have certain conditions like water, warmth, air and hormones to break dormancy and germinate to form a seedling (which then grows into a plant). Another part of a plant having dormancy is the bud. The bud is a young, undeveloped shoot of a plant which on breaking dormancy can form a branch, a leaf or a flower depending on its position in the plant. The breaking of dormancy of a bud also requires certain plant hormones.

Quiescence refers to the condition of the seed when it fails to germinate because the environmental conditions normally required for growth are not present. The dormant or inactive seed must attain water, warmth, air, hormones etc. to come out of dormancy. It can then only germinate to form a seedling and subsequently grow into a plant.

Dormancy refers to a resting, inactive condition of the seed when it fails to germinate even though the environmental conditions usually considered favorable for active growth are present. While Quiescence refers to the condition of the seed when it fails to germinate because the environmental conditions normally required for growth are not present.

2.5 PHOTOPERIODISM:

The duration of sunlight also regulates the germination of seeds and flowering in plants. The length of the day (in hours) during which the sunlight is available to the plants is called photoperiod. The effect of photoperiod on the germination of seeds and flowering in plants is called photoperiodism. Photoperiod acts as a stimulus for plants. The plants respond to this stimulus with the help of a special kind of pigment present in them in very small amounts. The special pigment is called phytochrome which is a blue-green pigment. Seeds of many plants germinate only if they are exposed to particular day length of light showing involvement of phytochrome pigment, e.g., seeds of lettuce. Photoperiod exerts significant effect in the flowering of plants. Garner and Allard (1920) recognized three classes of plants according to their photoperiodic responses (Fig. 2.13) such as:

(a) Short day plants

(b) Long day plants, and

(c) Day neutral plants.

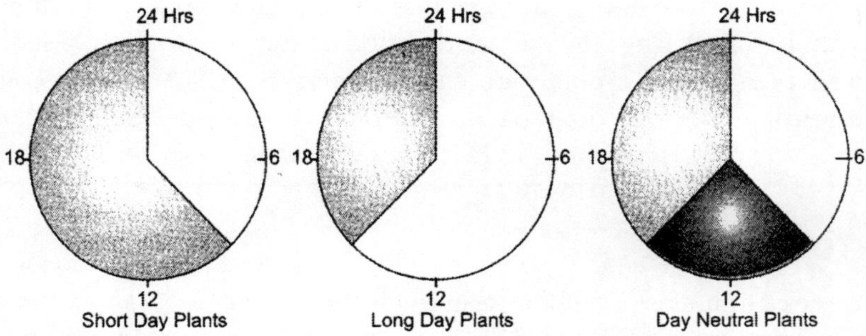

Fig. 2.13 Photoperidism : Requirement of light and dark periods during 24 hours for flowering in various type of plants

(a) Short Day Plants (SD Plants):

A short day plant is one that flowers on photoperiods shorter than the critical day length e.g. tobacco plant. The tobacco flower only after exposure to a number of short photoperiods (less than 12 hrs light and more than 12 hrs dark in 24 hrs cycle), which exist only in autumn season. In reverse situation (more than 12 hrs light and less than 12 hrs dark in 24 hrs cycle), these plants fail to flower. This happens with the help of phytochrome pigment.

(b) Long Day Plants (LD Plants):

A long day plant is one that flowers on photoperiods longer than critical day length e.g. wheat plants. Long-day plants flower when the night length falls below their critical photoperiod. These plants typically flower in the northern hemisphere during late spring or early summer as days are getting longer. In the northern hemisphere, the longest day of the year (summer solstice) is on or about 21 June. After that date, days grow shorter (i.e. nights grow longer) until 21 December (the winter solstice). This situation is reversed in the southern hemisphere (i.e., longest day is 21 December and shortest day is 21 June).

Some long-day obligate plants are such as Carnation (Dianthus), Henbane (*Hyoscyamus*) and Oat (*Avena*).

Some long-day facultative plants are such as Pea (*Pisum sativum*), Barley (*Hordeum vulgare*), Lettuce (*Lactuca sativa*) and Wheat (*Triticum aestivum*).

(c) Day Neutral Plants (DN Plants):

These plants do not require specific photoperiods to flower. These, in fact, flower in all photoperiods ranging from a few hours to 24 hrs of uninterrupted light, e.g. cucumber, maize, tomato etc. Thus, plant hormones (phytohormones) and phytochrome pigment together are involved in the control and coordination of plant responses. Some short-day facultative plants are such as Kenaf (*Hibiscus cannabinus*), Marijuana (*Cannabis*), Cotton (*Gossypium*), Rice (*Oryza*), Jowar (*Sorghum bicolor*), Green Gram (*Vigna radiata*) and Soybeans (*Glycine max*).

The length of the day (in hours) during which the sunlight is available to the plants is called photoperiod. The effect of photoperiod on the germination of seeds and flowering in plants is called photoperiodism. Garner and Allard (1920) recognized three classes of plants according to their photoperiodic responses as (i) Short day plants, (ii) Long day plants, and (ii) Day neutral plants.

2.6 PHYTOCHROMES:

Phytochromes are a class of photoreceptor in plants, bacteria and fungi use to detect light. They are sensitive to light in the red and far-red region of the visible spectrum and can be classed as either Type I, which are activated by far-red light, or Type II that are activated by red light. Recent advances have suggested that phytochromes also act as temperature sensors, as warmer temperatures enhance their de-activation. Phytochromes control many aspects of plant development. They regulate the germination of seeds (photoblasty), the synthesis of chlorophyll, the elongation of seedlings, the size, shape and number and movement of leaves and the timing of flowering in adult plants. Phytochromes are widely expressed across many tissues and developmental stages. Other plant photoreceptors include cryptochromes and phototropins, which respond to blue and UV-A light and UVR8, which is sensitive to UV-B light.

2.7 VERNALIZATION:

Vernalization is the induction of a plant's flowering process by exposure to the prolonged cold of winter, or by an artificial equivalent. After vernalization, plants have acquired the ability to flower, but they may require additional seasonal cues or weeks of growth before they will actually flower. Typical vernalization temperatures are between 5 and 10 ^{0}C (40 and 50 ^{0}F). For many perennial plants, such as fruit tree species, a period of cold is needed first to induce dormancy and then later, after the requisite period of time, re-emerge from that dormancy prior to flowering. Many monocarpic winter annuals and biennials, including some ecotypes of *Arabidopsis thaliana* and winter cereals such as wheat, must go through a prolonged period of cold before flowering occurs. It is possible to devernalize a plant by exposure to sometimes low and high temperatures subsequent to vernalization. For example, commercial onion growers store sets at low temperatures, but devernalize them before planting, because they want the plant's energy to go into enlarging its bulb (underground stem), not making flowers.

2.8 COORDINATION IN ANIMALS:

In animals, the message, communicated in the form of nerve impulses, from receptors (sensory neurons) to central nervous system and from latter to the effectors (muscles and glands) is very quick. The nervous coordination in animals, however, has certain limitations. For instance, **(i)** Nerve impulses can reach only those animal cells which are connected by nervous tissue, and **(ii)** Such cells, after generation and transmission of nerve impulses, take some time to reset their mechanisms before a new impulse is generated and transmitted. In other words, cells cannot continuously generate and transmit electrical impulses. This is the reason why most multicellular organisms use another means of communication between cells, commonly termed chemical communication. The stimulated cells release a chemical directly into the blood. Other body tissue cells have the means to detect these chemical using special molecules or receptors present either on their surfaces or inside their cytoplasm. The message is then transmitted and these chemicals produce their effects. Chemical communication is, however, slow but it can potentially reach all the cells of the body regardless of nervous connections. These chemicals are called hormones and are secreted by various endocrine glands. The latter constitute endocrine system.

Hormonal information, like nervous system, is meant for internal communication and regulation of animal body functions. However, there are some basic differences between the two controlling systems. Still, the two systems (nervous system and endocrine system) operate in a coordinated way on many occasions. Many important functions of the endocrine system are, in fact, under the control of nervous system. Therefore, the two systems are often collectively termed as neuroendocrine system.

Table 2.3 Differences between Nervous and Hormonal Information

Sr. No.	Nervous Information	Hormonal Information
1.	It is sent as an electrical impulse along axons, and as a chemical across synapse.	It is sent as a chemical messenger via blood stream.
2.	Information travels rapidly, in millisec-onds.	Information travels slowly.
3.	Information is directed to specific receptors one or a few nerve fibres, gland cells or other neurons, i.e., it is addressed by name.	Information is spread throughout the body by blood, from which the target cells or organs pick it up, i.e., it is addressed to 'whom it may concern'.
4.	It gets response immediately.	It gets response usually slowly.
5.	Its effects are short-lived.	Its effects are generally more prolonged.

Unicellular organisms respond to environmental stimuli by moving towards or away from the source of stimulus. Such responses of unicellular organisms (Amoeba) are termed as taxes (singular taxis). If they move towards the source of stimulus, their response is termed positive taxis and if they move away from the source of stimulus, their response is termed negative taxis. The responses of multicellular animals to stimuli, however, are not as simple. Evolution of multicellularity in animals necessitated the development of some system for the control and coordination of the activities of various cells of the body. Such a control and coordination, in fact, requires (a) Gathering information about changes in the external environment, **(b)** Transmitting this information to the internal cells located away from the body surface, and (c) Exchange of information between the cells situated away from each other. In lower multicellular animals, the coordination takes place through the nervous system. However, in higher animals, coordination takes place through two types of control systems: nervous system and endocrine system.

1. **Nervous System:**

The nervous system is composed of specialized cells called neurons (nerve cells which exercise control by sending electrical signals called nerve impulses. The nervous control is speedy and flexible but its effect is localized.

2. **Endocrine System:**

The endocrine system consists of specialized glands (endocrine glands) which bring about control by sending chemical messengers termed hormones. The hormonal control is usually slow acting and its effect is diffuse

2.9 NERVOUS SYSTEM IN ANIMALS:

All multicellular animals (except sponges) possess simple or complex nervous system. In all these animals, nervous systems is comprised of nervous tissue having specialized cells called neurons or nerve cells to respond to stimuli and coordinate animals' activities. Nerve cells are, in fact, the structural and functional units of nervous system. In higher multicellular animals, the nervous tissue consists of nerve cells or neurons, nerve fibres, bundle of nerve fibres forming nerves, packing cells (neuroglia), ependymal cells and neuro-secretory cells.

In animals, the message, communicated in the form of nerve impulses, from receptors (sensory neurons) to central nervous system and from latter to the effectors (muscles and glands) is very quick and has certain limitations. The two systems (Nervous System and Endocrine Glands System) are often collectively used which is termed as neuroendocrine system. The nervous system is composed of specialized cells called neurons (nerve cells which exercise control by sending electrical signals called nerve impulses. The nervous control is speedy and flexible but its effect is localized.

2.9.1 Structure of Neuron:

Neuron is the structural and functional unit of nervous system. It has a special structure to receive, conduct and transmit impulses; it varies greatly in size and shape. Neurons are, in fact, the largest cells present in the human body, sometimes reaching 90-100 cm. A neuron consists of three noticeable parts such as **(i)** Cell body **(ii)** Dendrites (Singular dendron), and **(iii)** Axon (Fig. 2.14).

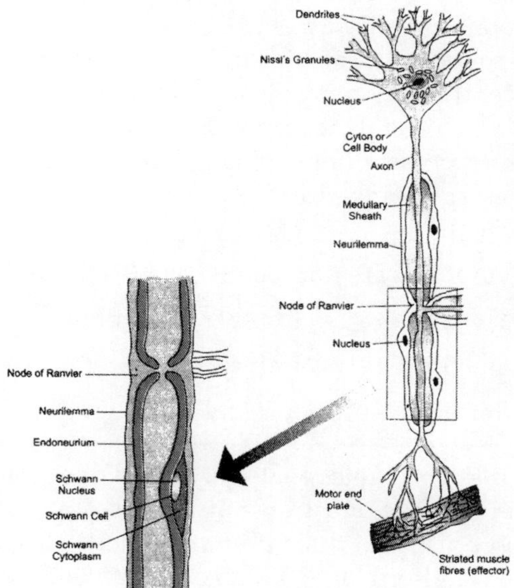

Fig. 2.14 Structure of a Neuron or Nerve cell

(i) Cell Body:

The cell body of a neuron is also called cyton or soma. It is broad, rounded, pyriform or stellate part of the neuron. It has abundant cytoplasm, called neuroplasm and a relatively large, spherical nucleus. The cytoplasm has mitochondria, golgi bodies, neurofibrils, neurotubules, and special granules called nissl's granules. These nissl's granules are characteristic of nerve cells. Centrioles are, however, absent in neurons. Cell body is concerned with metabolic maintenance and growth. It also receives nerve impulses from dendrites and transmits them to axon.

(ii) Dendrites :

These are several short, tapering, much branched protoplasmic processes stretching out from the cell body of a neuron. Dendrites are the parts of neuron where sensation (information) is acquired. The information then travels as an electric impulse towards the cell body. Dendrites contain Nissl's granules and neurofibrils.

(iii) Axon:

It is a single, very long, cylindrical protoplasmic process (nerve fibre) of uniform diameter arising from the cell body. Axon is highly branched at its terminal end. The terminal branches are called terminal arborization. Axon terminals are often knob-like and these may end in nerve fibres (forming neuromuscular junction) or glands or form synapses with dendrites of other neurons. The axon conducts nerve impulses away from the cell body. Axon is covered with one or two sheaths. Sheathed axon is termed nerve fibre. The cell membrane of the axon is called axolemma and its cytoplasm is termed axoplasm. It lacks Nissl's granules. However, neurofibrils are present. The single sheath present over the axon is made of Schwann Cells and called as Neurilemma. The axon may have an additional insulating and protective sheath of myelin around it. It is present between neurilemma and axon. Nerve fibres or axons having myelin sheath are termed myelinated nerve fibres and those without this sheath are termed non-myelinated nerve fibres. Myelinated nerve fibres conduct impulses more efficiently than non-myelinated nerve fibres. At intervals, myelinated nerve fibres possess unmyelinated areas called as Nodes of Ranvier. Neurons transmit messages in the form of nerve impulses. They have following special properties such as:

- They do not divide, and are not repaired, when injured.
- They die if deprived of oxygen for over five minutes.
- They use only glucose as a respiratory substrate.
- From shortly after birth, new neurons do not develop.

Neuron is the structural and functional unit of nervous system. It has a special structure to receive, conduct and transmit impulses; it varies greatly in size and shape. A neuron consists of three prominent parts such as (i) Cell body (ii) Dendrites (Singular Dendron), and (iii) Axon. The neurons are of three types such as (i) Sensory (Receptor) Neurons, (ii) Motor (Effector) Neurons, and (iii) Relaying (Connector) Neurons.

Table 2.4 Differences between Axons and Dendrites

Sr. No.	Axons	Dendrites
1.	These are long, uniform processes.	These are short, tapering processes.
2.	A single axon arises from the discharging end of a cell body.	Number of dendrites arises directly from the receiving surface of cell body.
3.	These do not possess Nissl's granules. However, neurofibrils are present.	These contain Nissl's granules and neurofibrils.
4.	Terminal branches of each axon enlarged to form knobs.	The ends taper and do not have knobs at the tips of branches.
5.	These carry nerve impulses away from the cell body.	These acquire sensations and send nerve impulses to the cell body.
6	Neurolemma is present	A covering of neurolemma is absent

2.9.2 Arrangement of Neurons:

The neurons lay end-to-end in chains to transmit nerve impulses in the animal body. Each neuron receives an impulse through its dendrites and passes it on to the next neuron in the chain through its axon via cell body.

Synapse: The neurons are not connected. There occurs a very minute gap between terminal portion of axon of one neuron and the dendron of other neuron. This minute gap is called synapse (Fig. 2.15). At the synapse, axon terminal comes in close proximity to the dendron terminal of next neuron. Axon terminal is expanded to from presynaptic knob. On the other hand, the dendrite terminal forms post-synaptic depression. In between the two, lies a narrow fluid filled space called synaptic cleft. Nerve impulse is a self-propagated electrochemical current that travels from one neuron to another neuron for the passage of message.

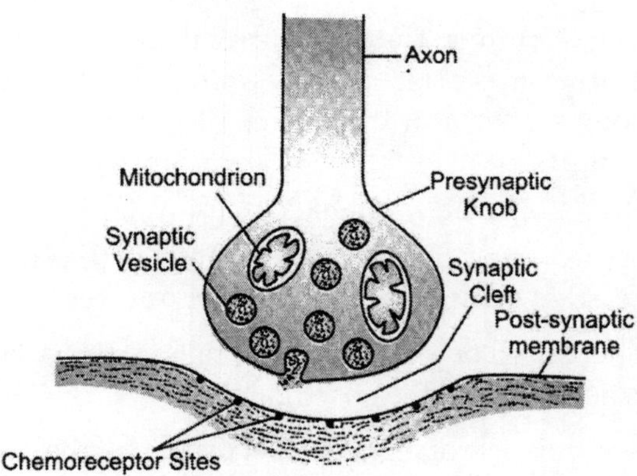

Fig. 2.15 Synapse : Impulse transmission

Route of Nerve Impulse:

The receptor/s, i.e., sensory neuron/s present in the body receives the stimulus. The stimulus generates a message in the form of an electrochemical impulse which then travels from the receptor (sensory neuron) to another neuron (Fig 2.16).

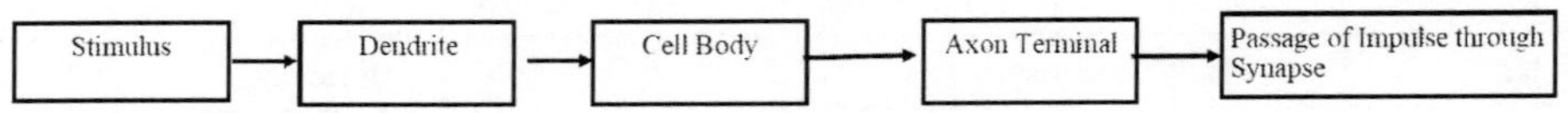

Fig. 2.16 Pathway of Nerve Impulse

As the nerve impulse reaches the presynaptic knob, the synaptic vesicles get stimulated to release a chemical called neurotransmitter in the synaptic cleft. The neurotransmitter molecules diffuse across the gap (synapse) to come in contact with the chemoreceptor sites in the post-synaptic membrane. In this way, nerve impulse passes across the minute gap to stimulate the dendron of other neuron. The synapse acts as a one-way valve to conduct nerve impulse in one direction only. This is so because chemical substance, called neurotransmitter, is secreted by synaptic vesicles only on one side of the gap, i.e. on axon's side. The neurotransmitter carries message across the synapse and passes it to the dendron of the other neuron. Dendrite of a neuron acquires the information. The information then travels as nerve impulse from dendron to cell body, axon and finally to dendron of other neuron through a synapse. In this way, impulses travel across the neurons only in one direction, i.e. from axon of one neuron to Dendron of other neuron through a synapse.

2.9.3 Types of Neurons:

The neurons are of three types:

(i) Sensory (Receptor) Neurons (ii) Motor (Effector) Neurons, and (iii) Relaying (Connector) Neurons.

(i) **Sensory (Receptor) Neurons:** These often occur in sense, and receive stimuli by their dendrites. The sensory neurons transmit impulses towards the central nervous system (brain and spinal cord) through their axons.

(ii) **Motor (Effector) Neurons:** The dendrites of these neurons synapse with axons of sensory neurons in central nervous system. They transmit impulses from central nervous system towards effectors (muscles or glands). The latter respond to stimuli.

(iii) **Relaying (Connector) Neurons:** These occur in the central nervous system (brain and spinal cord). These serve as links between sensory and motor neurons for distant transmission of nerve impulses.

A receptor is a nerve cell or group of nerve cells which is sensitive to a specific stimulus or to specific change in the environment.

An effector is some muscle or gland in specific part of the body which produces suitable response(s).

Sensory Receptors:

All higher chordate animals (vertebrates) receive a variety of external stimuli through specialized neurons termed sensory receptors or sensory neurons. These sensory receptors may be simple in structure or most complex sense organs. These animals have different sensory receptors/sense organs at specific regions of the body for receiving different stimuli. Every type of sensory receptor resembles in mode of working.

- The receptors collect information about changes in the environment around the animal in the form of stimuli.
- The stimuli set up appropriate electrical impulses in the sensory receptors (sensory neurons) which pass the information in the form of electrical impulses to the brain or spinal cord through nerves.
- The impulses that reach the brain from sensory neurons are termed sensations.
- The brain or spinal cord transmits motor impulses to appropriate effectors (muscles or glands) via motor neurons.
- In the brain, the sensations reanalyzed and interpreted as perceptions. Some of the reactions are also produced in the spinal cord without involving the brain.
- The effectors then produce suitable responses.

Sr. No.	Stimulus	Name of Receptors	Location of Receptors
1.	Heat, Cold	Thermoreceptors,	Skin
2	Smell	Olfactoreceptors	Nose
3.	Taste	Gustatoreceptors	Tongue
4.	Touch	Tangoreceptors	Skin
5.	Sound	Phonoreceptors	Ears

The sensory receptors, thus, merely receive the stimuli and set up similar nerve impulses. They cannot interpret the impulses generated in them by the stimuli. This power of interpretation lies with the brain. Actually, we feel, taste, smell, see and hear with the help of specific functional areas of the brain. Of course, brain without sensory receptors cannot detect environmental changes. Similarly, the sensory receptors alone are of no use. Hence, what actually matters is where the nerve impulses reach and not what generates them. Brain of higher vertebrates (birds and mammals) has specific area which provide areas which provide these animals learning abilities as well as store earlier experiences (memory) leading to efficient coordination.

2.9.4 Reflex Actions, Involuntary and Voluntary Actions:

Specific change in the environment evokes an appropriate response in the form of movement (action) in all living organisms. Such movements in the organisms are carefully controlled. Animals perform three types of actions such as:

- Reflex Actions

- Involuntary Actions
- Voluntary Actions

(a) Reflex Actions:

Reflex action was discovered by Marshal Hall (1833). These are nerve mediated, automatic involuntary actions that occur without the will of an animal. A reflex action may be defined as a spontaneous, automatic and mechanical response to a stimulus, action on a specific receptor, without the will of an animal.

Need for Reflex Action:

The receptors, present on various body parts, receive stimuli and transmit information in the form of nerve impulses to the central nervous system. Latter has various coordination centers where the information is processed. The processing is done on the basis of information and experiences already stored. The message is then sent by coordination centers through motor neurons to appropriate regions of the body which accordingly respond and react. The whole process takes some time. However, in certain situations, sensation requires immediate response as time used for processing may cause harm to the body. In such situations, reflex actions occur.

Examples: Blinking of eyes, movement of diaphragm during respiration, withdrawal of hand or foot every time it is suddenly pinched or pricked with a needle or touched by a hot object, coughing, yawning, sneezing etc. are all reflex actions (Fig 2.16 and 2.17). A gentle strike below the knee cap, while sitting with freely hanging legs, kicks the leg forward (knee-jerk reflex) is another example of reflex action. Similarly, watering of mouth on seeing the food, when hungry, is also a reflex action.

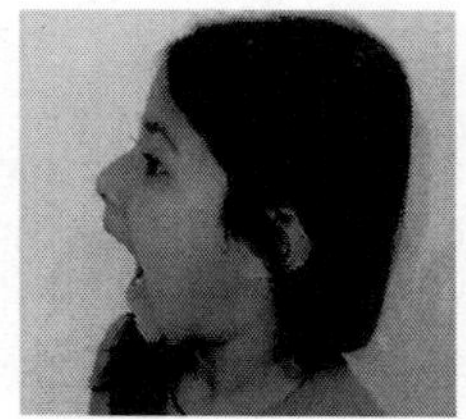

(A) Touch to hot object

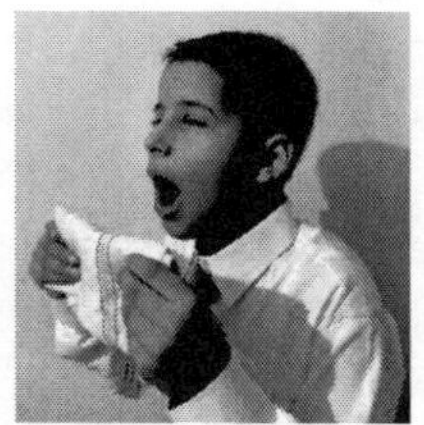

(B) Sneezing

(C) Coughing

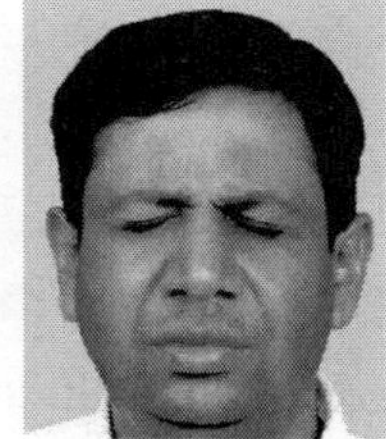

(D) Blinking of Eyes

Fig. 2.17 Various Examples of Reflex Action

Reflex Arc:

The path taken by nerve impulse in a reflex action is called reflex arc. It consists of five parts: Receptor (ii) Sensory Nerve (iii) Central Nervous System (iv) Motor Nerve and (v) Effector organ (Fig. 2.18).

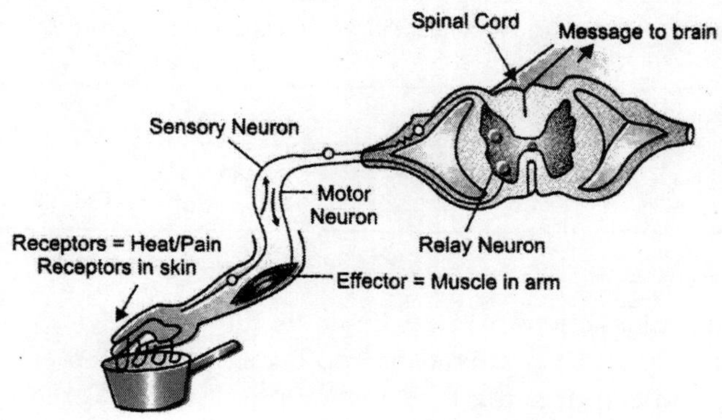

Fig. 2.18 Reflex Arc consists of all five parts

(i) **Receptor:** It is a specific group of cells or organ, the neurons of which receives a stimulus and set up a sensory impulse.

(ii) **Sensory (Afferent) Nerve:** It carries sensory impulse from the receptor to the central nervous system.

(iii) **Central Nervous System:** It is spinal cord or brain. Its neurons analyze and interpret the sensory impulse and set up an appropriate motor impulse. Accordingly, reflexes are termed spinal reflexes or cerebral reflexes.

(iv) **Motor (Efferent) Nerve:** It carries motor impulse from the central nervous system to specific per instructions received from central nervous system.

(v) **Effector organ:** It is a muscle, gland or organ. The effector organ is activated by motor impulse to provide a suitable response to the stimulus.

Significance of Reflex Action:

- It checks overloading and overtaxing of brain.
- It results in quick response to otherwise stimuli without the processing done by coordinating centres of central nervous system.
- It has a survival value.

Reflex action is nerve mediated automatic involuntary actions that occur without the will of an animal. A reflex action may be defined as a spontaneous, automatic and mechanical response to a stimulus, action on a specific receptor, without the will of an animal.

Table 2.5 Differences between Reflex Action and Walking

Sr. No.	Reflex Action	Walking
1.	It is inborn (inherited) and present in the individual since birth.	It is not inherited and is acquired through learning.
2.	It is automatic and occurs without the will of individual.	It is under the control of cerebellum part of brain and occurs with the will of individual.
3.	It cannot be changed.	It can be change.
4.	It has survival and protective value.	It has various functions other than survival and protective value

(b) Involuntary Actions:

Involuntary muscular actions are performed by the animal without its will. These occur automatically and the animal has no choice in it. Such actions are meant for controlling and coordinating the functioning of internal organs. Many of the involuntary action are controlled by the midbrain and hind brain. Regular beating of heart, blood pressure, movements of diaphragm during normal respiration, peristaltic movements in the oesophagus, salivation, vomiting, movement of the internal viscera etc. are all involuntary actions and are controlled by hind brain.

(c) Voluntary Actions:

Voluntary muscular actions are performed by the animal with its will. In each voluntary action, the animal exercise its choice so that the same stimulus may receive different times depending upon the situation. For instance, on seeing a snake in the way, one way to run away on first occasion or call for help on second occasion or try to kill it to save himself on the third occasion. All such actions are voluntary actions that are controlled by cerebellum part of hind brain. Similarly, walking in a straight line, riding a bicycle, picking up a pencil are also voluntary actions controlled by cerebellum. This part of the brain is responsible for precision of voluntary actions and maintains the posture and balance of the body. Had we not been thinking about each of these events, these events would have failed to happen.

Table 2.6 Differences between Voluntary and Involuntary Action

Sr. No.	Features	Voluntary Action	Involuntary Action
1.	Nature	Conscious thought like make decision about making action) Free will Consciously control skeletal muscles	Does not involve thought Not under control of the will No control on the activities
2.	Role	Respond with the benefit of experience	Respond quickly to avoid danger

3.	Controlled by	Forebrain (Cerebrum)	Hind brain (Cranial reflex action)
4.	Speed of Action	Slow response	Rapid response
5.	Response to the same stimulus	The same stimulus may produce various responses e.g. Decide to eat the food or not during hunger	The same stimulus always results in the same response e.g. the knee jerk reflex
6	Examples	If we want to ask question, we arise our hands	Involving as Skeletal muscle (Knee Jerk), Smooth muscles (Peristalsis) and Cardiac muscles (Pumping of the heart)

Involuntary muscular actions are performed by the animal without its will. These occur automatically and the animal has no choice in it. Such actions are meant for controlling and coordinating the functioning of internal organs. Voluntary muscular actions are performed by the animal with its will. In each voluntary action, the animal exercise its choice so that the same stimulus may receive different times depending upon the situation.

How does the nervous tissue cause action?

In the above account, we have studied about nervous tissue, mechanism of collection of information by receptors, its transmission to CNS and then to the effectors (muscles) which do the final act. In other words, actions/movements are finally performed by muscles. The question is how do animal muscles move? Let us now explain this process. The terminal part of axon of motor neuron synapses with the membrane of the muscle (motor end plate). When the nerve impulse from CNS reaches the muscle, the muscle fibres show movement. The muscle fibres, in fact, have special proteins that change their shape and their arrangement in response to nerve impulses. New arrangement of these proteins gives the muscle different shapes.

How the effectors (or muscles) cause action or movement?

When a motor nerve impulse sent by the spinal cord (or brain) reached the effector organs (which) are muscles), then the muscles cause action or movement (such as lifting the hand away from a hot plate). We will now describe how muscles are able to move in response to electrical nerve impulses and cause action. Muscles are made up of muscle cells. Muscle cells contain special proteins which can change their arrangement when stimulated by electrical impulses, causing the muscle cells to change shape and contract. When the muscle cells contract, the muscles also contract (and become shorter). When the muscles contract, they pull on the bones of the body part and make it move. For example, when electrical impulses sent by the spinal cord (or brain) stimulate the biceps muscle of the upper arm, they make biceps muscle to contract. And when the biceps muscle contracts, it pulls on a bone of the lower arm and makes it move (lifting the hand away from the hot plate).

Please note that the contraction of muscles (or muscle cells) caused by the action of electrical impulses is a reversible process.

2.10 HUMAN NERVOUS SYSTEM:

Human nervous system is the most complex. The nervous system can be divided into two main parts such as (Fig. 2.19).

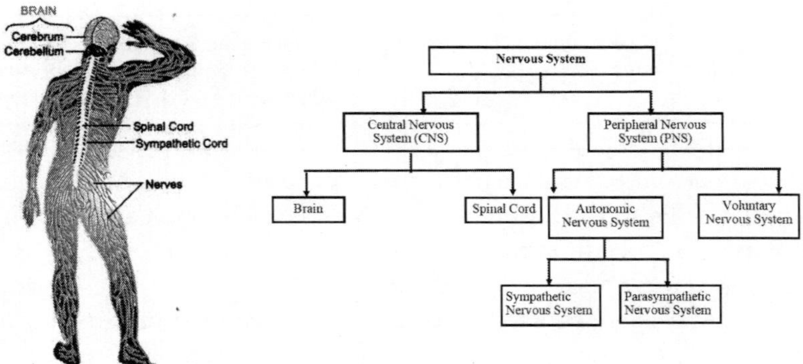

Fig. 2.19 Human Nervous System and its main parts

1. Central Nervous System (consisting of brain and spinal cord), and
2. Peripheral Nervous System (consisting of all the nerves of the body like cranial nerves, spinal nerves and visceral nerves). The peripheral nervous system can be further divided into two parts:

(i) Voluntary Nervous System (which is under voluntary control from the brain), and

(ii) Autonomic Nervous System (which operates automatically or involuntarily).

2.10.1. Central Nervous System (CNS).

It is hollow and lies on the mid dorsal part along the main (longitudinal) axis of the body. It is covered externally by part of axial skeleton. The CNS, in turn, consists of two parts such as (i) Brain or Encephalon (situated in the head) and (ii) Spinal Cord or Myelon (located in the neck and trunk) (Fig. 2.20)

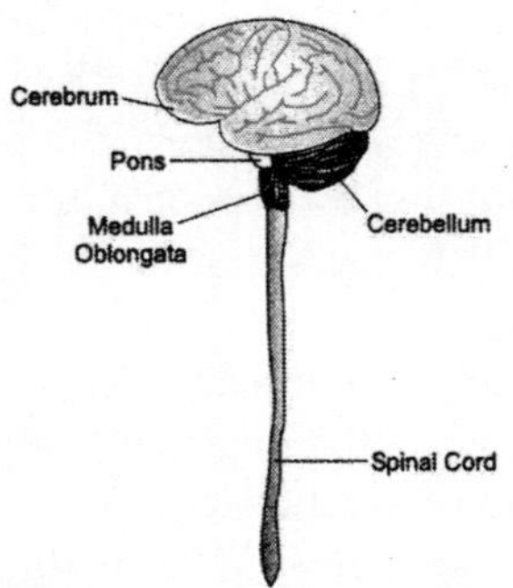

Fig. 2.20 Central Nervous System (CMS) of Human

2.10.1.1. Brain (Encephalon):

Brain is situated in the cranial cavity of the skull in the head region of the body. The bones of cranium or brain box protect this delicate organ from mechanical injury. The brain is the widest, and the uppermost part of the central nervous system. It is the highest coordinating centre in the body. Inside the box, the brain is contained in a fluid-filled balloon which provides further shock absorption. The study of brain in all aspects is called encephalogy.

Morphology:

The brain is soft, whitish organ. It weighs 1.2 to 1.4 kg and forms about 98% of the weight of the whole central nervous system. It has about 100 billion neurons or nerve cells. Brain is surrounded by three membranes known as meanings which provide protection to it. The space between these three meanings is filled with cerebro-spinal fluid which protects the brain from mechanical shocks. Brain is divisible into three main regions such as (Fig. 2.21).

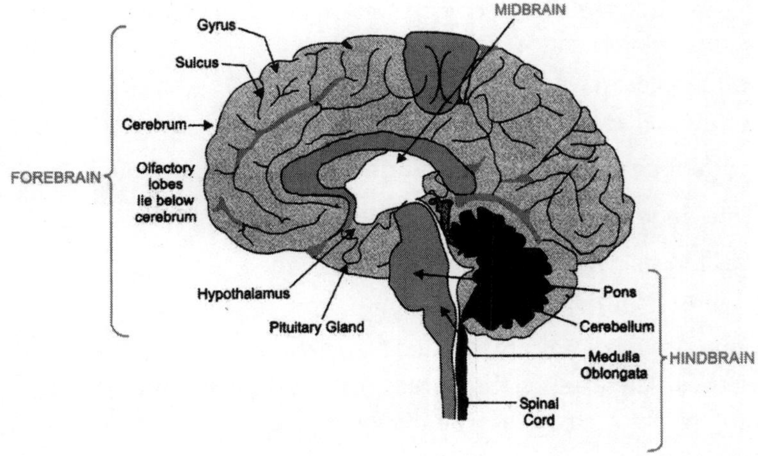

Fig. 2.21 The Human brain

(a) Forebrain or Prosencephalon

(b) Midbrain, and

(c) Hind brain

(a) Forebrain or Prosencephalon:

It forms the greater part of the brain. It further consists of three regions such as (i) Olfactory Lobes, (ii) Cerebral Hemispheres (Cerebrum) and (iii) Diencephalon.

(i) Olfactory Lobes:

These are a pair of widely separated club-shaped small structures. Olfactory Lobes are fully covered by the cerebral hemispheres and are, thus, visible only in the ventral view of the brain. Each olfactory lobe consists of an anterior olfactory bulb and a posterior narrow olfactory stalk. These lobes receive impulses from Olfacto-receptors (present in the olfactory epithelium in the nose) and relay sense of smell to the temporal region of the cerebrum.

(ii) Cerebral Hemispheres or Cerebrum:

It forms the largest part of the brain. It is also the most complex and specialized part of the brain. The two cerebral hemispheres lie side by side, being separated from each other by a deep longitudinal cerebral fissure. The surface of these cerebral hemispheres is greatly folded to accommodate large number of nerve cells. The folds are called gyri (singular gyrus) and the depressions between them are termed sulci (singular sulcus). Each cerebral hemisphere is divided by three deep gaps (fissures) into four lobes as another frontal lobe and lateral temporal lobe (Fig. 2.22). Different areas of cerebrum have different functions. For instance, it has sensory areas which receive impulses from the sense organs or sensory receptors, e.g., eyes, ears, tongue, nose and skin. Similarly, cerebrum also has motor areas which send instructions in the form of impulses to various effectors of the body (e.g., muscles). The latter respond to stimuli. There are specific regions in cerebrum for each kind of stimulus and its response. For instance-

- Frontal lobe is the region for speech, facial muscular activities as well as higher mental activities.
- Temporal lobe is the region for auditory reception (hearing).
- Occipital lobe is the region for visual reception (sight).
- Parietal lobe is the region for touch, taste, smell, temperature and conscious association.

Each lobe also has some areas called association areas which store information and experiences. These association areas control thinking, memory, learning and emotions. Internally, each cerebral hemisphere possesses a fluid-filled cavity called lateral ventricle.

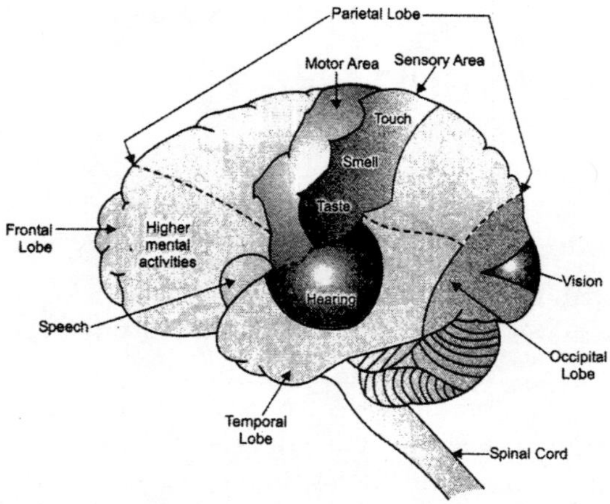

Fig. 2.22 Cerebrum showing the various function with specific region

(iii) Diencephalon:

It lies on the inferior side of the cerebrum and thus is visible in the ventral view of the brain. Its roof is called epithalamus, sides are called thalami and its floor is termed hypothalamus. Diencephalon has a narrow cavity called third ventricle. Hypothesis (Pituitary) is attached by a stalk or infundibulum to the hypothalamus region. Hypothalamus has control centres for hunger, thirst, fatigue, sleep, body temperature, sweating and emotions. It secretes neurohormones which regulate the secretions of anterior lobe of pituitary.

(b) Mid Brain (Mesencephalon):

It is significantly small region. It consists of two fibre tracts called crura cerebri and two swellings called superior and inferior colliculi on each side. The fibrous tracts, i.e., crura cerebri connect hind brain with the fore brain. The four swellings of both sides are together known as corpora quadrigemina. The two superior colliculi have centres for sight reflexes while the two inferior colliculi have centres for auditory reflexes. The mid brain controls reflex movements of:

(a) The head, neck and trunk in response to visual and auditory stimuli, and

(b) The eye muscles; changes in pupil size as well as shape of the eye lens.

(c) Hind Brain (Rhombencephalon):

The hind brain consists of three parts such as (a) Cerebellum, (b) Pons varolii and (c) Medulla oblongata. The cerebellum is the second largest part of the brain and constituting nearly 12.5% of hind brain. It has two large, lateral cerebellar hemispheres and a central vermis. Cerebellum maintains the posture, equilibrium and muscle tone. Pons varolii controls some aspects of respiration. Medulla oblongata is the posterior most part of the brain which lies below the cerebellum. It continues posteriorly into the spinal cord. It contains a fluid-filled cavity called fourth ventricle. Medulla oblongata controls (i) rate of heart beat (ii) breathing movements (iii) expansion and contraction of blood vessels to regulate blood pressure, and (iv) swallowing, coughing, sneezing and vomiting. Twelve pairs or cranial nerves arise from the brain

Table 2.7 Difference between Cerebrum and Cerebellum

Sr. No.	Cerebrum	Cerebellum
1.	It is the part of fore brain.	It is the part of hind brain.
2.	It is the largest part constituting 80% of the brain.	It is much smaller, constituting 12.5% of the brain.
3.	It forms the front, superior and lateral sides of the brain.	It lies in the posterior region of the brain.
4.	It has two parts called cerebral hemispheres.	It has three parts: two lateral cerebellar hemispheres and one central vermis.

5.	It has two cavities called lateral ventricles.	Cavity is nearly absent.
6.	Cerebrum is the set of intelligence and memory.	Cerebellum coordinates muscular activity.
7.	It controls movements, speech, sight, smell, taste, hearing, intelligence etc.	It maintains equilibrium of the body.
8	It is the centre of both voluntary and involuntary activities	It provides precision to voluntary activities

2.10.1.2 SPINAL CORD:

Spinal cord is a cylindrical structure and is about 45 cm long. It begins in continuation with the medulla oblongata of brain and extends downwards upto early part of lumbar region. It then extends to the end of vertebral column as fibrous connective called filum terminale. Internally, the spinal cord possesses a narrow, fluid-filled cavity called central canal. Spinal cord is enclosed in the vertebral column or backbone which protects it. Like brain, spinal cord too is surrounded by meninges. Thirty one pairs of spinal nerves arise from the spinal cord. Spinal cord performs two important functions such as (a) It conducts sensory and motor impulses to and from the brain, and (b) It acts as a centre for the reflex actions. Thus, it reduces brain's work.

2.10.2. Peripheral Nervous System:

It connects CNS with different parts of the body. It has two components, voluntary and involuntary nervous system.

- **Voluntary nervous system** is under the control of will. It consists of nerves that arise directly from CNS connecting different body parts for voluntary (conscious) control of the brain.
- **Involuntary peripheral nervous system (autonomic nervous system)** is not under the control of human will. It develops from branches of some cranial and spinal nerves called visceral nerves.

Peripheral nervous system, thus, consists of all the three types of nerves namely as (a) Cranial nerves, (b) Spinal nerves, and (c) Visceral nerves.

(a) Cranial Nerves:

Cranial nerves arise from the brain and spread to various parts of the head. They are 12 pairs in number. Cranial nerves I, II and VIII are sensory nerves ; cranial nerves III, IV, VI, XI and XII are motor nerves ; and cranial nerves V, VII, IX and X are mixed nerves (containing both sensory and motor nerve fibres).

(b) Spinal Nerves:

Thirty one pairs of spinal nerves from the spinal cord along most of its length and spread throughout the body (except head region). They are all mixed nerves as they carry both sensory and motor nerve fibres.

(c) Visceral Nerves:

Apart from regulating normal functions of the body, many activities of the internal organs such as heart, kidney, lungs, urinary bladder, blood vessels, glands etc. are controlled by specific set of nerves called visceral nerves which mostly arise from spinal cord but a few from the brain also. These form the autonomic nervous system.

2.10.2.1 Voluntary Nervous System:

Those actions which need thinking and which are performed by us knowingly are called voluntary actions. For example, speaking to a friend, writing a letter, dancing, cycling, kicking a football, standing in a room or sitting on a chair, are all voluntary actions. The voluntary nervous system helps us take voluntary actions which are under the conscious control of the brain. We will now give an example to understand the working of voluntary nervous system. Suppose we are walking down to school at a slow pace. After covering some distance, we look at our watch and find that we are getting late. So, we start walking very fast. We can do this because of our voluntary nervous system as follows:

(a) When our eyes see time on the watch, they send this information to the brain through the sensory nerves.

(b) The brain analyses this information and decides that since there is risk of being late to school, so we should walk faster.

(c) The brain sends the instructions to walk faster to the muscles of our legs through the motor nerves.

(d) The muscles of the legs act accordingly and make us walk faster.

This is an example of voluntary action and the decision to take this voluntary action has been made by the voluntary nervous system.

2.10.2.2 Involuntary or Autonomic Nervous System:

The term 'autos' means 'self' and 'nomos' means 'governing', so 'autonomic nervous system' means 'self-governing nervous system'. Thus, it is the system which operates automatically or involuntarily. It includes all those responses against stimuli which are not under the control of animal (involuntary activities). Visceral nerves of autonomic nerves system control the activities of internal organs. Autonomic nervous system is also termed as visceral nervous system. **The autonomic nervous system (ANS) is subdivided into two parts as (i) Sympathetic nervous system (ii) Parasympathetic nervous system.** The autonomic nervous system is that part of the peripheral nervous system which controls the activities of the organs inside our body automatically even without our thinking about them. The autonomic nervous system is a specific network of nerves in the body which controls the processes like breathing, heartbeat, digestion, sweating, etc., that maintain our life and keep us alive. The nerves of the autonomic nervous system are attached to the smooth muscles of the various internal organs of the human body like head, heart, blood vessels, alimentary canal, lungs, kidneys, urinary bladder, glands and skin, etc. Thus, the autonomic nervous system controls and regulates the functions of the internal organs of our body involuntarily (on its own). The smooth muscles of various internal organs receive both sympathetic and parasympathetic nerve fibres. Both the systems have involuntary opposing effects. If one system exerts stimulatory effect on an organ the other system exerts inhibitory effect on the organ (Table 2.8).

Table 2.8 Effects of Sympathetic and Parasympathetic System

Sr. No.	Organ	Sympathetic system	Parasympathetic system
1.	Heart	Increases contraction and rate of heart beat.	Decreases contraction and rate of heart beat.
2.	Blood vessels	Constriction of blood vessels.	Dilation of blood vessels.
3.	Lungs	Dilates bronchi and bronchioles	Constricts bronchi and bronchioles.
4.	Eyes	Dilates pupils.	Constricts pupils.
5.	Stomach	Inhibits secretion of gastric juice in stomach.	Stimulates secretion of gastric juice in stomach.
6.	Salivary glands	Inhibits secretion of saliva in the buccal cavity.	Stimulates secretion of saliva in the buccal cavity.
7.	Urinary bladder	Relaxation	Contraction.
8.	Hair	Erect	Oblique

Protective Coverings:

Both brain and spinal cord are protected from mechanical injury and shock by bony cases around them. Brain is protected by cranium (brain box) while spinal cord is protected by vertebral column. There are also present additional protective coverings called meanings (singular meninx) between the brain or spinal cord and their respective bony cases. The spaces between the meanings are filled by a fluid called cerebrospinal fluid. **Cerebrospinal fluid** is a clear, colourless, slightly alkaline fluid present in the ventricles of the brain, central canal of spinal cord and spaces between the meninges. It protects the CNS from shocks and keeps it moist. Electroencephalograph is an instrument which records the electrical activity of the brain in the form of a graph of electric potentials generated with time. Such a record is called electroencephalogram (EEG). The electroencephalogram (EEG) of a patient is done by placing two electrodes of electroencephalograph instrument on the scalp of the patient. Then a record of four different types of waves (alpha, beta, delta and theta) is produced on the graph paper. These waves vary in their frequency. These waves give the characteristic activity of the brain of a person. The EEG of a patient is useful to diagnose brain ailment.

2.11 HORMONES:

The term 'hormone' was introduced by William M. Bayliss and Ernest H. Starling. Hormones (Gk. Hormein = to excite) are informational molecules secreted by specialized ductless endocrine glands in specific part of the body in response to changes in external or internal environment. These chemical messengers are carried by blood to another part where they stimulate or inhibit specific physiological processes for the good of the body as a whole. Thus, hormones are chemical messengers secreted by ductless endocrine glands that regulate the biological processes in the living organisms. The various characteristics of hormones are given below:

- The hormones are secreted in small amounts by the endocrine glands.
- The hormones are poured directly into the blood and carried throughout the body by blood circulatory system.
- The hormones have their effect at the sites different from the sites where they are made. So, they are also called chemical messengers.
- The hormones act on specific tissues or organs (called target organs).
- The hormones coordinate the activities of the body and also its growth.

2.11.1 Characteristics of Hormones:

General characteristics of hormones are listed below:

- These are secreted by living cells to act as chemical messengers.
- They are organic compounds of diverse nature.
- They are secreted in small amounts by the cells of endocrine glands.
- They are poured directly into the blood and carried to different parts of the body.
- They act on specific tissues or organs called target organs.
- They act away from the site of production.
- The hormones may accelerate or inhibit specific physiological processes. They neither act as building material nor provide energy.
- They are often used up in their regulatory action.
- These are generally slow in action.

2.11.2 Release of Hormones:

Hormones are released from the endocrine glands by three types of stimuli such as:

(i) **Specific metabolite:** The presence of a specific metabolite in the blood elicits the hormone to deal with it. For instance, excess of glucose in the blood causes the release of insulin hormone from the pancreas (endocrine gland). Insulin lowers the level of glucose in the blood.

(ii) **Other hormone:** The presence of a specific hormone in the blood induces the release of another hormone. For example, certain hormones released from the anterior pituitary gland (endocrine gland) stimulate specific endocrine glands to release their hormones.

(iii) Neuronal impulse: Neurons of autonomic system stimulate hormone released from some endocrine glands. For instance, adrenaline (epinephrine) and noradrenaline (noerepinephrine) hormones are released from the adrenal medulla (endocrine gland) on the arrival of nerve impulses during anxiety, stress and danger.

2.11.3 Glands:

A cell, a tissue, or an organ which secretes certain useful chemical compounds required for particular functions is called a gland. Animals have three types of glans such as (1) Exocrine glands (2) Endocrine glands, and (3) Heterocrine glands

1. Exocrine Glands:

These glands have ducts for discharging their secretions on to the body surface or into the cavities in the body (Fig. 2.23). Sweat and sebaceous glands in the skin, salivary glands around the buccal cavity, gastric glands in the stomach wall, and liver are some examples of exocrine glands.

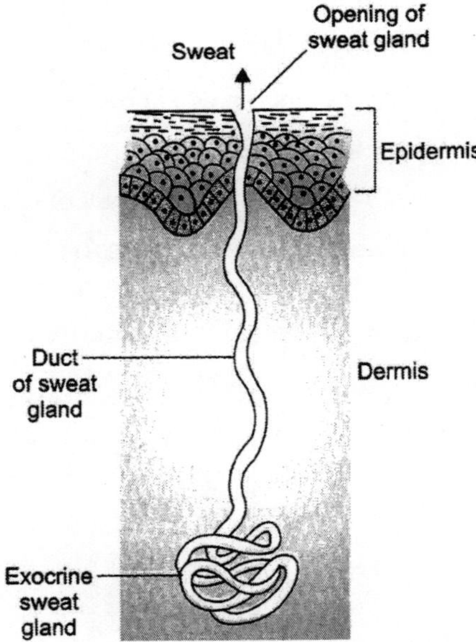

Fig. 2.23 Exocrine Gland showing discharge of their secretions

2. Endocrine Glands:

These glands lack ducts and pass their secretions into the surrounding blood for transport to the site of action (Fig. 2.24). They are also called ductless glands or glands of internal secretion (Endon = within; krinien = to separate). Their secretions are known as hormones or internal secretions. Pituitary, thyroid, parathyroid glands, adrenal glands etc. are examples of endocrine glands.

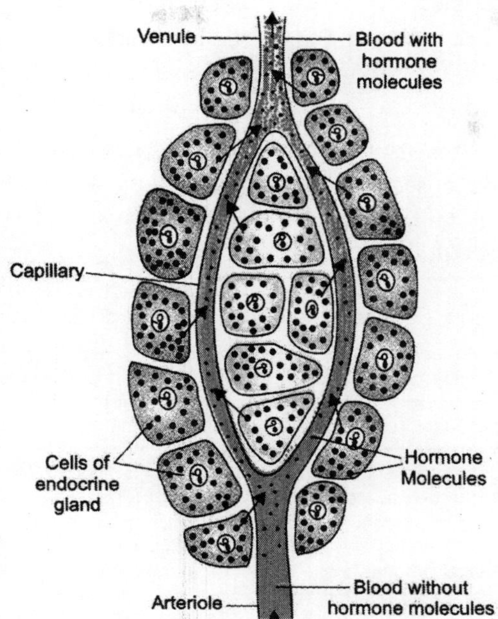

Fig. 2.24 Endocrine gland showing discharge of their secretions

3. Heterocrine Glands:

These glands consist of both exocrine tissue and endocrine tissue. The exocrine tissue sends its secretion or product by way of a duct; the endocrine tissue discharges its secretion into the blood. Pancreas and gonads (testes and ovaries) are examples of Heterocrine glands. The endocrine portion of pancreas juice containing digestive enzymes (trypsin, lipase and amylase) into the pancreatic duct that leads to the alimentary canal. Similarly, endocrine portion of the testis (male gonad) secretes testosterone hormone while the exocrine part releases sperms (male gametes) into the duct. Ovaries (female gonad) too have endocrine as well as exocrine tissues; the former secretes estrogen and progesterone hormones, and the latter release ova (female gametes) into the duct.

Hormones are chemical substances secreted in very small amounts by specialized tissues in the body called endocrine glands. These hormones coordinate the activities of living organisms and also their growth. Animals have three types of glands as Exocrine glands, Endocrine glands and Heterocrine glands.

- Exocrine glands have ducts for discharging their secretions on to the body surface or into the cavities in the body as sweat and sebaceous glands in the skin, salivary glands around the buccal cavity, gastric glands in the stomach wall, and liver etc.

- Endocrine glands lack ducts and pass their secretions into the surrounding blood for transport to the site of action such as Pituitary, thyroid, parathyroid glands, adrenal glands etc. are examples of endocrine glands.
- Heterocrine Glands consist of both exocrine tissue and endocrine tissue. The exocrine tissue sends its secretion or product by way of a duct; the endocrine tissue discharges its secretion into the blood. Pancreas and gonads (testes and ovaries) are examples of Heterocrine glands.

Table 2.9: Difference between Exocrine Glands and Endocrine Glands

Sr. No.	Exocrine Glands	Endocrine Glands
1.	These glands possess ducts.	These glands lack ducts.
2.	These glands pour their secretions to the target directly or through ducts.	These glands pour their secretions into blood.
3.	Secretions are various enzymes, mucus, excretory substance, lubricants etc.	Secretions consist of hormones (chemical messengers).
4.	These glands do not occur in isolation.	These glands usually occur in isolation.
5.	The target sites are adjacent to the glands.	The target sites are far away from the glands.

2.11.4 Human Endocrine Glands and their Hormonal Secretions:

The endocrine glands vary in embryonic origin and are isolated from one another. Thus, they do not form an organ-system comparable to the other organ-systems of the body such as digestive system, respiratory system, circulatory system, excretory system, reproductive system etc. Still, they are together referred to as an endocrine system because they interact. A hormone from one gland may stimulate or inhibit another endocrine gland. Hormones from the hypothalamus regulate the secretion of hormones form the pituitary gland; and the hormones from the pituitary gland, in turn, influence many other endocrine glands (e.g., thyroid gland, adrenal gland, testes, ovaries etc.). Human body possesses large number of endocrine glands which elaborate, store and release chemical messengers called hormones. These endocrine glands are: hypothalamus, pituitary (or hypophysis), thyroid gland, parathyroid glands, pancreas, adrenal glands, pineal gland, thymus, testes (in males) and ovaries (in females). These glands are located in different parts of the body (Fig. 2.25).

(1) Hypothalamus:

It is composed of nervous tissue and situated at the base of the brain. The neurosecretory cells of the hypothalamus secrete several neurohormones called Releasing Hormones (RH) and Inhibiting Factors or Hormones (IF or IH). These neurohormones are carried to the pituitary gland to stimulate or inhibit the secretion of anterior pituitary hormones. Hypothalamus also elaborates two neurohormones such as (i) oxytocin and (ii) vasopressin (also called antidiuretic hormone or ADH).

These are subsequently stored in posterior lobe of pituitary. A brief summary of hypothalamic hormones and response of pituitary is given in (Table 2.10).

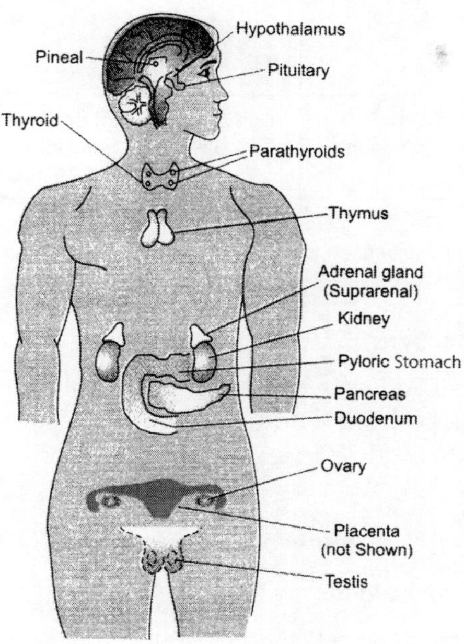

Fig. 2.25 Human Endocrine Glands

Table 2.10 Summary of Hypothalamic Hormones and Pituitary Response

Sr. No.	Hypothalamic Hormone	Target Organ	Response of Pituitary
1.	Thyrotrophin releasing hormone (T-RH)	Thyroid	Anterior Lobe Thyroid stimulating hormone (TSH) secretion
2.	Adrenocorticotrophin releasing hormone (A-RH)	Adrenal cortex	Adrenocorticotrophic hormone (ACTH) secretion
3.	Follicle stimulating hormone-releasing hormone (FSH-RH)	Ovary/Testis	Follicle stimulating hormone (FSH) secretion
4.	Luteinizing releasing hormone (LH-RH)	Ovary/Testis	Luteinizing hormone (LH) secretion
5.	Growth hormone-releasing hormone (GH-RH)	Most tissues	Growth hormone (GH) or Somatotrophic hormone (STH) secretion
6.	Growth hormone release inhibiting hormone (GH-RIH) or Somatostatin	-	Growth hormone secretion inhibited

7.	Prolactin-release hormone (P-RH)	Mammary glands	Prolactin hormone (PH) or Luteotrophic hormone (LTH) secretion
8.	Prolactin release inhibiting hormone (P-RIH	-	Prolactin secretion inhibited
9.	Melanocyte stimulating hormone-releasing hormone (MSH-RH)	Skin pigment cells	Intermediate Lobe Melanocyte stimulating hormone (MSH) secretion
10.	Melanocyte stimulating hormone release inhibiting hormone (MSH-RIH)	-	Melanocyte stimulating hormone secretion inhibited

(2) Pituitary Gland (Hypophysis):

Pituitary gland is a small, red-grey, pea-shaped gland attached to the hypothalamus of the brain by a stalk or infundibulum, and present just below the brain. The pituitary gland consists of three lobes such as (i) Anterior lobe, (ii) Intermediate lobe, and (iii) Posterior lobe.

All the three lobes of the pituitary secrete separate hormones.

(i) Anterior Lobe of Pituitary Gland:

This lobe of Pituitary gland secrets 6 types of hormones such as:

(a) **Follicle Stimulating Hormone (FSH):** It stimulates sperm formation in the male and growth of ovarian follicles in the female.

(b) **Luteinizing Hormone (LH):** In the male, it induces the leydig cells of the testes to produce male sex hormones named androgens (e.g. testosterone). The androgens make the male genital system to become fully grown and functional. In the female, the luteinizing hormone causes ovulation and secretion of female sex hormones as Estrogen from the maturing ovarian follicle and Progesterone by the corpus luteum formed in the empty ovarian follicle.

Follicle stimulating hormone and luteinizing hormone are together referred to as Gonadotrophic hormones (GTHs) or Gonadotrophins.

(c) **Somatotrophic or Growth Hormone (STH or GH):** It stimulates growth and development of all tissues by accelerating protein synthesis and cell division, and by retaining calcium in the body.

(d) **Prolactin Hormone (PRL):** It enhances mammary gland development and milk production in females.

(e) **Adrenocorticotrophic Hormone (ACTH):** It stimulates the adrenal cortex to grow and secrete its hormones.

(f) **Thyroid Stimulating Hormone (TSH):** It stimulates growth of thyroid gland and production of thyroid hormones.

Disorders: Improper secretion of GH hormone produces three important disorders.

- **Dwarfism:** It is caused due to the deficiency of growth hormone from early age.
- **Gigantism:** It is caused due to excess secretion of growth hormone from childhood.
- **Acromegaly:** It is caused due to excess secretion of growth hormone after adolescence.

(ii) Intermediate Lobe of Pituitary Gland:

It secretes a single hormone named melanocyte stimulating hormone (MSH). The hormone stimulates the synthesis of black pigment melanin in the skin, and also causes dispersal of melanin granules in the pigment cells (melanocytes) thereby darkening the colour in certain fishes and amphibians.

(iii) Posterior Lobe of Pituitary Gland:

It stores and releases two hormones:

(a) **Oxytocin (OT):** It induces contractions of smooth muscles of the uterus during the birth of the young one, and myoepithelial cells of the mammary glands to cause release of milk during sucking by the infant. Because of its role, oxytocin is called 'birth hormone' and 'milk ejecting hormone'.

(b) **Vasopression**: It is also called antidiuretic hormone (ADH). It decreases the loss of water in the urine by increasing the reabsorption of water in the distal convoluted tubules, collecting tubules and collecting ducts in the kidneys. Vasopression also stimulates the contraction of smooth muscles of the arterioles, thereby enhancing the arterial blood pressure, hence its name vasopressin.

Disorder of Pituitary Gland

- **Dwarfism:** Dwarfism is caused due to deficiency of growth hormone from early age.
- **Gigantism:** It is giant size of the young's with very tall skeleton and proportionally large muscle and viscera. It is caused due to excess secretion of growth hormone from childhood.
- **Acromegaly:** It is caused due to excess secretion of growth hormone after adolescence. Acromegaly in adults leads to overgrowth of the jaw bones and bowing of the spine (backbone).
- **Diabetes insipidus**: Deficiency of ADH reduces reabsorption of water and increases urine output, causing excessive thirst. This disorder is called diabetes insipidus. No glucose is lost in the urine of such a patient.

(3) Thyroid Gland:

It is the largest endocrine gland. It is a bi-lobed or H-shaped in structure and situated in the neck region. The thyroid gland secretes three hormones:

(a) Thyroxine (T_4)

(b) Triiodothyronine (T_3), and

(c) Calcitonin

Both thyroxine and triiodothyronine are iodinated forms of the amino acid tyrosine. These two hormones, therefore, need iodine for their synthesis. These two hormones have four functions such as:

- They maintain the basal metabolic rate (BMR) of the body by regulating the rate of oxidation of carbohydrates, fats and proteins and production of energy in our body.
- They promote growth of body tissues and development of mental faculties.
- They stimulate tissue differentiation. Therefore, they affect metamorphosis of a tadpole into an adult frog.
- Thyroxine also controls the working of kidneys.

Disorders:

Improper secretion of the above thyroid hormones causes the following disorders:

- **Grave's Disease (Exophthalmic Goitre):** It is caused by hyper secretion (over secretion) of thyroid hormones due to enlargement of thyroid. Excess of thyroid hormones increases metabolic rate and accelerates oxidation. This results in quick consumption of food, leaving nothing for storage and causing emaciation (excessive leanness).
- **Cretinism:** It is caused by hypo secretion (under secretion) of thyroid hormones in infants.
- **Myxodema:** It is caused by deficiency of thyroid hormones in adults. It is more common in women than in men.
- **Simple Goitre (Iodine deficiency goitre or Endemic goitre):** It is the enlargement of thyroid gland accompanied with cretinism or myxoedema. It is caused due to dietary deficiency of iodine. The disease is common in hilly areas. It causes enlargement of thyroid gland. Swollen neck is one of the symptoms of this disorder. Addition of iodides to the table salt prevents the disorder. In our country, common salt is iodized to provide required iodine to the thyroid gland. Over activity of thyroid is known as hyperthyroidism, and under activity as hypothyroidism. Calcitonin regulates the concentration of calcium and phosphorus in the blood.

(4) Parathyroid Glands:

These are four small, flat, oval glands situated on the posterior surface of the thyroid gland, two in each lobe of thyroid. They secrete parathormone (PTH). It is also called Collip's hormone. When the blood calcium level is below normal, then parathormone is released. Parathormone from parathyroid glands and calcitonin from thyroid gland act antagonistically to regulate the calcium-phosphorus balance in the blood.

(5) Adrenal Glands:

These are a pair of glands situated on upper side of each kidney. Therefore, they are also called Suprarenal Glands. Each endocrine gland has two distinct portions such as (i) Outer Adrenal Cortex and (ii) Inner Adrenal Medulla. Both these parts secrete different hormones.

(i) Adrenal Cortex:

It is firm, outer pale yellowish pink region essential for life. Adrenal cortex secretes three groups of steroid hormones:

- Mineralocorticoids (e.g., aldosterone). They help in mineral metabolism, and control the sodium and potassium ratio in the body fluids. In short, mineralocorticoids maintain water and electrolyte balance, and blood volume in the body.
- Glucocorticoids (e.g., cortisol). They regulate the metabolism of carbohydrates, proteins and fats. They also have anti-inflammatory and anti-allergic effects.
- Sex corticoids. They stimulate the development of secondary sexual characters both in males and females.

(ii) Adrenal Medulla:

It is an internal, soft, dark reddish brown tissue of adrenal gland. It enables the animal to face physical and emotional stress.Adrenal medulla secretes two hormones:

(i) Adrenaline (Epinephrine)

(ii) Noradrenaline (Norepinephrine).

Adrenaline is also termed as emergency hormone. In normal situations, these hormones are secreted in small amounts. However, when a person faces stress or danger, these are secreted in large amounts to prepare the body to face emergency situations. These increase the rate of heart beat to supply more oxygen to muscles. Also, blood supply to the digestive system and skin is reduced due to contraction of muscles around small arteries in these organs to direct more blood to our muscles. The breathing rate also increases due to the contraction of the diaphragm and rib muscles. Also, blood pressure, basal metabolic rate and sugar level in blood is increased. Because of the above roles of these hormones in 'fight or flight reaction', the adrenals are known as the glands of emergency.

(6) Pancreas:

Pancreas lies below the stomach in the loop of duodenum part of small intestine. It is an elongated, yellowish Heterocrine gland. Its exocrine part consists largely of lobules that secrete pancreatic juice which contains digestive enzymes (e.g., trypsin, amylase and lipase) for the digestion of proteins, carbohydrates and fats respectively. Interspersed at random among the lobules are endocrine cells, called islets of Langerhans which produce hormones. Endocrine part of pancreas (Islets of Langerhans) secretes two hormones such as (a) Insulin, and (b) Glucagon

(a) **Insulin** is secreted by -cells of islets of Langerhans. Insulin helps to lower the blood glucose level. Deficiency of insulin hormone in the body causes the disease known as diabetes mellitus.

(b) **Glucagon** is secreted by the -cells of islets of Langerhans. This hormone increases the blood glucose level.

Deficiency of Insulin:

- **Diabetes mellitus:** Deficiency of insulin hormone in the body causes a disease known as diabetes mellitus. In this disease, the patient excretes sugar (glucose) in urine, feels excessive thirst and also does excessive urination.

Table 2.11 Differences between Diabetes mellitus and Diabetes insipidus

Sr. No.	Diabetes Mellitus	Diabetes Insipidus
1.	It occurs due to the deficiency of insulin hormone.	It occurs due to the deficiency of ADH (antidiuretic hormone).
2.	Blood sugar level is high.	Blood sugar level is normal.
3.	Sugar is excreted in urine.	Sugar is not excreted in urine.
4.	Urine has normal concentration	Urine is dilute

(7) Gonads:

(i) **Testes:** In males, pair of testes is extra abdominal in position and located in scrotum. Testes secrete male hormones (e.g. testosterone). Groups of interstitial cells (also called Leydig cells) are present in connective tissue between the sperm-producing seminiferous tubules. The testosterone is secreted by these Leydig cells under the influence of Luteinizing hormone (LH) from the anterior lobe of pituitary gland. Testosterone serves the following functions:

- It stimulates the growth and development of male accessory glands, e.g., prostate, penis etc.
- It stimulates the development of secondary sexual characters in males at puberty, e.g., moustache, beard, hair on the body, deepening of voice, broadening of shoulders, enlarged and stronger bones and muscles.

- It stimulates formation of sperms (spermatogenesis) in the seminiferous tubules of testes.

(ii) Ovaries: Unlike testes in males, a pair of ovaries lies in the abdomen in females. They secrete three female sex hormones such as **(a)** Estrogen **(b)** Progesterone, and **(c** Relaxin

(a) Estrogen: Estrogen is secreted by the specific cells of Graffian follicles. It serves following functions:

- It stimulates the formation of ova (oogenesis) in the ovary.
- It stimulates the development of accessory sex characters such as enlargement of breasts during puberty, broadening of pelvis, growth of pubic and axillary hair, onset of menstrual cycle.

(b) Progesterone: It is secreted by the corpus luteum. The latter is formed by rupturing of mature Graffian follicle during ovulation. Progesterone plays following important roles:

- Maintenance of pregnancy by suspending ovulation.
- It stimulates thickening of uterine epithelium (i.e., changes in uterus) during menstrual cycle.

(c) Relaxin: It is produced by the corpus luteum at the end of the gestation period. It helps in easy birth of the young one.

(8) Pineal Gland:

It is a very small, reddish- grey, vascular, solid body lying between the two cerebral hemispheres of the brain. It secretes melatonin hormone. It regulates the working of gonads.

(9) Thymus Gland:

It is situated in the upper chest near the front side of the heart. Thymus is soft, pinkish, bi-lobed structure and is a prominent gland in the young child. However, it gradually atrophies in the adult. It secretes thymosin hormone which stimulates the development and differentiation of lymphocytes and thereby increasing resistance to infection.

Table 2.12 Summary of Major Hormones

Sr. No.	Gland	Position	Hormones	Functions
1.	Hypothalamus (Head Region)	Base of dien-cephalon	(a) Realizing hormones (RH)	Stimulate anterior/intermediate pituitary to secrete hormones.
			(b) Inhibiting hormones (IH)	Inhibit the secretion of hormones from anterior/intermediate pituitary gland

2.	Pituitary Gland(master Endocrine gland)	It has three lobes. It is attached to the lower surface of the brain.		
	(i) Anterior lobe	Beneath brain	(a) Follicle stimulating hormone (FSH)	Stimulates spermatogenesis (Sperms formation) in male. In females, it stimulates the follicle cells in the ovaries to develop into mature eggs and also stimulates them to produce Oestrogen.
			(b) Luteinizing hormone (LH)	Induces testosterone (Sex hormone) secretion in males. Ovulation and secretion of estrogen and progesterone (sex hormones) in female.
			(c) Growth Hormone (GH) or Somatotrophic hormone (STH)	Regulates protein metabolism and growth of bones. Deficiency causes dwarfism, excess produces gigantism in young and acromegaly in adults.
			(d) Prolactin Hormone (PRL)	Enhances mammary gland development and milk production in females.
			(e) Adrenocorticotrophic hormone (ACTH)	Stimulates the adrenal cortex to secrete cortisol hormone.
			(f) Thyroid stimulating hormone (TSH)	Controls the growth and functioning of the thyroid gland. Stimulates the thyroid gland to produce thyroxine.
	(ii) Intermediate Lobe	Front of larynx	(a) Melanocyte stimulating hormone (MSH	Stimulates the synthesis of melanin in the skin, and controls skin colour.
	(iii) Posterior Lobe	Front of larynx	(a) Oxytocin	Stimulates contraction of smooth muscles at the time of child birth. It also helps in milk ejection (lactation) from the mammary glands.
			(b) Vasopression or Antidiuretic Hormone (ADH)	Regulates water and electrolyte balance in body fluids. Reduces loss of water in urine.
3.	Thyroid	In the back of thyroid On top of kidneys	(a) Thyroxine or T4	Regulates oxidation of food. Excess causes exophthalmic goitre.
			(b) Triiodothyronine or T3	Deficiency produces cretinism in children, myxodema in adults.

			(c) Calcitonin (CT)	Controls calcium and phosphorus balance. Calcitonin lowers calcium level when by suppressing release of calcium ions from the bones. calcium level is high in blood.
4.	Parathyroid		(a) Para hormone (PTH)	Regulates calcium and phosphate levels in the blood. When blood calcium level is below normal, it mobilizes the release of calcium into the blood from bones. It has an action opposite to that of calcitonin on calcium metabolism.
5	Adrenals (i) Cortex		(a) Glucocorticoids e.g., Cortisol	Regulates carbohydrate, fat and protein metabolism in the body and the level of blood sugar. Regulates heart rate and blood pressure.
			(b) Mineralocorticoids e.g., Aldosterone	Controls salt and water metabolism. Its deficiency causes Addison's disease
			(c) Sex corticoids	Stimulates the development of secondary sexual characters both in males and females.
	(ii) Medulla		(a) Adrenaline	Whip up metabolism in emergency by increasing heartbeat, blood pressure, respiratory rate, blood sugar level etc. Both of these hormones together control emotions, fear, anger, blood pressure, heartbeat, respiration and relaxation of smooth muscles.
			(b) Noradrenaline	
6.	Pancreas	Below stomach	(a) Insulin	Regulates the conversion of glucose to glycogen; i.e., it lowers blood glucose level. Controls carbohydrate metabolism, its deficiency causes diabetes mellitus.
			(b) Glucagon	Regulates the conversion of glycogen back to glucose i.e., it increases blood glucose level. Release of sugar from liver.

7.	Gonads (a) Testes	In scrotal sacs	(a) Testosterone	Regulates the growth, development and functioning of accessory sex organs and controls the secondary sexual characteristics in males such as enlargement of penis and scrotum, growth of facial and public hair, and enlargement of larynx that causes deepening of voice. Develops male reproductive organs and accessory sexual character.
	(b) Ovaries	Lower abdomen	(a) Estrogen or Estradiol	Develops sexual genital organs and accessory sexual characters Thus oestrogens are responsible for development of secondary sexual characteristics in females like mammary gland, voice, hair pattern, etc.
			(b) Progesterone	Stops egg production, attaches-foetus to uterine wall.
			(c) Relaxin	Widening of pelvis at birth.
8	Pineal	It lies between the two cerebral hemispheres of the brain.	(a) Melatonin	Lightens colour, regulates working of gonads. Regulates circadian rhythm.
9.	Thymus	Near heart	(b) Thymosin	Regulates growth. Stimulates proliferation of lymphocytes (WBCs).

2.11.5 Homeostasis and Feed Back:

Homeostasis means keeping the internal chemical environment of the body constant. Hormones help to maintain homeostasis by their integrated action and feedback control. Feedback control is mostly negative, and rarely positive.

(a) Negative Feed Back Control: In a negative feedback control, synthesis of a hormone slows or halts when its level in the blood rises above normal. Two specific examples are cited to explain the negative feedback control.

- **Blood-thyroxine Homeostasis:** In response to some external stimulus, hypothalamus produces a thyrotophin- releasing hormone (T-RH). The latter then stimulates the anterior lobe of pituitary to secrete thyroid stimulating hormone or TSH. The TSH, in turn, stimulated the thyroid gland to produce hypothalamus and anterior lobe of pituitary which then secrete less 'releaser' hormone and thyroid stimulating hormone respectively (Fig. 2.26). A rise in the level of thyroid stimulating hormone (TSH) in the blood may also exert negative feedback on the hypothalamus and retard the secretion of thyrotophin-releasing hormone (T-RH). This restores the normal blood-thyroxine level.

If blood-thyroxine level falls below normal, the negative feedback effect on the hypothalamus and anterior lobe of pituitary disappears. These two endocrine glands then secrete more thyrotophin-releasing hormone (T-RH) and thyroid stimulating hormone (TSH) respectively. This stimulates are thyroid gland to secrete more thyroxine. As a result, normal blood-thyroxine level is restored.

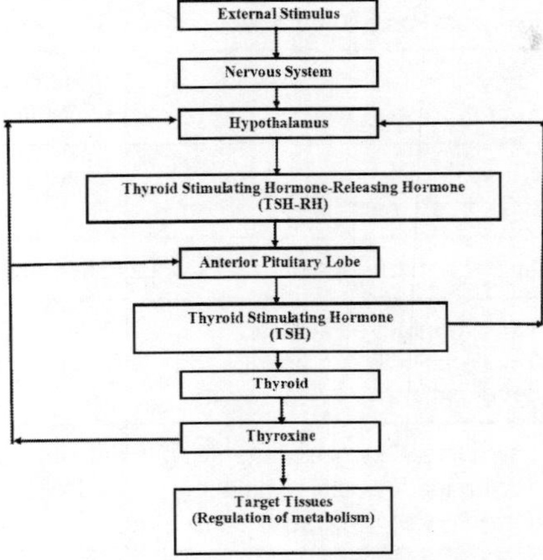

Fig. 2.26 A case of negative feedback mechanism of Homeostasis by hormones

- **Blood Glucose Homeostasis:** When we eat a carbohydrate-rich meal, blood sugar level is increased. It stimulated pancreas gland to secrete insulin. The latter stimulates the target cells to take up extra glucose which is either utilized in cell respiration or is stored as glycogen. In this way, blood-glucose level is brought back to normal (Fig. 2.27). If blood-glucose level falls below normal, insulin secretion by pancreas decreases. This checks further fall in the blood-glucose level. In this manner, insulin maintains blood-glucose homeostasis.

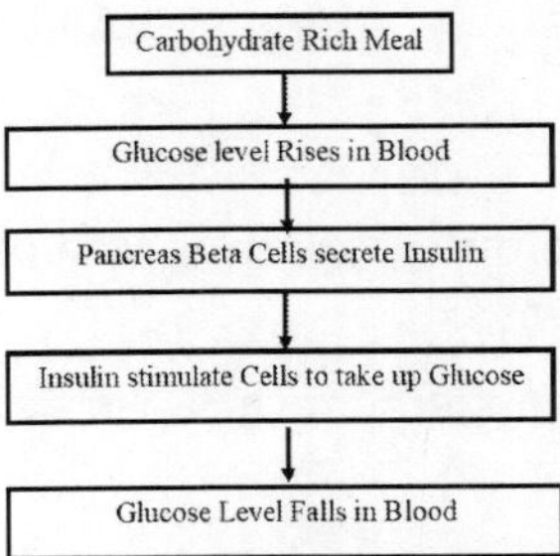

Fig. 2.27 Negative feedback control of Blood Glucose Level

(b) **Positive Feed Back Control:** In the positive feedback control, accumulating biochemical substances increase its own production. For example, at the onset of labour pain in female before child birth, uterine contraction stimulated the release of oxytocin hormone from posterior lobe of pituitary gland. The latter intensifies uterine contractions. The contractions further stimulate the production of oxytocin. The cycle of increase stops suddenly after birth of the baby.

Table 2.13 Differences between Animal hormones and Plant Hormones

Sr. No.	Parameters	Animal Hormones	Plant Hormones
1.	Number	They are more in number	They are lower in number
2.	Glands	They are produced by endocrine glands	Specific glands are absent for their secretion
3.	Synergistic Action	No synergistic action is found in case of animals	Synergistic action is more common
4.	Specificity	They are more specific	They are less specific
	Functions	They regulate all kinds of functions including growth and development	They regulate growth and develop-ment
	Examples	Thyroxine, Oxytocin, Lutenizing Hormones, Follicle stimulating hormones etc.	Auxin, Gibberellic Acid, Cytokinins, ABA, Ethylene etc.

REMEMBER SOCKETS

- All the living organisms (plants and animals) respond and react to changes in the environment around them. The changes in the environment to which the organisms respond and react are called stimuli (singular-stimulus).
- Plants do not have nervous system, muscles and sense organs like animals. They respond and react to various environmental stimuli such as light, water, gravity, touch, and chemicals etc. by the action of plant hormones.
- Plant movements are largely brought about by some definite internal and external stimuli.
- Both, plants and animals react to various stimuli around them. But the method of reacting to stimuli is not similar in plants and animals. They react to stimuli in different ways.
- Plants generally show movements at a very slow rate. The movements in plants are not as apparent as in case of animals.
- The response of organisms to a stimulus is usually in the form of some movement of their body part.
- Induced plant movements are not directional movements, and classified as nastic movements and tropic movements. Nastic movements are induced by external stimuli such as light, temperature, touch. Whereas, Seismonastic movements occur in response to touch (shock).

- Tropic movements or tropisms are directional movements or orientations of specific part of a plant in response to external stimuli. These are very slow and the specific part of the plant shows response in the form of growth by the action of plant growth hormones. Light, force of gravity, chemical substances and water are common environmental stimuli to which the plants respond.
- Tropic movements are induced growth movements of curvature that occur due to differential growth. The responses of plants to these stimuli are termed as (a) Phototropism (b) Chemotropism (c) Hydrotropism and (d) Geotropism.
- Plant hormones or Phytohormones are naturally occurring organic chemical substances secreted in plants in minute quantities.
- Plant hormones control and coordinate various activities of plants by controlling one or the other aspect of growth of plant, so known as plant growth substances also.
- Plant hormones or phytohormones regulate various functions in plants such as Breaking the dormancy of seeds, Growth of root, stem and leaves, Stomatal movement), Flowering of plants, Ripening of fruits, and Phototropism, geotropism, chemotropism, hydrotropism, thigmotropism and nastic movements
- Phytohormones are divided into groups-growth promoters (Auxins, Gibberellins, Cytokinins and Ethylene) and growth inhibitors (Abscisic Acid or ABA).
- The main functions of plant hormones include promotion of dormancy, breaking of dormancy, stomatal movements, flowering of plants, growth or root, stem and leaves, growth and ripening of fruits, apical dominance, parthenocarpy, tropic and nastic movements in plants etc.
- The duration of sunlight (photoperiod) also regulates the germination of seeds and flowering in plants. This effect of photoperiod on the germination of seeds and flowering in plants is called photoperiodism.
- Plants respond to photoperiod with the help of a phytochrome pigment.
- Dormancy refers to a resting, inactive condition of the seed when it fails to germinate even though the environmental conditions usually considered favourable for active growth are present.
- Quiescence refers to the condition of the seed when it fails to germinate because the environmental conditions normally required for growth are not present.
- Photoperiod acts as a stimulus for plants. The plants respond to this stimulus with the help of a special kind of pigment present in them in very small amounts. The special pigment is called phytochrome which is a blue-green pigment.

- A short day plant (SD) plant is one that flowers on photoperiods shorter than the critical day length e.g. tobacco plant.
- A long day plant (LD) plant is one that flowers on photoperiods longer than critical day length e.g. wheat plants.
- Day Neutral Plants (DN Plants) do not require specific photoperiods to flower. These, in fact, flower in all photoperiods ranging from a few hours to 24 hrs of uninterrupted light, e.g. cucumber, maize, tomato etc.
- Phytochromes are a class of photoreceptor in plants, bacteria and fungi use to detect light. They are sensitive to light in the red and far-red region of the visible spectrum and can be classed as either Type I, which are activated by far-red light, or Type II that are activated by red light.
- Vernalization is the induction of a plant's flowering process by exposure to the prolonged cold of winter, or by an artificial equivalent.
- Nervous tissue forms the nervous system in animals.
- Nervous system consists of neurons, nerve fibres, nerves, packing cells (neuroglia), ependymal cells and neurosecretory cells.
- Neurons are the structural and functional units of nervous system.
- Neuron has a special structure to receive, conduct and transmit impulses; it varies greatly in size and shape.
- Neurons are, in fact, the largest cells present in the human body, sometimes reaching 90-100 cm.
- Each neuron consists of three noticeable parts such as (i) Cell body (ii) Dendrites (Singular dendron), and (iii) Axon
- Nerve cells have Nissl's granules.
- Neurons lie end-to-end in chains to transmit nerve impulses in the body. They are not connected. There occurs a minute gap between terminal portion of axon of one neuron and the dendron of other neuron. This minute gap is called a synapse.
- The synapse acts as a one-way valve to conduct the impulse in one direction only with the help of neurotransmitter.
- Nerve impulse travels from axon of one neuron to Dendron of other neuron through a synapse.
- A receptor is a sensory nerve cell or group of sensory nerve cells which is sensitive to a specific stimulus or to a specific change in the environment.
- The animals have different sensory receptors for receiving different stimuli such as Photocreceptors for light, Thermoreceptors for heat and cold, Gustatoreceptors for taste Phonoreceptors for sound, Olfactoreceptors for smell and Tangoreceptors for touch.

- The receptors transmit nerve impulses to central nervous system. The latter transmits motor impulses to appropriate effectors, which produce suitable responses.
- Reflex action was discovered by Marshal Hall (1833).
- Reflex actions are nerve mediated, automatic involuntary actions that occur without the will of an animal.
- Reflex action is defined as a spontaneous, automatic and mechanical response to a stimulus, action on a specific receptor, without the will of an animal.
- Reflex arc is the basis of reflex action.
- Blinking of eyes, Knee-jerk reflex, coughing, yawning, sneezing, movement of diaphragm during respiration etc, are common reflex actions.
- Reflex actions involve spinal cord and hence are also termed as spinal reflexes. Some reflexes that involve brain are termed cerebral reflexes.
- Involuntary actions are performed by the animal without its will. These occur automatically and the animal has no choice in it. Most of these actions are controlled by hind brain.
- Voluntary actions are performed by the animal with its will. These are controlled by cerebellum part of hind brain. In each voluntary action, the animal exercise its choice so that the same stimulus may receive different times depending upon the situation.
- Human nervous system is divided into (i) central nervous system (brain and spinal cord), and (ii) peripheral nervous system (cranial) nerves, spinal nerves and visceral nerves.
- Brain consists of three regions such as (i) fore brain, (ii) mid brain, (iii) hind brain.
- Fore brain have three regions such as (a) olfactory lobes, (b) cerebral hemispheres (cerebrum) and (c) diencephalon.
- Mid brain controls reflex movements of the head, neck and trunk in response to visual and auditory stimuli, and of the eye muscles, changes in pupil size and shape of eye lens.
- Hind brain has three parts such as (a) cerebellum, (b) pons varolii and (c) medulla oblongata.
- Cerebellum maintains posture, equilibrium and muscle tone. Pons controls some aspects of respiration.
- Medulla oblongata controls swallowing, coughing, sneezing, vomiting, breathing, movements, rate of heart beat, expansion and contraction of blood vessels to regulate blood pressure etc.
- Spinal cord is enclosed in vertebral column. 31 pairs of spinal nerves arise from it.

- Spinal cord conducts sensory and motor impulses to and from the brain, and acts as a centre for the reflex actions.
- Autonomic nervous system (ANS) controls involuntary activities of the body.
- ANS is subdivided into sympathetic and parasympathetic and parasympathetic nervous systems. These have opposing effects.
- Endocrine system is meant for internal communication and regulation of the animal body.
- Animals have three types of glands such as (a) exocrine glands (b) endocrine glands and (c) Heterocrine glands.
- The term 'hormone' was introduced by William M. Bayliss and Ernest H. Starling.
- Hormones (Gk. Hormein = to excite) are informational molecules secreted by specialized ductless endocrine glands in specific part of the body in response to changes in external or internal environment.
- A cell, a tissue, or an organ which secretes certain useful chemical compounds required for particular functions is called a gland.
- Animals have three types of glands such as (i) Exocrine glands (ii) Endocrine glands, and (iii) Heterocrine glands
- The study of endocrine glands and the role of their secretions is called endocrinology.
- Endocrine glands lack ducts and pass their secretions (hormones) into the surrounding blood for transport to the site of action.
- Hormones are informational molecules secreted by specialized endocrine glands. These chemical messengers are carried by blood to target organs where they stimulate or inhibit specific physiological processes for the good of the body as a whole.
- Hormones are released from the endocrine glands by three types of stimuli such as (i) Specific metabolite: (ii) Other hormone and (iii) Neuronal impulse
- Human body possesses large number of endocrine glands such as (a) hypothalamus, (b) pituitary gland (c) thyroid gland (d) parathyroid glands (e) pancreas (f) adrenal glands (g) pineal gland (h) thymus (i) Gonads (testes in males and ovaries in females).
- Endocrine glands are located in different parts of the body e.g. head, neck and trunk regions.
- Hypothalamus is situated at the base of the brain. The neurosecretary cells of the hypothalamus secrete several neurohormones called releasing Hormones (RH) and inhibiting hormones or factors (IH or IF).
- Pituitary or hypophysis is present just below the brain. Pituitary gland consists of three lobes such as (a) anterior lobe, (b) intermediate lobe, and (c) posterior lobe.

- Each lobe of Pitutary gland produces different hormones.
- Thyroid is the largest endocrine gland present in the neck region.
- Thyroid gland secretes the hormones such as (a) thyroxine (T_4), (b) triiodothyronine (T_3) and (c) calcitonin with specific actions.
- Parathyroid glands are four small glands situated on the posterior surface of thyroid gland, two in each lobe of thyroid.
- Parathyroid gland secretes parathormone (PTH or Collip's hormone).
- PTH from parathyroid glands and calcitonin from thyroid gland act antagonistically to regulate the calcium-phosphorus balance in the blood.
- Adrenal glands are a pair of glands situated on upper side of each kidney. Therefore, they are also called suprarenals.
- Each adrenal gland has two distinct endocrine regions as (a) outer adrenal cortex and (b) inner adrenal medulla.
- Adrenal cortex secretes three groups of steroid hormones such as (a) mineralocorticoids (e.g., aldosterone), glucocorticoids (e.g., cortisol), and (b) sexcorticoids.
- Mineralocorticoids maintain water and electrolyte balance and blood volume in the body.
- Glucocorticoids regulate the metabolism of carbohydrates, proteins and fats.
- Sex corticoids stimulate the development of secondary sexual characters both in males and females.
- Adrenal medulla secretes two hormones such as (a) adrenaline (epinephrine) and (b) noradrenaline (norepinephrine).
- In normal conditions, these hormones (adrenal medulla) are secreted in small amounts. However, in stress or danger situations, they are secreted in large amounts to prepare the body to face emergency situations.
- Pancreas lies below the stomach.
- Endocrine part of pancreas is called islets of Langerhans. It secretes to hormones such as (i) insulin and (ii) glucagon
- Adrenaline is also termed as emergency hormone.
- Insulin is secreted by β-cells of islets of Langerhans.
- Insulin helps to lower the blood glucose level.
- Glucagon is secreted by the α-cells of islets of Langerhans.
- Glucagon hormone increases the blood glucose level.
- Gigantism is caused due to excess secretion of growth hormone from childhood.

- Acromegaly is caused due to excess secretion of growth hormone after adolescence. Acromegaly in adults leads to overgrowth of the jaw bones and bowing of the spine (backbone).
- Diabetes insipidus: Deficiency of ADH reduces reabsorption of water and increases urine output, causing excessive thirst. This disorder is called diabetes insipidus. No glucose is lost in the urine of such a patient.
- Grave's Disease (Exophthalmic Goitre) is caused by hyper secretion (over secretion) of thyroid hormones due to enlargement of thyroid. Excess of thyroid hormones increases metabolic rate and accelerates oxidation.
- Cretinism is caused by hypo secretion (under secretion) of thyroid hormones in infants.
- Myxoedema is caused by deficiency of thyroid hormones in adults. It is more common in women than in men.
- Simple Goitre (Iodine deficiency goitre or Endemic goitre) is the enlargement of thyroid gland accompanied with cretinism or myxoedema. It is caused due to dietary deficiency of iodine. The disease is common in hilly areas. It causes enlargement of thyroid gland. Swollen neck is one of the symptoms of this disorder.
- Parathormone from parathyroid glands and calcitonin from thyroid gland act antagonistically to regulate the calcium-phosphorus balance in the blood.
- Relaxin is produced by the corpus luteum at the end of the gestation period. It helps in easy birth of the young one.
- Pineal Gland is a very small, reddish- grey, vascular, solid body lying between the two cerebral hemispheres of the brain.
- Pineal Gland secretes melatonin hormone. It regulates the working of gonads.
- Thymus Gland secretes thymosin hormone which stimulates the development and differentiation of lymphocytes and thereby increasing resistance to infection.
- Homeostasis means keeping the internal chemical environment of the body constant. Hormones help to maintain homeostasis by their integrated action and feedback control. Feedback control is mostly negative, and rarely positive.
- In a negative feedback control, synthesis of a hormone slows or halts when its level in the blood rises above normal. Two specific examples are cited to explain the negative feedback control such as Blood-thyroxine Homeostasis and Blood Glucose Homeostasis
- In the positive feedback control, accumulating biochemical substances increase its own production.

VERY SHORT ANSWER TYPE QUESTIONS:

1. Which part of the plant shows positive geotropism and why?
2. What is parthenocarpy?
3. Which one of the following actions is an example of non-directional movement? (a) Movement in the touch-sensitive plant and (b) Promoting cell division in plants
4. Name the hormone responsible for (a) Phototropic movement of plant and (b) Promoting cell division in plants.
5. Define the term phytohormones? Name any two phytohormones.
6. What is the meaning of movements of curvature?
7. Define the term nastic or seismonastic movements.
8. Define the term thermonastic movements.
9. Define the term vernalization.
10. Give the example of phototropism movement.
11. Give the example of geotropism movement.
12. Give the example of phytohormones.
13. Define the term growth promoters.
14. Define the term dormancy.
15. Which plant hormone is responsible for (a) promote cell enlargement and cell differentiation in plants.(b) to induce parthenocarpy in many plants (c) to promote cell division in plants, and (d) for ripening of fruits
16. Which plant hormone is responsible for abscission and senescence in leaves?
17. Give the example of short day plants.
18. Give the example long day plants.
19. Give the example day neutral plants.
20. Name three disorders which occur due to improper secretion of growth hormone from anterior lobe of pituitary.
21. Name the endocrine gland which secretes neurohormones.
22. Why endocrine glands release their secretions directly into the blood?
23. Why pituitary gland is called master gland?
24. Which is the posterior most part of the brain? Name the cavity present in it.
25. Give the function of any one hormone secreted by the anterior lobe of pituitary
26. Distinguish between voluntary and involuntary actions of our body.

SHORT ANSWER TYPE QUESTIONS:

1. Which animal or plant hormone is associated with the following?
   (i) Decreases sugar level of blood
   (ii) Develop secondary sexual characters at puberty in boys
   (iii) Inhibits growth of plants
   (iv) Ripening of fruits
   (v) Dwarfism
   (vi) Goitre
2. What are plant hormones? Give one example of a plant hormone that promotes growth.
3. Discuss different tropic movements.
4. Difference between tropic movements and nastic movements
5. Demonstrate the response of plants to any one light, gravity and chemicals
6. Write a short note on (a) Hydrotropism (b) Photoperiodism (c) Seismonastic Movements (d) Photonastic Movements (e) Growth Promoters (f) Growth Inhibitors (g) Dormancy (h) Quiescence (i) Short Day Plants (j) Touch me not plant (k) Vernalization (l) Vernalization and Photoperiodism
7. Give the functions of Plant Hormones.
8. Role of Auxins in Phototropism.
9. Give protective features of the nervous system.
10. Give the difference between sensory and motor nerves.
11. Compare hormonal and nervous mechanism for control and coordination.
12. Differentiate between:
    (a) Spinal nerve and cranial nerve.
    (b) Dendrite and axon
    (c) Central nervous system and peripheral nervous system.
    (c) Nervous and hormonal information
13. Write down the short note on (a) Dendrites (b) Axon (c) Synapse (d) Homeostatic

LONG ANSWER TYPE QUESTIONS:

1. Define the term plant movement. Give their types and explain the induced plant movements.
2. Define the term plant movement. Give their types and explain the spontaneous plant movements.

3. Explain the types of tropism movement with suitable examples.
4. Define the term plant hormones. Give their types and explain any two with functions.
5. Explain the types of photoperiodism plants with suitable examples.
6. Give an account on coordination in animals in brief.
7. Give an account on structure of neuron with diagram.
8. Write an account on the parts of human nervous system.
9. Define the term endocrine glands, give their types and explain any two with functions.
10. Explain the terms Reflex Actions, Involuntary and Voluntary Actions in details.
11. Define the term endocrine gland, give their types and explain the pituitary gland with disorders.
12. Prepare a table of the endocrine gland with their hormones, functions and disorders.
13. List the functions of testosterone and estrogen.
14. Describe various parts of central nervous system in human beings.
15. Define reflex arc and draw its with labeled diagram.
16. Nervous and hormonal systems together perform the function of control and coordination in human beings. Justify the statement.
17. Name various plant hormones. Also give their physiological effects on plant growth and development.
18. How does control and coordination take place in plants?

MULTIPLE CHOICE QUESTIONS:

1. Any change in the environment to which an organism responds is called

   (a) Tropism (b) Tonocity (c) Stimulus (d) Response

2. The directional movement or orientation of a plant part in response to light is termed.

   (a) Photoperiodism (b) Thigmotaxis

   (c) Chemotropism (d) Phototactic

3. Name the plant hormone which is termed growth inhibitor.

   (a) Auxin (b) Ethylene (c) Gibberellin (d) Abscisic acid

4. Seismonastic movements are shown by which plant?

   (a) Cucumber plant (b) Indian telegraph plant

   (c) Rose plant (d) 'Touch-me-not' plant

5. In plants, growth occurs by
   (a) Bark (b) Permanent tissue
   (c) Ground substance (d) Meristematic tissue
6. Leaf fall can be induced by
   (a) Abscisic acid (b) Cytokinins (c) Florigens (d) Auxins
7. Opening and closing of flowers represent a kind of
   (a) Geotropism movement (b) Autonomic movement
   (c) Nastic movement (d) Nutation
8. Production of seedless fruits is referred to as
   (a) Exocarpy (b) Endocarpy
   (c) Parthenogenesis (d) Parthenocarpy
9. Bending of stem towards the sunlight in plants is due to Unequal distribution of auxins
   (a) Uniform occurrence of gibberellins
   (b) Unequal distribution of auxins
   (c) Unequal distribution of cytokinins and gibberellins
   (d) Inhibition of cytokinins synthesis.
10. The movement of shoot towards light is Phototropism
   (a) Hydrotropism (b) Phototropism
   (c) Geotropism (d) Chemotropism
11. The main function of Abscisic acid in plants is to promote growth of stem
   (a) Promote cell division (b) Inhibit growth
   (c) Promote growth of stem (d) Increase the length of cells
12. Which of the following is not associated with growth of plant?
   (a) Gibberellins (b) Abscisic acid (c) Auxin (d) Cytokinins
13. The growth of tendril in pea plants is due to Rapid cell division in tendrillar cells in contact with the support
   (a) Effect of gravity
   (b) Effect of light
   (c) Rapid cell division in tendrillar cells in contact with the support
   (d) Rapid cell divisions in tendrillar cells that are away from the support
14. The movement of sunflower in accordance with the path of sun is due to
   (a) Geotropism (b) Phototropism
   (c) Hydrotropism (d) Chemotropism

15. Which of the following is not a reflex action?
    (a) Weeping (b) Blinking of eyes (c) Coughing (d) Knee-jerk
16. Which statement is not true about thyroxin?
    (a) It regulates carbohydrates, protein and fat metabolism in the body.
    (b) Iron is essential for the synthesis of thyroxin
    (c) Thyroxin is also called thyroid hormone
    (d) Thyroid gland requires iodine to synthesize thyroxin.
17. The term hormone was coined
    (a) Tansley (b) Krebs
    (c) Ernst Hackel (d) Bayliss and Starling
18. Endocrine glands are
    (a) Ductless gland (b) Ducted glands
    (c) Both a and b (d) None of these
19. Which hormone is known as emergency hormone?
    (a) Progesterone(b) Aldosterone (c) Adrenaline (d) Insulin
20. Melatonin hormone regulates the working of
    (a) Sugar level (b) Gonads
    (c) Calcium Level (d) Heart beats
21. Which is secreted by β-cells of islets of Langerhans.
    (a) Progesterone (b) Aldosterone
    (c) Insulin (d) Glucagon
22. Which is secreted by the α-cells of islets of Langerhans.
    (a) Progesterone (b) Aldosterone
    (c) Insulin (d) Glucagon
23. Which disease is caused by deficiency of thyroid hormones in adults?
    (a) Myxedema (b) Acromegaly (c) Goiter (d) Skin Allergy
24. Which disease is caused due to excess secretion of growth hormone after adolescence?
    (a) Myxedema (b) Acromegaly (c) Goiter (d) Skin Allergy
25. Name the hormone which is commonly termed as birth hormone and milk ejecting hormone.
    (a) Vasopressin (b) Oxytocin
    (c) Latinizing hormone (d) Thyroxine

26. The fight or flight response is developed by hormones secreted by
    (a) Adrenal medulla (b) Hypothalamus
    (c) Pancreas (d) Adrenal cortex
27. The main function of parathyroid in body is to
    (a) Form calcium compounds
    (b) Built teeth
    (c) Build bones
    (d) Maintain proper amount of calcium and phosphorus in the blood
28. The hormone that helps in the reabsorption of sodium and water in the body is
    (a) Progesterone (b) Aldosterone
    (c) Testosterone (d) All of these
29. Master gland in the body is
    (a) Adrenal (b) Thyroid
    (c) Pituitary (d) Islets of Langerhans
30. Which gland stimulates proliferation of lymphocytes (WBCs)?
    (a) Adrenal (b) Thyroid (c) Pituitary (d) Thymus
31. Which hormones regulates the glucose level
    (a) Insulin (b) Glucagon
    (c) Both (a) and (b) (d) None of these
32. Which hormone is produced by the corpus luteum at the end of the gestation period?
    (a) Insulin (b) Relaxin (c) Lactin (d) None of these
33. Which part of human brain is more developed in comparison of others?
    (a) Optic lobes (b) Cerebrum
    (c) Medulla oblongata (d) Cerebellum
34. Which of the following protects the brain from shocks?
    (a) Dura mater (b) Pons
    (c) Arachnoid membrane (d) Cerebrospinal fluid
35. Cavities of brain are called
    (a) Coelom (d) Lumen (c) Auricles (d) Ventricles
36. Which part of the brain regulates the body temperature?
    (a) Thalami (b) Hypothalamus
    (c) Medulla oblongata (d) Pituitary

37. Which is the correct sequence of the components of a reflex arc?
    (a) Receptors → sensory neuron → spinal cord → motor neuron → muscles
    (b) Receptors → spinal cord → sensory neuron → motor neuron → muscles
    (c) Receptors → motor neuron → spinal cord → sensory neuron → muscles
    (d) Receptors → muscles → sensory neuron → motor neuron → spinal cord
38. Cerebral hemispheres are the centres of
    (a) Thinking and memory (b) Posture and equilibrium
    (c) Reflex actions (d) All of these
39. Medulla oblongata of hindbrain controls
    (a) Vision (b) Equilibrium
    (c) Thinking (d) Rate of heart beat
40. Example of negative feedback control such as
    (a) Blood-thyroxine Homeostasis (b) Blood glucose Homeostasis
    (c) Both a and b (d) None of these

ANSWER KEY

1. (c)	2. (a)	3. (d)	4. (d)	5. (d)	6. (a)	7. (c)	8. (d)
9. (b)	10. (b)	11. (a)	12. (b)	13. (c)	14. (b)	15. (a)	16. (a)
17. (d)	18. (b)	19. (c)	20. (b)	21. (c)	22. (d)	23. (a)	24. (b)
25. (b)	26. (a)	27. (d)	28. (b)	29. (c)	30. (d)	31. (c)	32. (b)
33. (b)	34. (d)	35. (d)	36. (b)	37. (a)	38. (c)	39. (a)	40. (b)

3

CHAPTER

REPRODUCTION

Reproduction in Animals and Plants (Asexual and Sexual), Reproductive Health-need and Methods of Family Planning, Safe Sex Vs HIV/AIDS, Child bearing and Women's Health.

3.1 INTRODUCTION AND DEFINITION:

Reproduction is not an essential process for the survival of an individual, it is essential for multiplication and perpetuation of the species. Reproduction may be defined as the production of new generation of individuals of the same species that are physically independent of their parents. It ensures the continuation of life, perpetuation of species, replacement of dead individuals and introduction and transfer of variation.

3.1.1. Why do organisms reproduce?

Reproduction is an energy spending process which is not at all essential to maintain the life of an individual organism. Even then, the organisms reproduce. They reproduce to increase in number. All the members produced from the parents look like parents as well as they look very much like themselves. Thus, reproduction results in the increase of members which look alike. The organisms which look similar to each other belong to the same species. Moreover, reproduction provides group immortality by replacing the dead individuals with new ones for the survival of the species on this earth.

3.1.2. Why do the organisms belonging to same species look similar to each other?

The reproducing organisms create new individuals which are similar to each other and look exactly similar to their parents. Do you know why? They look similar body designs, they produce similar blue prints. This is possible only when a mechanism of production of similar copies of blue print is involved in the process of reproduction.

That is true; the reproduction involves a mechanism, called DNA (Deoxyribo-nucleic Acid) copying or DNA replication which is capable of producing similar multiple of blue prints. As we know that the nucleus contains chromosomes which carry information for inheritance of features from parent to the next generation in the form of DNA molecules. DNA copying is an essential phenomenon of reproduction through which the organisms pass on their body features to their offspring's. Similar copies of DNA molecules carry similar molecular composition and, in other words, similar genes for production of similar features. The molecule of DNA in the cell nucleus possesses information's in the form of genes to synthesize specific kind of proteins. If this information is changed, consequently the production of the kind of proteins will also be changed. The proteins act as enzymes to carry out specific functions. Thus, a change in the structure of DNA may cause a change in the information that ultimately leads to change in the kind of protein formed from this DNA sequence. Change in the structure of protein may result in the formation of a different kind of enzyme that leads to the production of a different kind of feature. It is being clear here that creation of a DNA copy is the basic event in the process of reproduction. Cells usually employ chemicals to build copies of their DNA. This process results in the production of two similar copies of DNA in a reproducing cell. The two copies of the DNA molecule, which are similar with each other, come to the DNA molecule, which are similar with each other, come to two small daughter cells. Therefore, DNA copying is always accompanied by the creation of another cellular apparatus. Finally, the DNA copies are separated into two similar cells, each with its own cellular apparatus. Finally, a cell divides to give rise to two daughter cells.

These two daughter cells thus formed, are exactly similar with each other due to their similar genetic makeup. There are, however, chances that they may not be exactly similar. They may be identical or may be slightly different with each other. It all depends on how accurately the process of DNA copying has occurred. In fact, no biochemical reaction is absolutely reliable. Therefore, it is always expected that process of copying of DNA is not truly identical each time. It may cause a slight difference during each DNA copying. As a result, the two copies of DNA may be similar, but not identical to the original. Some of these copies of DNA may possess such drastic variation from the original that their survival become difficult. Such individuals having distinct variations to survive and simply die. On the other hand, some variations in the DNA copies may create such individuals which can survive but with a slight variation in its feature.

Table 3.1 Life span of some organisms

Organisms	Lifespan	Organisms	Lifespan
Tortoise	100-150 years	Polar beer	40-50 years
Humans	80-110 years	Dog	4 years
Elephant	80-90 years	Mouse	25-30 years
Crocodile	60-80 years	Butterfly	1-2 weeks

3.1.3. The Importance of Variation:

We all know that a particular kind of vegetation grows in a particular kind of habitat or a particular niche. For example, xerophytes have specific adaptations and survive in a particular kind of ecological niche (i.e., desert). We also know that the ecological niches keep on changing with time. Usually the DNA copying, which is an essential feature of reproduction, remains consistent generations after generations. This helps the offspring's to adapt the particular kind of niche in the ecosystem. Both niches and the populations of the organisms inhabiting in them are in perfect harmony so that they continue to survive as long as both remain unchanged. Thus, stability of populations of species in particular niche is maintained by the process of reproduction.

However, there are chances that niches can change due to reasons beyond the control of the organisms. The changes may be in the form of increased temperature on earth, varying water levels, meteorite hits, or any other ecological disbalances. Under such conditions any alteration in the particular kind of niche is surely going to hit the consistent population. The population of species may be wiped out due to sudden alteration of niche. However, a few individuals of the population of a species may survive the changed niches which have undergone variation in their features due to variations in DNA copying mechanism. As stated earlier, the biochemical reactions, undergoing inside the body cells, are not always absolutely reliable. There are chances of changes in DNA copying resulting in the production of such organisms which are suited in the changed niches. If a population of reproducing organisms achieves such inheritable changes which help them to survive under changed niches then the organisms have chances to survive. In case, if the organism has not changed and the niche has altered drastically then the organism is surely going to be wiped out. For example, if a population of bacteria is growing in temperate water and has developed an environmental harmony. Suppose the temperature of water is somehow increased due to global warming then the bacteria will definitely die. However, only those will survive which have developed resistance to heat. Variations are thus useful for the survival of species over time.

Reproduction may be defined as the production of new generation of individuals of the same species that are physically independent of their parents. It ensures the continuation of life, perpetuation of species, replacement of dead individuals and introduction and transfer of variation. It involves a mechanism, called DNA (Deoxyribo-nucleic Acid) copying or DNA replication which is capable of producing similar multiple of blue prints.

3.2 TYPES OF REPRODUCTION:

There are many different ways in which new organisms are produced from their parents. Some organisms like *Amoeba* just split into two parts to produce new Amoebae; some organisms like hydra and yeast grow out of the parent's body in the form of a bud; some organisms like birds and snakes hatch out of the eggs laid by their parents; whereas some organisms like human babies, cubs, kittens and puppies are born from their mother (Fig. 3.1). This means that each species of organisms reproduces in a different way. All the different ways of reproduction can be divided into two main groups such as asexual reproduction and sexual reproduction.

(A) Mother with her baby

(B) Cat with babies

(C) Dog with babies

(D) Cow with calf

Fig. 3.1 The process of reproduction ensures continuity of life

1. Asexual Reproduction:

In asexual reproduction, the offspring arises from a single parent. The production of new organisms from a single parent without the involvement of sex cells (or gametes) is called asexual reproduction. In asexual reproduction, a part of the parent organism separates off and grows into a new organism. Thus, in asexual reproduction, only one parent is needed to produce a new organism. But no sex cells are involved in asexual reproduction. Asexual reproduction takes place mainly in those organisms whose bodies have a simple structure. So, the simple animals, simple plants and micro-organisms (like bacteria) reproduce by asexual reproduction methods. Actually, asexual reproduction takes place in unicellular animals and plants, micro-organisms (bacteria) and simple multicellular animals (*Hydra* and *Planaria*) and some multicellular plants (Bryophyllum and Rose plants, etc.).

2. Sexual Reproduction

In sexual reproduction, the offspring arises from two parents of different sexes as a male sex and a female sex. The male parent contains male sex cells or male gametes and the female parent contain female sex cells or female gametes. The production of a new organism from two parents by making use of their sex cells (or gametes) is called sexual reproduction. In sexual reproduction, the sex cell of one parent fuses with the sex cell of the other parent to form a 'zygote'. This zygote then grows and develops to form a new organism. Thus, in sexual reproduction, two parents are needed to produce a new organism. Our father is a male and our mother is a female parent. The male and female parents have special organs in them which produce male sex cells and female sex cells, respectively (which are required in sexual reproduction). The humans, fish, frogs, hens, cats, dogs, cows, horses, deer, rabbits, lions and tigers all reproduce by the method of sexual reproduction. Most of the flowering plants also reproduced by sexual reproduction. As we will study after a while, some organisms use both the methods (asexual and sexual) for reproduction whereas other organisms use only one of these methods for reproduction.

The basic difference between asexual and sexual reproduction is that only one parent is needed in asexual reproduction whereas two parents are needed in sexual reproduction. Another difference is that no sex cells (or gametes) are involved in asexual reproduction but sex cells (or gametes) take part in sexual reproduction.

3.2.1. Asexual Reproduction:

In the asexual reproduction method, certain body cells of the parent organism undergo repeated mitotic cell divisions to form two (or more) new organisms of the same kind. Asexual reproduction takes place by six different methods such as Fission, Budding, Spore formation, Regeneration, Fragmentation and Vegetative propagation.

3.2.1.1. Fission:

Many single-celled organisms like protozoa and bacteria just split (or break) into two identical halves during cell division, leading to the creation of new organism. This is called fission. In biology, fission is the process of reproduction in unicellular organism such as protozoa (like *Amoeba, Paramecium, Leishmania* etc.) and many bacteria. In the process of fission, a unicellular organism splits (or divides) to form two (or more) new organisms. Depending on whether the parent organism splits to form two new organisms or more than two organisms, fission is of two types such as (i) binary fission and (ii) multiple fission.

(i) Binary Fission:

Binary fission is an asexual method of reproduction of organism. The word 'binary' means 'two' and the word 'fission' means 'splitting'. So, the term 'binary fission' means 'splitting into two'. In binary fission, the parent organism splits (or divides) to form two new organisms. When this happens, the parent organism ceases to exist and two new organisms come into existence. The unicellular organisms like *Amoeba, Paramecium, Leishmania*, Bacteria, etc., reproduce by binary fission.

Amoeba reproduces by binary fission by dividing its body into two parts. When the Amoeba cell has reached its maximum size of growth, then first the nucleus of Amoeba lengthens and divides into two parts. After that the cytoplasm of Amoeba divides into two parts, one part around each nucleus. In this way, one parent Amoeba divides do form two smaller *Amoebae* (called daughter *Amoebae*) (Fig. 3.2). One Amoeba produces two *Amoebae*. The two daughter *Amoebae* produced here grow to their full size by eating food and then divide again to produce four *Amoebae*, and so on. In the unicellular organisms such as Amoeba, the splitting of the parent cell during fission (or cell division) can take place in any plane.

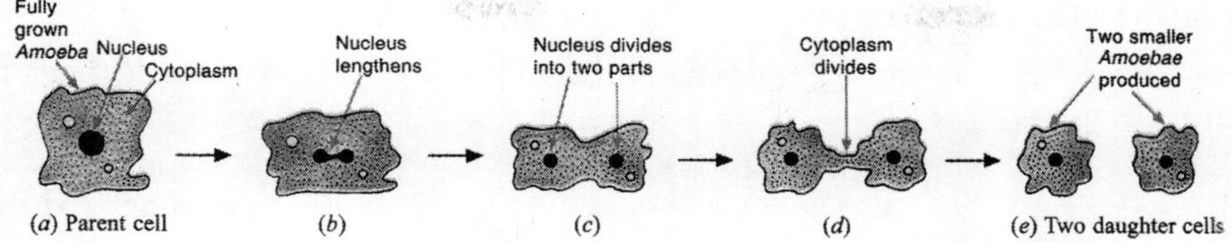

Fig. 3.2 Binary fission in *Amoebae*

Paramecium is a unicellular animal having short thread-like structures called cilia over its surface. *Paramecium* also reproduces by the method of binary fission. A fully grown *Paramecium* divides its body into two parts to form two smaller Paramecia (Fig. 3.3A, B). This happens by the division of nucleus followed by the division of cytoplasm.

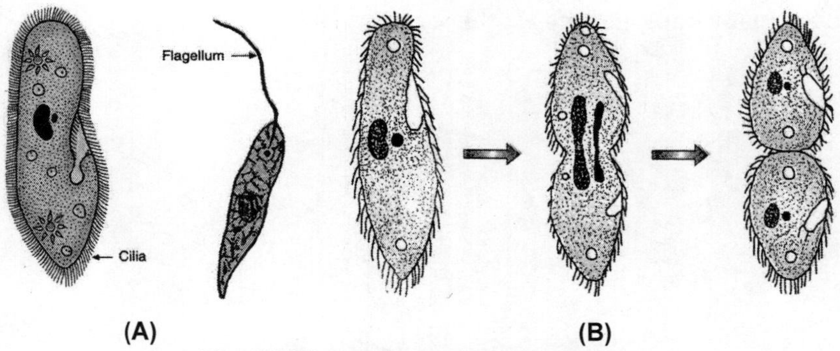

Fig. 3.3 Binary fission in *Paramecium*

Leishmania is a unicellular animal (which is a protozoan). It is a parasite which causes the disease known as kala-azar (or black fever). Kala-azar is also known as leishmarinasis. *Leishmania* has a greater degree of organization in its body, having a whip-like structure called flagellum at its one end. *Leishmania* reproduces by the process of binary fission (Fig. 3.4). In *Leishmania*, the splitting of parent cell during fission (cell division) takes place in a definite plane (longitudinally) with respect to flagellum at its end. In this respect *Leishmania* differs from *Amoeba* (in which fission can take place in any plane).

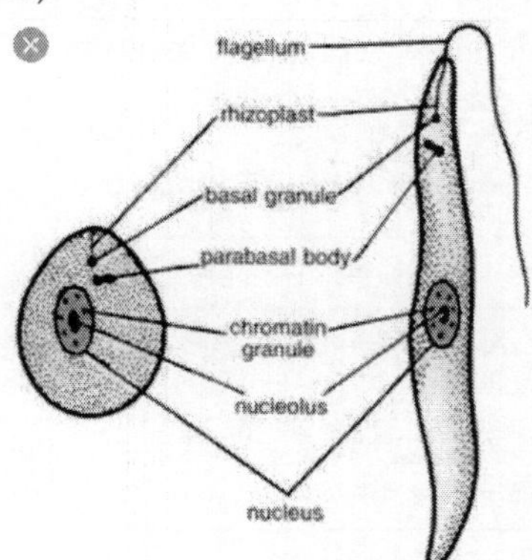

Fig. 3.4 Binary fission in *Leishmania*

(ii) Multiple Fission:

The term multiple means many or several. So, multiple fission means 'splitting into many' or 'splitting into several'. In multiple fission, the parent organism splits or divides to form many new organisms at the same time. Sometimes (particularly during unfavourable conditions), a cyst or protective wall is formed around the cell of a single-celled organism like that of Plasmodium (Fig. 3.5). Inside the cyst, the nucleus of cell splits/divides several times to form many smaller nuclei called daughter nuclei. Little bits of cytoplasm collect around each daughter nuclei and thin membranes are formed from a single parent cell within the cyst. In fact, as many daughter cells are formed as the number of daughter nuclei produced by the divisions of the parent nucleus. When the favourable conditions arrive, the cyst breaks open and many daughter cells present in it are released, each forming a new organism. In this way, a single celled parent undergoes in multiple fission to reproduce many daughter cells at the same time.

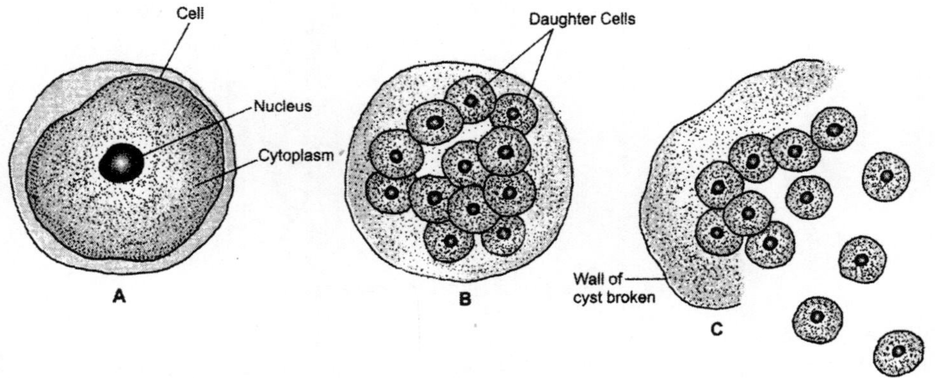

Fig. 3.5 Reproduction by Multiple fission

Plasmodium is a protozoan (a microscopic, single-celled animal) which reproduces by the asexual method called as multiple fission. About 1000 daughter cells are produced by the multiple fission of one Plasmodium cell. *Plasmodium* is the malarial parasite which causes malaria disease in human beings. Malaria parasite plasmodium is carried by female Anopheles mosquitoes from one person to another thereby spreading the malaria.

Table 3.1 Differences between Binary and Multiple Fission

Sr. No.	Characteristics	Binary Fission	Multiple Fission
1.	Production of individuals	It results in the production of two individuals.	It results in the production of several individuals.
2.	Conditions	It occurs under favourable conditions.	It occurs both under favourable and unfavourable conditions

3.	Number of Nuclei	Nucleus of the parent cell divides only once to form two daughter nuclei.	Nucleus of the parent cell undergoes repeated divisions to form a number of daughter nuclei.
4.	Cytoplasm division	Cytoplasm divides after each nuclear division.	Cytoplasm does not divide after each nuclear division.
5.	Part of parent body	No part of parent body is left unused.	A part of the body (covering and residual cytoplasm) is left unused.

3.2.1.2. Budding:

Budding is an asexual method of reproduction. In budding, a small part of the body of the parent organism grows out as a 'bud' which then detaches and becomes a new organism. The asexual reproduction by budding is observed in hydra and yeast.

Hydra is a simple multicellular animal, and reproduces by the process of budding. In hydra, first a small outgrowth called 'bud' is formed on the side of its body by the repeated mitotic divisions of its cells. This bud then grows gradually to form a small hydra by developing a mouth and tentacles. Finally the tiny new hydra detaches itself from the body of parent hydra and lives as a separate organism. In this way, the parent hydra turns into a new hydra (Fig. 3.6). Thus, hydra reproduces asexually by growing buds from its body. This is called budding. Please note that the bud formed in a hydra is not a single cell. It is a group of cells.

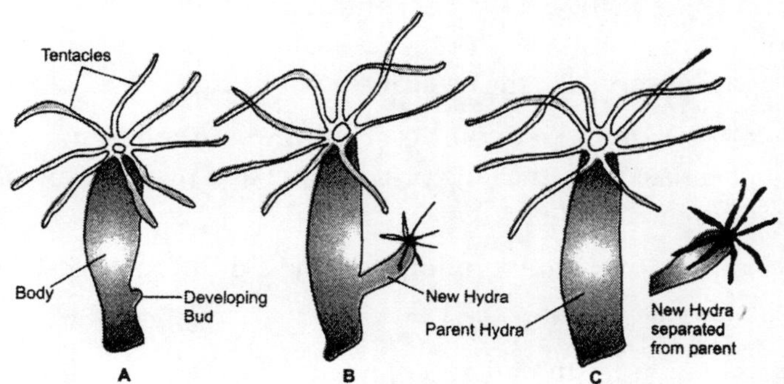

Fig. 3.6 *Hydra* reproducing asexually by budding

Yeast is tiny, unicellular, non-green plant (which is a fungus). Yeast reproduces by budding. In yeast, first a bud appears on the outside of the cell wall. The nucleus of parent yeast cell then divides into two parts and one part of the nucleus moves into the bud. Ultimately, the bud separates off from the parent yeast cell and forms a new yeast cell (or new yeast plant). The budding in yeast, however, often takes place so fast that the first buds start forming their own buds and all of them remain attached to the parent yeast cell forming a chain of yeast cells (Fig. 3.7). After some time, all the yeast cells of the chain separate from one another and form individual yeast plants.

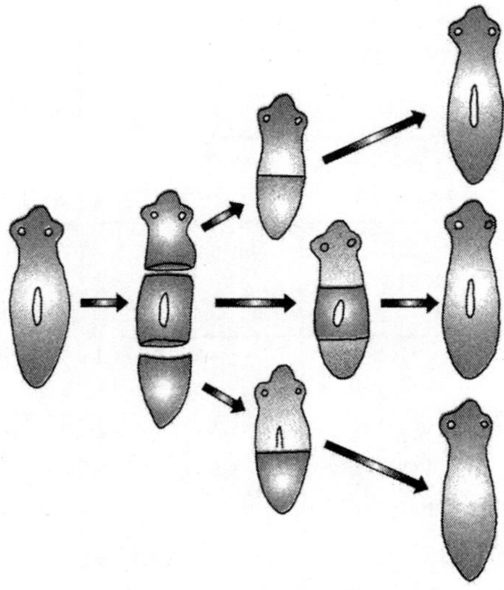

Fig. 3.7 Budding in yeast

Experiment to show Asexual Reproduction in Yeast by Budding:

- Take 100ml of water in a conical flask and dissolve 10gm of sugar in it.
- Add 5gm of yeast powder (or yeast granules) to this sugar solution and stir it well with a glass rod.
- Put a cotton plug in the neck of the conical flask. This conical flask containing sugar solution and yeast mixture is kept aside in a warm place for 3 to 5 days. When froth is observed in the flask, the yeast culture is ready for examination.
- Take out a small quantity of the yeast culture solution from near the bottom of the conical flask with the help of a dropper and place a drop of this culture solution on a clean slide.
- Add a very little of iodine solution over the culture solution drop to stain it.
- Place a cover slip over the slide.
- Keep the slide under the microscope and observe it first under low power and then under the high power of the microscope.
- Note the formation of buds on the yeast cells and how they separate from the parent cell.

In some organisms like sponges and corals, the buds remain attached to the parent organism permanently. These buds then grow and produce buds of their own. In this way, a colony of sponges or corals is formed.

3.2.1.3. Spore Formation:

Spores are the microscopic asexual reproductive bodies which are covered by a hard protective coat. This coat enables them to survive under unfavourable conditions like lack of food, lack of water and extreme temperatures. But when the conditions are favourable (food and water is available, and temperature is suitable), then the spores grow to produce new plants. Thus, spores are a kind of seeds of plants. These spores are very light and keep floating in air all around us. They are so small that we cannot see them with naked eyes. The spore formation method of asexual reproduction is used by unicellular organisms as well as by multicellular organisms. For example, bacteria are the unicellular organisms which reproduce by spore formation whereas fungi such as *Rhizopus* (bread mould), *Aspergillus, Mucor* and *Penicillium* (3.8 A, B, C, D) and non-flowering plants such as ferns and mosses are multicellular organisms which reproduce by spore formation method.

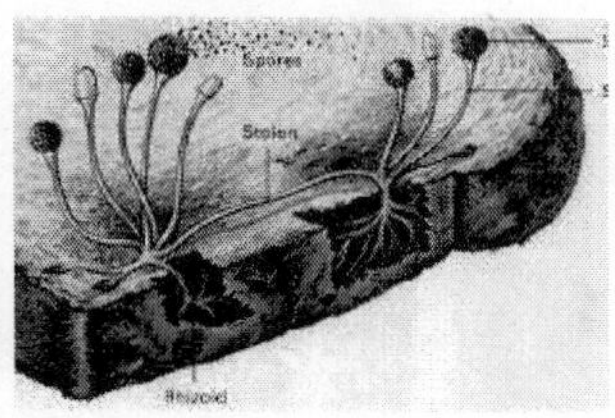

(A) *Rhizopus* fungs (Bread mould) growing on a slice of bread

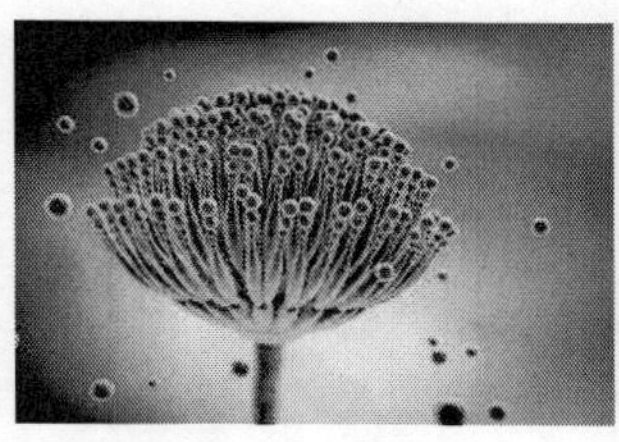

(B) *Aspergillus*

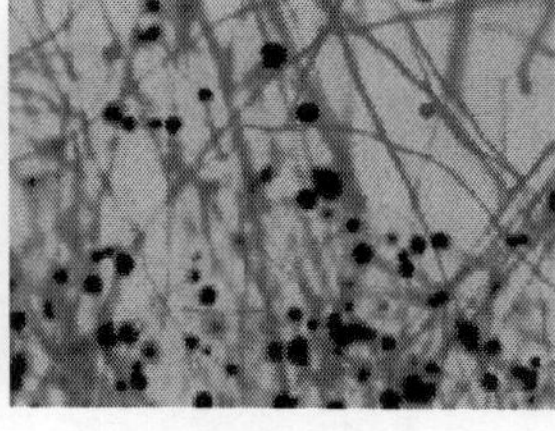

(C) *Mucor*

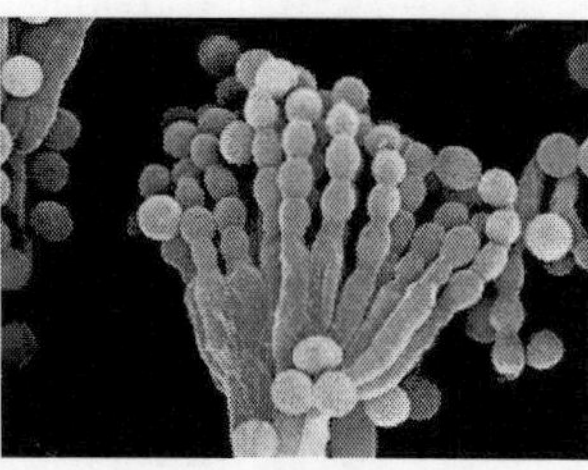

(D) *Penicillium*

Fig. 3.8 Spore formation in various fungus

The reproduction by spore formation takes place in plants. In spore formation, the parent plant produces hundreds of microscopic reproductive units called spores. When the spore case of the plant bursts, then the spores spread into air. When these air-borne spores land on food or soil under favourable conditions (like damp and warm), they germinate and produce new plants. Most of the fungi (like *Rhizopus, Mucor,* etc.), bacteria and non-flowering plants such as ferns and mosses reproduce by the method of spore formation. The common bread mould is a fungus plant whose scientific name is *Rhizopus*. The common bread mould or *Rhizopus* fungus reproduces by the method of spore formation.

The tiny spores of 'bread mould' (a fungus plant) are almost always present in the air. If we keep a moist slice of bread aside for a few days, then the spores of bread mould plant present in air settle on the moist bread and germinate to form new fungus plants. The bread mould plants look like a white cottony mass covering the bread slice which later on turns black. If we observe the surface of this slice of bread through a magnifying glass, then the bread mould plant express grow. The common bread mould plant consists of fine, thread-like projections called hyphae and thin stems having knob-like structures called sporangia. Each knob-like structure (or sporangium) contains hundreds of minute spores enclosed in a spore case. When the spore case bursts, the tiny spores are dispersed in air. These spores are the asexual reproductive units which can produce more bread mould plants under suitable conditions. Actually, it was one such air-borne spore which grew on the moist slice of bread kept aside by us for a few days. If we remove one sporangium from the bread mould, keep it on a slide, put a coverslip over it and observe spores through a microscope.

3.2.1.4. Regeneration:

In some organisms (plants as well as animals) small cut parts of their body can grow or regenerate to form whole new organisms complete in all respects. The process of getting back a full organism from its body parts is called regeneration. The simple animals like Hydra and Planaria show regeneration.

Planaria is a flatworm which is found in freshwater ponds and slow-moving streams. Planaria possesses great power of regeneration. If the body of Planaria somehow gets cut into a number of pieces, then each body piece can regenerate into a complete Planaria by growing all the missing parts (Fig 3.9). Planaria worm somehow gets cut into three pieces. After a certain time, each cut piece of the body of Planaria worm grows into a complete Planaria worm. In this way, three Planaria worms are produced from just one Planaria worm. Similarly, if the body of a Hydra gets cut into a number of pieces, then each body piece of Hydra can grow into a complete Hydra. (Please note that regeneration is not exactly the same as reproduction because most simple animals would not depend on being cut into pieces to be able to reproduce).

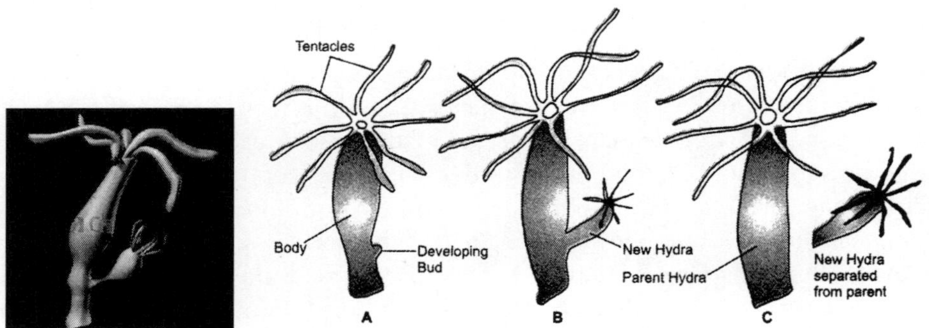

Fig. 3.9 Regeneration in *Planaria*

The regeneration of an organism from its cut body part occurs by the process of growth and development. This happens as follows: The cells of cut body part of the organism divide rapidly to make a 'ball of cells'. The cells present in the 'ball of cells' move to their proper places within the ball where they have to form various organs and body parts of the organism. The cells then change their shapes (or become specialized) to form different types of tissues. These different tissues form various organs and body parts of the organism. In this way a complete organism is regenerated.

A summarized account of the different modes of asexual reproduction is given in Figure 3.10.

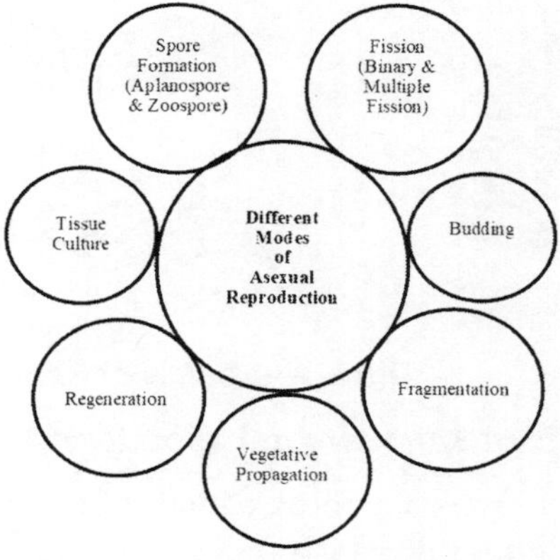

Fig. 3.10 Different modes of Asexual Reproduction

3.2.1.5. Fragmentation:

The breaking up of the body of a simple multicellular organism into two or more pieces on maturing, each of which subsequently grows to form a complete new organism, is called fragmentation. The breaking up of the body of an organism in fragmentation to form new organism occurs naturally (on its own) when the parent organism matures. Fragmentation is an asexual method of reproduction. The reproduction by fragmentation method can occur in simple multicellular plants as well as animals. The organisms like *Spirogyra* (Fig 3.11) and *Sea anemones* (Fig 3.12) can reproduce by the method of fragmentation.

Fig. 3.11 Fragmentation in *Spirogyra*

Fig. 3.12 Fragmentation in *Sea anemones*

Spirogyra is a green, filamentous alga plant which is found in ponds, lakes and slow moving streams. *Spirogyra* filament simply breaks into two or more fragments on maturation, and each fragment then grow into a new *Spirogyra*. This breakup of the filament of a mature Spirogyra on its own brings about asexual reproduction. Thus, *Spirogyra* reproduces by the asexual method of fragmentation. A mature

Spirogyra filament is undergoing fragmentation to produce the three new *Spirogyra*. All the *Spirogyra* will mature in due course of time and break again to produce even more *Spirogyra*. This process of reproduction goes on and on (Fig 3.13 and 3.14).

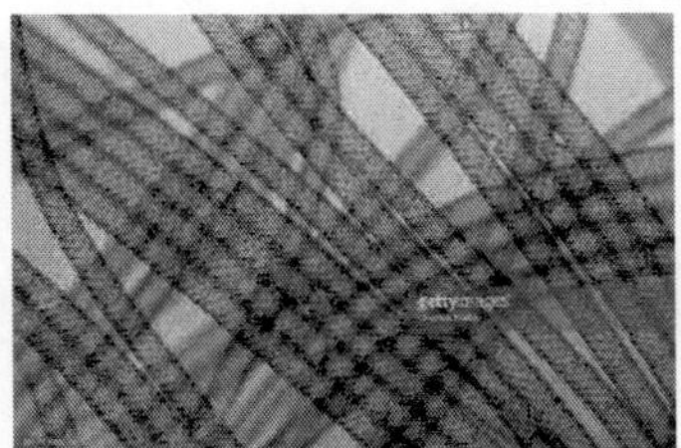

Fig. 3.13 *Spirogyra*

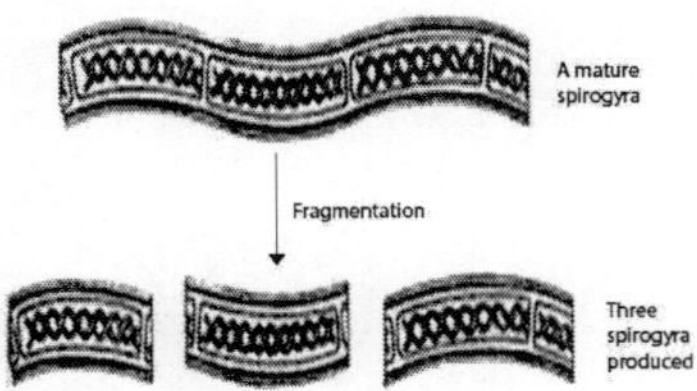

Fig. 3.14 Process of Reproduction in *Spirogyra*

Illustration on Study of *Spirogyra* in Laboratory:

- Collect some water from a pond or lake which appears dark green and contains long filament or thread-type structures.
- Take out the green coloured mass from the pond water sample and separate its threads or filaments by using two needles.
- Place one filament on a clean slide, put a drop of glycerin over it and cover it with a coverslip.
- Keep this slide under the microscope and see it first under the low power and then under the high power of microscope.
- Observe the detailed structure of the green filament of *Spirogyra* and draw a diagram accordingly.

Please note that the main difference between fission and fragmentation is that in fission, a unicellular organism breaks up to form two or more daughter organisms, whereas in fragmentation, a multicellular organism breaks up to form two or more daughter organisms.

Each species of organisms reproduces in a different way. All the different ways of reproduction are grouped under two broad categories such as (a) Asexual reproduction and (b) Sexual reproduction. Asexual reproduction is the production of offsprings by a single parent without meiosis, formation of gametes, fertilization and transfer of genetic materials between individuals. In asexual reproduction, a part of the parent organism separates off and grows into a new organism. Only one parent is needed to produce a new organism in asexual reproduction. But no sex cells are involved in asexual reproduction. Asexual reproduction takes place in unicellular animals and plants, micro-organisms (like bacteria) and simple multicellular animals (like *Hydra* and *Planaria*) and some multicellular plants (like *Bryophyllum* and Rose plants etc.).

The common methods of asexual reproduction are (i) Fission, (ii) Budding, (iii) Spore formation, (iv) Regeneration, (v) Vegetative propagation (vi) Micro propagation *in vitro* and (vii) Parthenogenesis. Binary fission is an asexual method of reproduction of organism. In binary fission, the parent organism splits (or divides) to form two new organisms. Budding is an asexual method of reproduction. In budding, a small part of the body of the parent organism grows out as a 'bud' which then detaches and becomes a new organism. The asexual reproduction by budding is observed in hydra and yeast. Spores are asexual reproductive bodies which are covered by a hard protective coat. This coat enables them to survive under unfavourable conditions like lack of food, lack of water and extreme temperatures. But when the conditions are favourable, then the spores grow to produce new plants. Thus, spores are a kind of seeds of plants.

3.2.1.6. Vegetative Propagation:

Vegetative propagation is an asexual method of reproduction. The reproduction by vegetative propagation occurs only in plants. In vegetative propagation, new plants are obtained from the parts of old plants (like stems, roots and leaves), without the help of any reproductive organs. Vegetative propagation usually involves the growth and development of one (or more) buds present on the old part of the plant to form a new plant. These buds are in the dormant state (inactive state) in the old part of the plant. When provided suitable conditions (like moisture, optimum temperature etc.), these buds grow to form new plants. It is a common observation that green grass plants spring up in dry fields after the rains.

This happens due to vegetative propagation as follows:

Buds are present on the stems as well as the leaves of the *Bryophyllum* plant which can develop into new plants. So, *Bryophyllum* plants can be reproduced by vegetative propagation by using either a piece of its stem or its leaves. For example, if we plant a broken piece of the stem of a *Bryophyllum* plant in the ground, we will get a new *Bryophyllum* plant growing from it in a week's time. Even the leaves of a *Bryophyllum* plant can produce new plants. The leaves of a *Bryophyllum* plant have special type of buds in their margins (or edges). These buds may get detached from the leaves, fall to the ground and then grow to produce new *Bryophyllum* plants (Fig 3.15).

Fig. 3.15 Vegetative propagation in *Bryophyllum*

Money plant can also be grown by vegetative propagation (Fig 3.16) by using a piece of its stem as follows:

Fig. 3.16 Vegetative propagation in Money plant

- Cut a piece of stem of money plant in such a way that it contains at least one leaf on it (The point on stem where a leaf is attached is called a node).
- Dip one end of this stem in water. After few days we will find that new roots appear at the point where leaf was attached.
- The piece of stem will gradually grow into a new money plant.

Please note that if we cut the stem of money plant in between two leaves, then it will not grow into a new plant. This is because it does not have a growing point in it. Potato tuber is an underground stem of the potato plant. Potato tuber can be used for the vegetative reproduction of potato plants. Each potato tuber can produce more than one plant.

This happens as follows:

A potato tuber has many buds (called eyes) on its body. These buds act as organs for vegetative reproduction (Fig 3.17). When a potato tuber is planted in the soil then the buds of the potato tuber start growing to form a new potato plants, and potato growing from only two buds of the potato tuber. Other buds can also do the same.

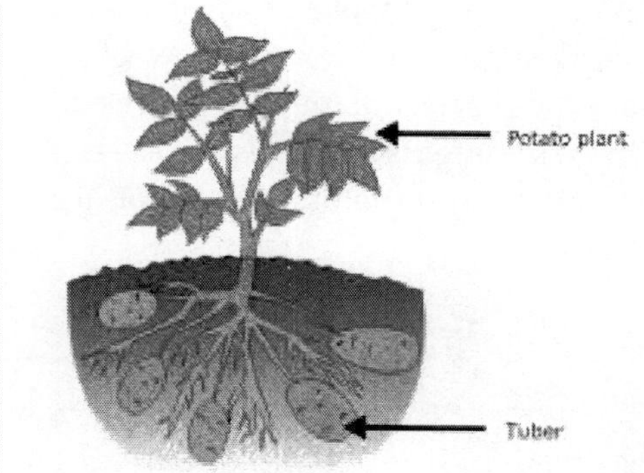

Fig. 3.17 Potato plant with tubers

It may be noted here that it is not necessary to plant the whole potato tuber in the ground to produce new potato plants. We can even plant 'cut pieces' of a potato tuber in the ground to obtain new potato plants. But all these cut pieces of potato tuber should have a bud or eye on them. So, if we cut a potato tuber into a number of pieces in such a way that every piece has a on it and plant them in the ground, then each cut piece of potato tuber will produce a new potato plant in due course of time. Each potato plant produces more than one tuber and each tuber has more than one bud, which produce more than one new plant). Due to this, the vegetative propagation method of producing potato plants by tubers is much faster than the production of potato plants from seeds.

Vegetative Propagation of Potatoes:

- Take a potato and cut it into small pieces in such a way that some pieces contain a bud or eye in them.
- Place the potato pieces having buds on wet cotton kept in a tray.
- Keep the tray aside for a few days (but sprinkle water on the cotton daily to keep it wet).
- We will see that green shoots and roots appear from the buds of the potato pieces. These are the new potato plants. If, however, we take potato pieces without buds in this experiment, then no new potato plants will grow from them.

Artificial Propagation of Plants:

We can grow many plants from one plant by using the man-made methods. The process of growing many plants from one plant by man-made methods is called artificial propagation of plants. A number of methods of artificial propagation of plants are used in agriculture (for raising crops), and horticulture (cultivation of vegetables, fruits and flowers). The three common methods for the artificial propagation of plants are such as Cuttings, Layering, and Grafting.

1. Cuttings:

A small part of a plant which is removed by making a cut with a sharp knife is called a 'cutting'. A cutting may be a piece of stem, root or even a leaf. While making a cutting, care should to taken to see that there are some buds on it. Plants like Rose, Grapes, *Bougainvillea, Chrysanthemum,* Sugarcane, Banana and Cactus etc. can be grown by means of cuttings (Fig 3.18, 3.19, 3.20 and 3.21).

Fig. 3.18 Rose

Fig. 3.19 Grapes

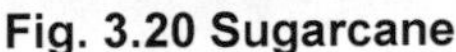

Fig. 3.20 Sugarcane

Fig. 3.21 Banana

2. Layering:

In this method, a branch of the plant is pulled towards the ground and a part of it is covered with moist soil leaving the tip of the branch exposed above the ground. After some time, new roots develop from the part of the branch buried in the soil. The branch is then cut off from the parent plant. The part of the branch which has developed roots grows to become a new plant (just like the parent plant). There are two methods of layering as (a) Mount layering in Jasmine plant (Chameli), grapevine, rose and lemon. Many plants like strawberry and raspberry are propagated by the natural layering method (Fig 3.22A, B). (b) Air layering in lichi pomegranate, lemon and mango (Fig 3.22C,D).

(A) Jasmine Plant

(B) Grapes

(C) Air layering in Mango

(D) Litchi

Fig. 3.22 Propagation in various plants through layering method

3. Grafting:

Grafting is a method in which the cut stems of two different plants ,one with roots (root stock)and other without roots(scion) are joined together in such a way that the two stems join and grow as a single plant. This new plant will have the characteristics of both the original plants. Grafting is used to breed fruit trees and flowering bushes. Apple, mango, peach, apricot and pear trees are often grafted.

Artificial vegetative propagation is usually used for the propagation (or reproduction) of those plants which produce either very few seeds or do not produce viable seeds. Some examples of such plants which are reproduced by artificial vegetative propagation methods are such as Banana, Pineapple, Orange, Grape, Apple and Rose etc. (Fig. 3.23, 3.24).

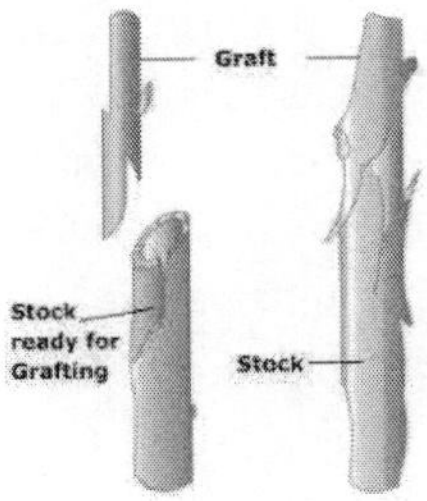

Fig. 3.23 The Grafting method for the artificial propagation of plants or trees

Fig. 3.24 Grafting in Apple tree

Advantages of Artificial Vegetative Propagation:

Artificial propagation of farm and garden plants has several advantages. Some of the important advantages of the artificial vegetative propagation of plants are given below:

1. The new plants produced by artificial vegetative propagation will be exactly like the parent plants. Any desirable features of the parent plant will be replicated in the new plants.
2. The fruit trees grown from seeds may take many years before they start to bear fruit. But the fruit trees grown from cuttings or by grafting start to bear fruits much earlier (only after a few growing seasons).
3. The plants grown by vegetative propagation usually need less attention in their early years than the plants grown from seeds.
4. Many plants can be grown from just one parent plant by artificial propagation.
5. We can also get seedless plants by artificial propagation.

The cuttings, layering and grafting are the traditional methods for the artificial propagation of plants. For commercial purposes, they are being replaced by the modern methods of artificial propagation of plants involving tissue culture. In tissue culture, they just put a few plant cells (or plant tissue) in a growth medium with plant hormones in it and it grows into new plants.

Advantages of Vegetative Propagation:

It is a cheaper, easier and more rapid method of plant propagation.

Limitations of Vegetative Propagation:

- Plants produced by this technique possess less vigour.
- They are more prone to diseases.
- They show no genetic variations.

Tissue Culture:

The production of new plants from a small piece of plant tissue (or cells) removed from the growing tips of a plant in a suitable growth medium (culture solution) is called tissue culture. The growth medium (or culture solution) used for growing plant tissues is very important in this process because it contains various plant nutrients in the form of 'jelly' (called agar) and plant hormones which are necessary for plant growth. Agar is obtained by Gelidium algae, and used as solidifying agent only. The process of tissue culture for producing new plants is carried out as follows (Fig. 3.25A, B,C):

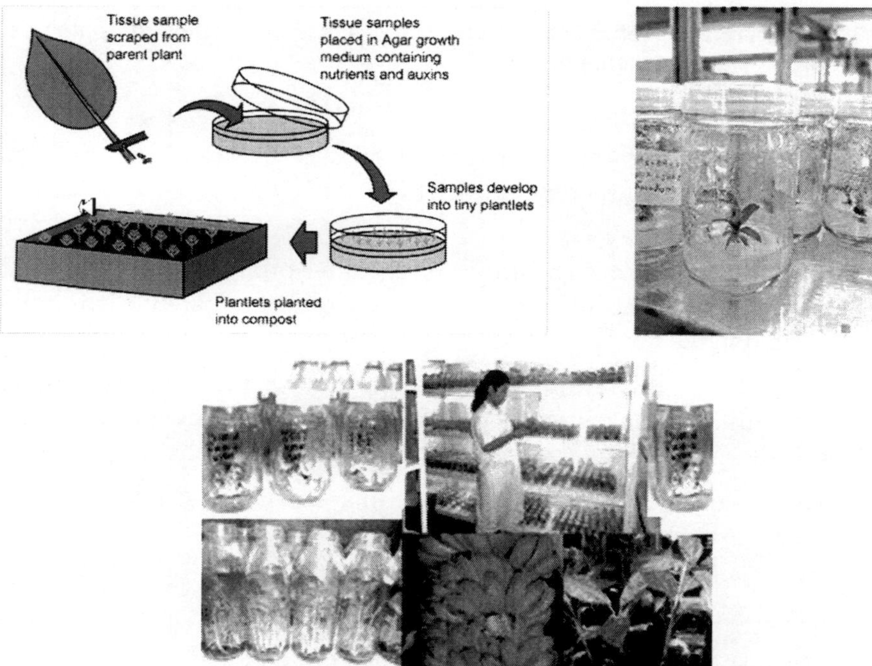

Fig. 3.25 Process of Tissue culturing

- A small piece of plant tissue is taken from the growing point of the plant (tip of the plant) and placed on a sterile jelly which contains nutrients and plant hormones. The hormones make the cells in the plant tissue divide rapidly producing many cells which form a shapeless lump of mass called callus.
- The callus is then transferred to another jelly containing suitable plant hormones which stimulate the callus to develop roots.

- The callus with developed roots is then put on a yet another jelly containing different hormones which stimulate the development of shoots.
- The callus having roots and shoots separates into tiny plantlets. In this way, many tiny plantlets are produced from just a few original plant cells (or tissue).
- The plantlets thus produced are transplanted into pots or soil where they can grow to form mature plants.

The tissue culture technique is being used increasingly for the production of ornamental plants like Orchids, Dahlia, Carnation, Chrysanthemum, etc. The production of plants by the method of tissue culture is also known as micro propagation (due to the extremely small amount of plant material used).

Advantages of Tissue Culture:

(a) Tissue culture is a very fast technique. Thousands of plantlets can be produced in a few weeks' time from a small amount of plant tissue.

(b) The new plants produced by tissue culture are disease free.

(c) Tissue culture can grow plants round the year, irrespective of weather or season.

(d) Very little space is needed for developing new plants by tissue culture.

> The process of growing many plants from one plant by man-made methods is called artificial propagation of plants. The three common methods for the artificial propagation of plants are such as (a) Cuttings, (b) Layering, and (c) Grafting.
>
> The production of new plants from a small piece of plant tissue (or cells) removed from the growing tips of a plant in a suitable growth medium (culture solution) is called tissue culture. The growth medium (or culture solution) used for growing plant tissues is very important in this process because it contains various plant nutrients in the form of 'jelly' (called agar) and plant hormones which are necessary for plant growth. The tissue culture technique is being used increasingly for the production of ornamental plants like Orchids, Dahlia, Carnation, Chrysanthemum, etc.

3.2.2 Sexual Reproduction:

Sexual reproduction may be defined as the production of offspring's (new individuals) by the fusion of two gametes (usually one from male parent and the other from female parent) to form a diploid zygote which develops into a mature organism. The act of fusion of gametes is called fertilization, which may be defined as fusion of two gametes to form a single cell (zygote). As stated earlier, the sexual mode of reproductions involves two individuals – one acting as male and the other acting as female. Sometimes, the individual is one but the sex organs are separate (as in many plants) which produce different gametes. In any case, the two gametes, belonging to different sex, fuse with each other to form a zygote leading to the development of a new generation. In most of the cases, formation of gametes involves meiosis so that the chromosome number becomes half in gametes. For example, if an organism is diploid, i.e., the somatic cells of the organism have 2n number of chromosome, the meiosis in gamete formation results in the production of haploid (n) gametes. Fertilization brings about fusion of nuclei of gametes. The two nuclei (belonging to two gametes) each having single set of chromosomes (haploid), fuse to form zygote that develop into an adult body of the organism.

In majority of animals including human beings, male and female individuals are separate, but majority of flowering plants and a few animals (like tapeworm, earthworm, garden snail, star fish, etc.) are bisexual or hermaphrodite, i.e., each individual has both male and female sex organs.

Higher animals possess gonads (primary sex organs). The gonads produce haploid gametes by meiotic (reduction) division. The testes, which are male gonads, produce sperms after meiosis in primary spermatocytes. The ovaries, which are female gonads, produce eggs or ova after meiosis in primary oocytes. Meiosis is an essential feature of sexual life cycle which provides mechanism for reducing the chromosomes by half. This ensures production of diploid number of chromosomes in zygote, when the gametes fuse.

3.2.2.1 External and Internal Fertilization:

In certain aquatic animals (such as fishes and amphibians), in terrestrial animals, fertilization is internal. In case of reptiles, birds and mammals, including human beings, the male gametes are inserted into the female's reproductive tract where fertilization takes place. Transfer of male gametes from testes into the female reproductive tract occurs at the time of mating or copulation. The organs involved in the process of mating are called copulatory organs or accessory sex organs.

Table 3.3. Differences between External and Internal Fertilization

Sr. No.	External Fertilization	Internal Fertilization
1.	It occurs outside the body of the female organism.	It occurs inside the body of the female organism.
2.	Both male and the female discharge their gametes outside their body.	Male discharges the gametes into the reproductive tract of the female. The female releases the gametes inside its reproductive tract.
3.	It is not a sure method of fertilization.	It is a sure method of fertilization.
4.	Embryo develops unprotected, outside the body of the female.	Embryo is well protected but may develop outside or inside the body of the female.

Advantages of Sexual Reproduction over Asexual Reproduction:

Sexual reproduction has several advantages over asexual reproduction because it involves:

- Fusion of male and female gametes coming from male and female organisms. Since the fusing gametes come from two different and sexually distinct individuals, the offspring's exhibit diversity of characters, and
- Meiosis during gametogenesis provides opportunities for new combination of genes. It plays a prominent role in the origin of new species and leads to variation required for evolution.

Gamete formation, inside the gonads of diploid organisms, involves meiosis or reduction division. The gamete mother cell is diploid (2n), i.e., it has two sets of chromosomes. This single diploid cell divides by meiosis to form 4 haploid daughter cells. Each daughter cell becomes a gamete, either male or female. Each gamete possesses single set of chromosomes. That means this division involves copying of the DNA as well as the cellular apparatus. There is a stage in such nuclear division where crossing over of chromosomes takes place. This is very important step which results a slight different composition of chromosomes in gametes. Fusion of these gametes results in the formation of slightly different individuals which show variations. The variations which lead to the appearance of such characters which fit to the changing environment result in the survival of the species, Chances of variation, therefore, are much more in sexual mode of reproduction as compared to asexual reproduction. Moreover, chances of the production of compatible generations are also more in sexual reproduction.

The variation generated due to DNA copying during gamete formation moves on to the next generation. This happens by the process of combination of DNA belonging to two different individuals. Again a new variation appears during DNA copying at the time of gamete formation in this generation that has already variations accumulated from previous generations. Thus, two different individuals in a population would have distinct patterns of accumulated variations. Since all the different variations occur in living organisms, it is assumed that these do not have any bad effects. Combining the distinct variations from two or more individuals during variations from two or more individuals during sexual reproduction would thus create new combinations of variants. The meiosis plays very important role in the life cycle of an organism. We are now aware of the fact that each sexual fusion results in the combination of DNA copies from two individuals doubling the DNA content in the next generation. If there is no meiosis, each new generation will result in having twice the amount of DNA that the previous generation had. If this process continues for few generations, very soon a situation will arise when the amount of DNA will not be accommodated inside the cell and there will be no other cellular contents except DNA. Furthermore, a time will come when there will be only DNA on earth and there will be no room for anything else (Fig 3.26). To avoid this situation, nature has solved the problem by creating a special kind of cell division during the formation of gametes. This cell division is called meiosis or reduction division. During meiosis, the chromosome number or DNA content of parents becomes half in their gametes. Again the number becomes same in the next generation when the two gametes carrying half chromosome number fuse with each other (Fig 3.27).

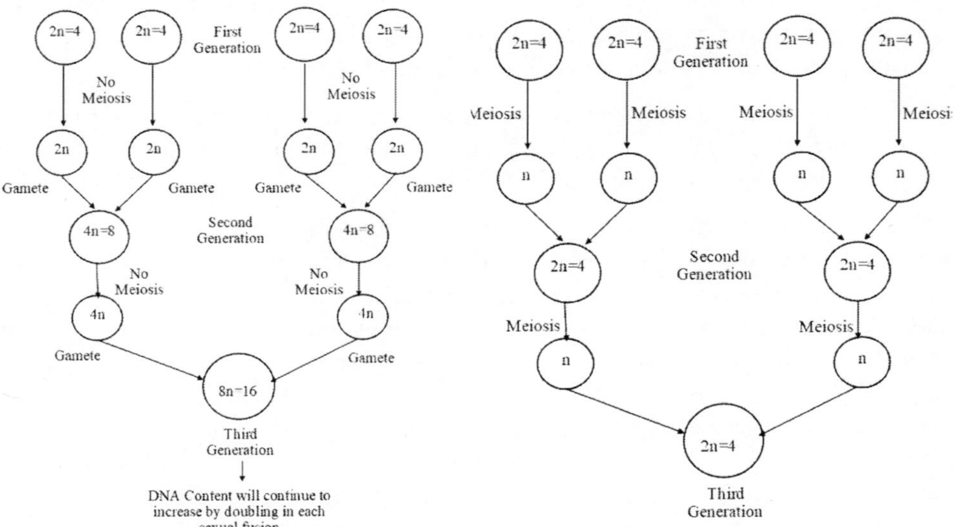

Fig. 3.26 Fusion of Gamets (without meiosis)

Fig. 3.27 Fusion of Gamets (with meiosis)

Sexual reproduction is the production of new individuals by the fusion of two gametes (usually one male and the other female) following the act of fusion and formation of zygote. Thus, two parents are needed to produce a new organism in sexual reproduction. The humans, fish, frogs, hens, cats, dogs, cows, horses, deer, rabbits, lions and tigers all reproduce by the method of sexual reproduction.

The basic difference between asexual and sexual reproduction is that only one parent is needed in asexual reproduction whereas two parents are needed in sexual reproduction. Another difference is that no sex cells (or gametes) are involved in asexual reproduction but sex cells take part in sexual reproduction.

In certain aquatic animals (such as fishes and amphibians), in terrestrial animals, fertilization is internal. In case of reptiles, birds and mammals, including human beings, the male gametes are inserted into the female's reproductive tract where fertilization takes place. Transfer of male gametes from testes into the female reproductive tract occurs at the time of mating or copulation. The organs involved in the process of mating are called copulatory organs or accessory sex organs.

3.2.2.2 Sexual Reproduction in Plants:

In flowering plants, all the steps of sexual reproduction occur within specialized reproductive organs, called the flowers. Thus, a flower may be defined as a specialized condensed reproductive shoot of flowering plants inserted with the essential reproductive parts.

Parts of a flower: It consists of four whorls of floral appendages attached on the receptacle. The receptacle is the top of the flower stalk (pedicel). The four whorls of floral appendages are such as (Fig. 3.28) :

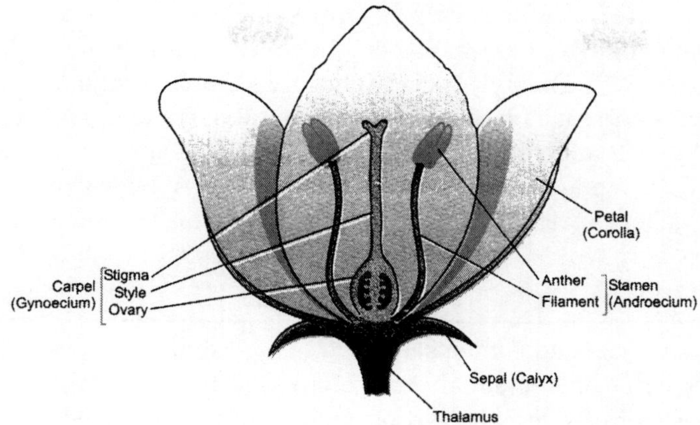

Fig. 3.28 A typical Flower consisting of Floral Parts

- Calyx or Sepals
- Corolla or Petals
- Androecium or Stamen
- Gynoecium (Pistil) or Carpels

1. Calyx or Sepals:

It is the outermost whorl of floral leaves, called sepals; Sepals are generally green in colour and protective in function. Occasionally they are brightly colored.

2. Corolla or Petals*:*

It is the collection of petals, Petals are generally large, showy and brightly coloured to attract the insect pollinators. Calyx and Corolla are non-essential parts of the flower because they are not directly involved in reproduction.

3. Androecium or Stamen:

It is the collection of stamens, the male reproductive organs of the flower. Each stamen consists of anther and filament. The filament is a long stalk that bears an anther at the top. Usually anthers are bi-lobed and contain four pollen sacs (micro sporangia). The pollen grains are made inside the pollen sacs. Pollen grains are haploid microspores formed inside the pollen sacs by meiosis (reduction division) is diploid microspore mother cells. Initially, each microspore is uni-nucleate, but soon it becomes 2-3 nucleate. The two male nuclei along with a little cytoplasm differentiate into two gametes. Thus, each pollen grain produces two male gametes, when germinates to produce pollen tube.

4. Gynoecium (Pistil) or Carpels*:*

It is the collection of carpels, the female reproductive organs of the flower. Each carpel has a swollen ovary, long style and a terminal stigma. If a flower has only one carpel, it is called monocarpellary. Generally, the flowers have more than one carpels (free or fuse). The fused carpels form a single unit consisting of a common ovary, style and stigma. The stigma receives pollen grains during pollination and supports their growth. The style bears stigma in a suitable position to receive pollen and also provides passage for the entry of pollen tube. The ovary is the swollen and hollow basal-part that contains ovules. Each ovule has an embryo sac that bears a haploid egg-the female gamete. In plants, the fusion of male gamete and female gamete (egg) takes place inside the embryosac, which is located inside the ovule. Fusion occurs only when the pollen grains are transferred to the stigma of same or another flower.

A flower is a specialized condensed reproductive shoot of flowering plants on which are inserted the essential reproductive parts. A typical angiospermous flower consists of (a) Calyx (whorl of sepals), (b) Corolla (whorl of petals), (c) Androecium (whorl of stamens), and (d) Gynoecium (whorl of carpels). Anthers produce pollen grains containing male gamete.

3.3 POLLINATION

The transfer of pollen grains from the opened anther of the stamen to the receptive stigma of the carpel is called pollination. Pollination is an important mechanism by which the pollen grains, produced inside the pollen sacs of anthers, are transported by various agencies to the stigma of female reproductive organ. It is an essential process that facilitates successful fertilization in plants. The pollen grains protect male gametes from drying out. The common agents responsible for successful pollination are such as wind, water, insects, birds, bats and many other living organisms.

3.3.1 Types of Pollination:

Pollination is of two types such as Self-pollination and Cross pollination

1. Self-Pollination:

It involves the transfer of pollen grains from the anther of a flower to the stigma of the same flower or to the stigma of another flower borne on the same plant.

2. Cross Pollination:

It involves the transfer of pollen grains from the flower of one plant to the stigma of the flower of another plant of the same species.

Table 3.4 Difference between Self and Cross Pollination

Sr. No.	Character	Self-Pollination	Cross Pollination
1	Occurrence	Occurs within a flower or between two flowers of the same plant.	Occurs between two flowers of two different plants of the same species.

2	Agent of pollination	No external agent of pollination required (usually).	External agents such as wind, water, insects and birds required.
3	Transfer of pollen grains	It is the transfer of pollen grains from anther to the stigma of the same flower.	It is the transfer of pollen grains from the anther of one flower to the stigma of another flower.
4	Requirement of external for the transfer of pollen grains	It does not require an external agent for the transfer of pollen grains.	It requires an external agency for the transfer of pollen grains from one flower to another.
5	Production of pollen grains	Produced in small numbers, thus no wastage of pollen grains occurs.	Produced in large numbers (usually), thus, wastage of pollen grains occurs.
6	Number of Pollen grains production	Such bisexual flowers produce small number of pollen grains.	Such flowers produce large number of pollen grains.
7	Nature of variations	It does not bring variations.	It brings about large variations.
8	Appearance of flowers	Flowers are not attractive (usually).	Flowers are attractive with coloured petals.
9	Fragrance and nectar	Flowers do not (usually) produce scent or nectar.	Flowers generally produce scent and nectar.
10	Nature of offspring's produced	Offspring's produced have genetic makeup identical to the parent plant; purity of race maintained, no variation occurs.	Offspring's produced may differ in genetic make-up, and variations occur.

The transfer of pollen grains from the opened anther of the stamen to the receptive stigma of the carpel is called pollination. The common agents responsible for successful pollination are such as wind, water, insects, birds, bats and many other living organisms.

Pollination is of two types such as Self-pollination and Cross pollination. Self-Pollination involves the transfer of pollen grains from the anther of a flower to the stigma of the same flower or to the stigma of another flower borne on the same plant. Cross Pollination involves the transfer of pollen grains from the flower of one plant to the stigma of the flower of another plant of the same species.

3.4 FERTILIZATION IN PLANTS:

Pollination results in the deposition of related pollen grains over the receptive stigma of the carpel. They absorb water, swell and then germinate to produce pollen tubes. One pollen tube grows into the stigma, passes through the style and then moves towards the ovarian cavity. Two non-motile male gametes are formed inside the tube during its growth through the style. After reaching the ovary, pollen tube enters the ovule through the micropyle. The tip of the tube finally pierces the egg apparatus end of the embryo sac. After penetration, the tip of pollen tube ruptures releasing two male gametes into the embryo sac. The mature embryo sac consists of an egg apparatus (one haploid egg and two synergids), two polar nuclei and three antipodal cells (Fig. 3.29). During the act of fertilization, one male gamete fuses with the egg to form the diploid zygote. The process is called syngamy or generative fertilization. The diploid zygote finally develops into embryo. The other male gamete fuses with the two polar endosperm nucleus. The process is called triple fusion or vegetative fertilization. This mechanism involving two acts of fertilization in an embryo sac is called double fertilization. After fertilization, zygote develops into embryo, ovule develops into seed and the ovary develops into the fruit. A summarized account of sexual reproduction in flowering plants is represented in Fig. 3.30.

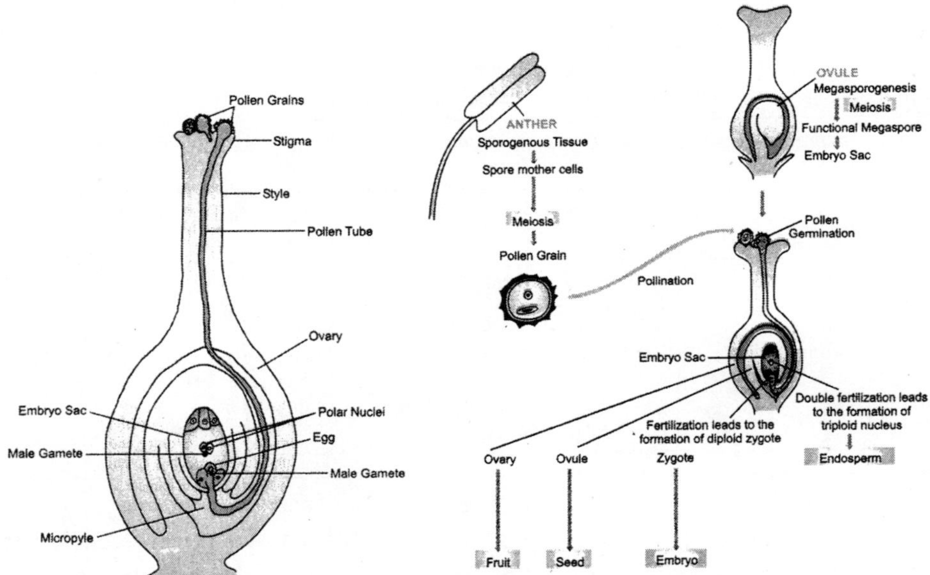

Fig. 3.29 Fertilization in the flowering plant

Fig. 3.30 Diagrammatically representation of sexual reproduction in flowering plants

3.5 REPRODUCTION IN HUMAN BEINGS

Human beings are unisexual and the human reproduction is highly evolved. There is a distinct sexual dimorphism, i.e., males are visibly different from females in physical buildup, external genital organs and accessory (external) sex characters. Thus, the structures associated with reproduction are different in males and females. The structures associated with male reproduction constitute the male reproductive system. Similarly, the structures associated with female reproduction constitute the female reproductive system. The reproductive systems of males and females consist of many organs which are distinguishable into primary and secondary sex organs.

3.5.1 Primary and Secondary Sex Organs

The primary sex organs are gonads, which produce gametes (or sex cells) and secrete sex hormones. The gonads of the male are called testes which produce male gametes (sperms) and the male hormone (testosterone). The gonads of the female are ovaries which produce female gametes (ova) and female hormones (estrogen and progesterone). The secondary sex organs include the genital ducts and glands which help in transportation of gametes and reproductive process. They do not produce gametes or secrete hormones.

3.5.2 Primary and Secondary Sexual Characters:

Primary sexual characters are those present at birth, whereas secondary sexual characters are those that develop at puberty. Puberty is the age of human males and females at which the reproductive organs become functional, gonads start producing gametes and sex hormones, and the boys and the girls become sexually mature. It is attained between the ages of 13 to 14 years in males and between 10 to 12 years in females. In males, puberty is triggered by the secretion of the hormone testosterone from the testes. The hormone brings about the development of secondary sexual characters during puberty and maintains these throughout adult life.

In females, puberty is triggered by the production of the hormone estrogen from the ovaries. This hormone brings about the growth and maturation of the reproductive tract and the development of secondary sexual characters. Puberty changes a girl child into women fit to bear child.

3.5.3 Secondary Sexual Characters in Human (Males):

The secondary sexual characters in human males include.

- Broadening of the shoulders and increased muscle development.
- Changes in behavior associated with courtship and mating.
- Enlargement of larynx and thickening of vocal cords producing deepening of the voice.
- Enlargement of penis and scrotum.
- Growth of pubic hair extra hair on the face, in the armpits and on the chest.
- Increase in height.

3.5.4 Secondary Sexual Characters in Human (Females):

The secondary sexual characters in human females include:

- Broadening of pelvis.
- Growth of breasts and external genitalia (vulva).
- Growth of pubic hair and extra hair in the armpits.
- Increase in the subcutaneous fat, particularly in thighs, shoulders, buttocks and face.
- Initiation of menstruation and ovulation.

Male Reproductive System:

It consists of the following organs such as Testes, Scrotum, Vas Deferens, Urethra and Penis (Fig.3.31 and 3.32):

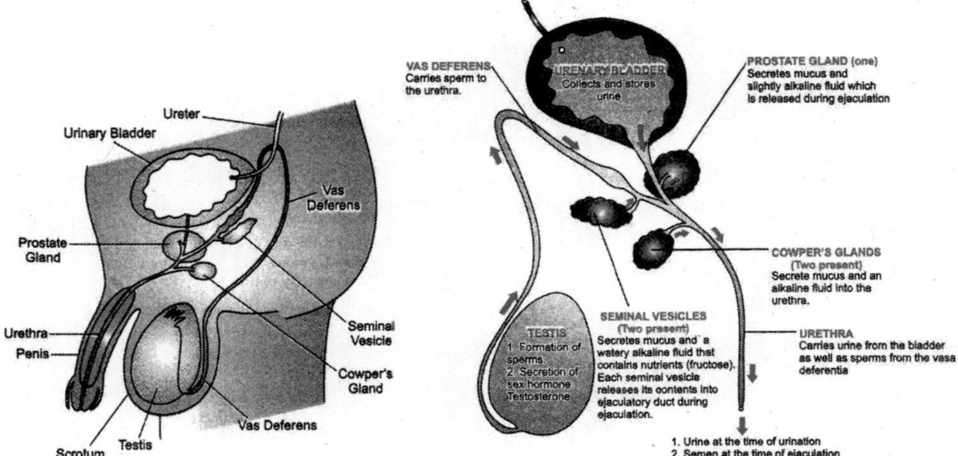

Fig. 3.31 Human Male's Reproduction system

Fig. 3.32 Path of sperm and role of different glands in Humans (Males)

1. **Testes:** The human male (Man) possesses two testes, which are the primary reproductive organs, lying outside the abdominal cavity. The two testes are the male gonads, which are the sites where gametes i.e., sperms, are made. The testes also produce the male sex hormone-testosterone. The testes of man produce sperms from puberty onwards, throughout his life.

2. **Scrotum:** The scrotum is a pouch of skin that hangs between the legs. It is divided internally into right and left scrotal sacs by a muscular partition. The two testes lie in respective scrotal sacs. The scrotum acts as a thermo-regulator and provides an optimal temperature for the formation of sperms. The sperms develop at a temperature 1-3°C lower than the normal body temperature. The life of sperms is greatly reduced if the temperature is higher. During winters, when temperature falls, the scrotum shrinks to bring the testes close to body to get warmth. During summers, when temperature rises, the scrotum becomes relaxed to lose heat.

3. **Vas deferens (plural Vasa differentia):** This is a straight tube, about 40 cm long, which carries the sperms to the seminal vesicles. The sperms are stored temporarily in the seminal vesicle, where mucus and a watery alkaline fluid containing the sugar-fructose, mix with the sperms.

4. **Urethra:** It is about 20 cm long tubes that arise from the urinary bladder to carry urine. It runs through the penis and opens to the outside through male genital pore. The contents of two seminal vesicles, i.e., sperms from vas

deferens also join the urethra. Thus urethra carries urine from the bladder, as well as sperms from the, vasa differentia, through the penis.

5. **Penis:** Penis is a long and thick muscular organ made up of mostly erectile tissue. The tip of penis consists of a soft and highly sensitive glans penis, which is covered by a loose, retractable fold of skin, called foreskin. At the time of sexual excitement, the erectile tissue gets filled with blood causing the penis to become erect. It is inserted into the vagina of the female where sperms are ejaculated for the purpose of reproduction.

In human males, there is only one opening for the urine and sperms to pass out of body.

3.5.5 Female Reproductive System:

Female reproductive system is more complex as compared to that of males. The complexity of sexual structures in females is due to the fact that it accepts sperms from male, supports their movements up to egg, fertilization and subsequent post fertilization changes. Female reproductive system (Fig 3.33) consists of the following organs such as (a) Ovaries (b) Fallopian tubes or Oviducts (c) Uterus and, (d) Vagina.

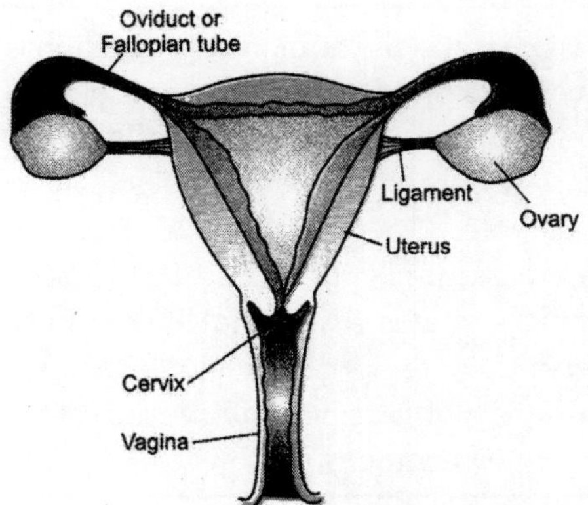

Fig. 3.33 Human Female's Reproductive System

3.5.5.1 Ovaries:

Each human female contains two almond shaped ovaries located in the lower part of abdominal cavity near the kidney. Each ovary is connected by a ligament to the uterus. The ovaries are primary sex organs (or female gonads), which perform the dual function such as (a) production of female gametes (egg or ova) and (b) Secretion of female sex hormones (estrogen and progesterone). Each ovary is composed of ovarian follicles, at various stages of development. Each follicle contains a large ovum surrounded by many layers of follicle cells. The production of ova starts at the age of puberty. Usually, one ovum is produced every month during the fertile years of a woman. After menopause, the ovaries become small and lose follicle.

3.5.5.2 Fallopian tubes (Oviducts):

A fallopian tube is about 10-12 cm long muscular tube which carries egg from the ovary to the uterus and also provides the appropriate environment for its fertilization. The funnel-shaped opening end of each fallopian tube lies near the posterior ends of each ovary. The other ends of the long convoluted tubes open into the uterus.

3.5.5.3 Uterus:

The uterus is a large, inverted pear-shaped, muscular structure that lies behind the bladder. If fertilization takes place, the embryo gets attached to the wall of uterus and grows there until birth.

3.5.5.4 Vagina:

This is a muscular tube about 7-10 cm long whose walls contain elastic tissue. It is well adapted to receive the male's penis during copulation. The vagina is also called "birth canal", as it allows passage of the baby at the time of child birth. In human females, the urinary opening (i.e., external opening of urethra) and vaginal openings are separate.

Human beings are unisexual and the human reproduction is highly evolved. Male reproductive system consists of testes, scrotum, vas deferens, urethra and penis. Female reproductive system consists of ovaries, fallopian tubes (oviducts) uterus and vagina. The primary sex organs in males are testes and in females are ovaries. Testes produce male gametes (Sperms) and male hormone (Testosterone). The sperms develop at a temperature 1-3°C lower than the normal body temperature. Ovaries produce female gametes (ova) and female hormones (Estrogen and Progesterone). The secondary sex organs include the genital ducts and glands which help in transportation of gametes and reproductive process. They do not produce gametes or secrete hormones.

3.6 SEXUAL CYCLE IN FEMALES

The fertility period of human female begins at the age of puberty (about 10-12 years of age), when one of the two ovaries starts releasing ovum (female gamete). The two ovaries do not release their ova at one time. They alternatively release their ova one after the other and the period between the releases of ova is generally 28-30 days. For example, if one ovum is released from left ovary on 1st October, the next ovum will be released from the right ovary on 28th to 30th October. The release of ova is controlled by a female hormone called as estrogen, which is not secreted constantly but in cycles. Each cycle lasts about 28-30 days, i.e., one ovum is released per cycle (Fig. 3.34). The control centres for production of gametes (ova) are the hypothalamus and the pituitary gland. The hypothalamus stimulates release of follicle stimulating hormone (FSH) and luteinizing hormone (LH) from the pituitary gland. These are referred to as gonadotrophic hormones because they control the ovary and other reproductive organs.

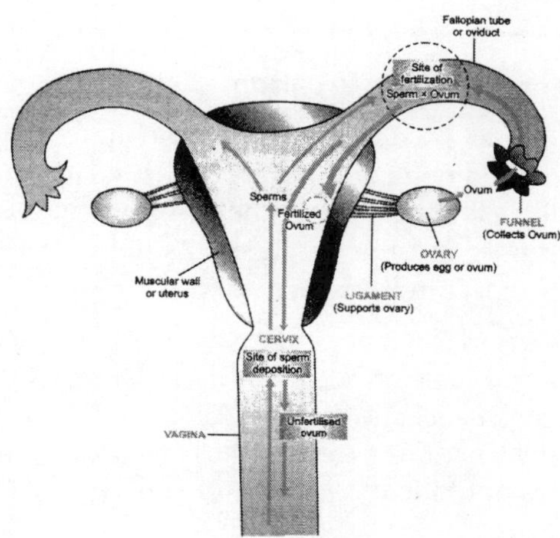

Fig. 3.34 Path of Sperms and Ova in Female Reproductive System

Differences between Zygote, Embryo and Foetus

A zygote is formed after fertilization. A zygote develops and becomes an embryo and finally, an embryo develops and becomes a foetus (Fig 3.35A, B and C).

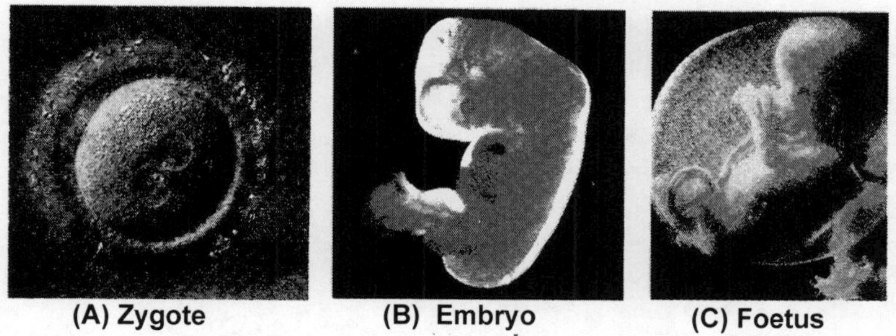

(A) Zygote **(B) Embryo** **(C) Foetus**

Fig. 3.35 Diagram showing differences among a human zygote, embryo and foetus

Table: Differences among Zygote, Embryo and Foetus

Sr. No.	Zygote	Embryo	Foetus
1.	A zygote is formed by the fusion of male and female gametes (sperm and egg)	An embryo is formed by the repeated cell division of a zygote.	A foetus is formed by the growth and development of an embryo.
2.	A zygote is the beginning of the formation of a baby.	An embryo is an unborn baby in the uterus in the early stages of development (up to 8 weeks after fertilization).	A foetus is an unborn baby in the uterus in the later stages of development (after 8 weeks till birth).

3.	A zygote is a single cell. It is smaller than a full stop (.)	An embryo is multicellular. The body features of growing baby in the embryo are not much developed.	A foetus is also multicellular. The body features of developing baby (like hands, legs, head, eyes and ears, etc.) can be identified.

3.6.1 Sexual Cycle in Females: Menstruation

When a girl child is born, her ovaries already contain many thousands of immature ova (or eggs) which are contained in immature follicles. When a girl reaches the age of puberty, then one follicle develops at a time to form a mature ovum (or egg). On maturing, the follicle bursts and the ovum shoots out of the ovary. This is called ovulation i e. release of an ovum from an ovary.

In a normal, healthy girl (or woman), ovulation takes place on the 14th day of the beginning of menstrual cycle of 28 days. This means that ovulation takes place in the middle of the menstrual cycle (because 14th gamete) once every 28 days from the age of puberty. That is, in girls ovulation starts when they attain puberty. Please note that ovulation does not take place every day after puberty but after a period of 28 days.

Before every ovulation, the inner lining of the uterus becomes thick and soft with lot of blood capillaries or blood vessels in it (Fig. 3.36A). These changes in the uterus are necessary because in case the ovum released by the ovary gets fertilized by the sperm, then the uterus has to keep this fertilized ovum for further development and supply it with food and oxygen, etc., so that it many grow into a baby in due course of time. If, however, a sperm is not available at the time of ovulation, then fertilization of ovum does not take place. Since the ovum is not fertilized, so the thick and soft uterus lining having lot of blood lining also breaks down. Since the thick and soft uterus lining contains a lot of blood vessels, so the breaking or disintegration of the uterus lining produces blood along with other tissues. This blood and other tissues come out of the vagina in the form of bleeding (Fig, 3.36B). Thus, the breakdown and removal of the inner, thick and soft lining of the uterus along with its blood vessels in the form of vaginal bleeding is called menstrual flow or menstruation.

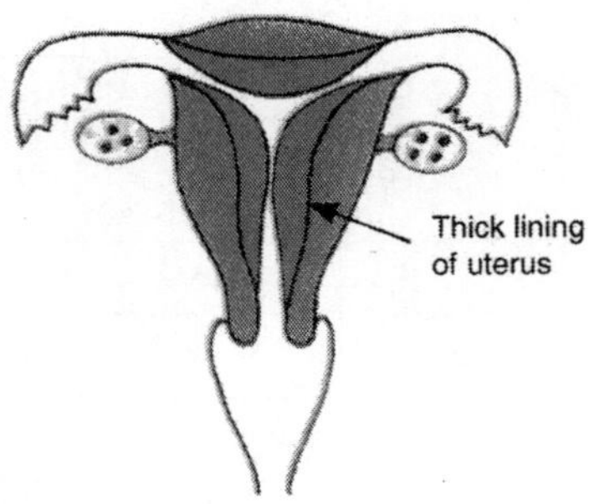

Fig. 3.36 A A thick lining grows in the uterus to receive the fertilized egg cell (if any)

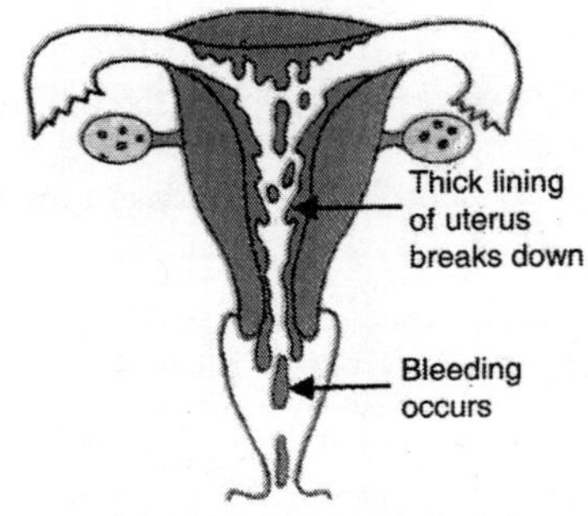

Fig. 3.36 B The thick pining of uterus breaks down leading to bleeding (In case the egg cell is not fertilized), It is called as menstruation cycle

Menstruation occurs if an ovum released by the ovary of a woman does not get fertilized due to non-availability of sperm at the time of ovulation. Since the process of menstruation in a woman occurs again and again after a fixed period of 28 days (to 30 days), so it is also known as menstrual cycle. Menstruation occurs every 28 days because ovulation (release of ovum or egg by ovary) occurs every 28 days. In everyday language, menstruation is called 'periods'. The menstrual cycle is described here with following points:

(a) When a girl reaches puberty at the age of about of 10 to 12 years, the sex hormones released into her blood cause some of the ova (or egg cells) in her ovaries to become mature (or ripe).

(b) Usually one mature ovum is released from the ovary into the oviduct once every 28 days. This is called ovulation.

(c) Before ovulation (or release of ovum), the inner lining of uterus becomes thick and spongy, and full of tiny blood vessels (or blood capillaries), and prepares itself to receive the fertilized ovum or egg (in case it gets fertilized by sperm).

(d) If the ovum does not get fertilized (due to non-availability of sperm in the female body) then the thick and soft inner lining of uterus is no longer needed and hence it breaks. So, the thick and soft inner lining of uterus along-with the blood vessels and the dead ovum comes out of the vagina in the form of a bleeding called menstruation.

(e) Menstruation usually occurs 14 days after ovulation and usually lasts for about 3 to 5 days.

(f) After menstruation is over, the inner lining of the uterus starts building up again so that it may become ready to receive the next ovum in case it gets fertilized.

If the ovum does not get fertilized even now, then menstruation takes place again. This cycle of menstruation is repeated again and again in women after every 28 days (till the time ovum gets fertilized). The menstrual cycle is controlled by hormones. Menstruation stops temporarily when the ovum gets fertilized and the woman gets pregnant. This is because in this case the thick and soft lining of the uterus containing lot of blood vessels is needed for the growth and development of the fertilized ovum or fertilized egg cell to form a baby. Menstruation restarts after the birth of the baby.

3.6.2 Menarche and Menopause:

The first menstruation (or menstrual flow) begins at puberty (when the girl or woman is around 10 to 12 years of age). The first occurrence of menstruation (or periods) at puberty is called menarche. Menarche is the beginning of the reproductive life of a girl (or woman). In other words, menarche is the time from which a girl (or woman) becomes capable of having a baby. Menstruation stops permanently when a woman reaches the age of about 45 to 50 years. With the permanent stoppage of menstruation, a woman loses her ability to bear child. She becomes infertile. The permanent stoppage of menstruation (or periods) in a woman is called menopause. Thus, the reproductive life of a woman starts at menarche and ends at menopause.

3.6.3 What is menstrual cycle?

The cycle of events taking place in female reproductive organs, i.e., in ovaries and uterus, under the control of sec hormones, in every 28 days and marked by bleeding or menstrual flow, is called menstrual cycle (Sexual cycle in Human females). As stated above, the ovaries and the uterus start exhibiting certain events in cyclic manner at the age of puberty. The events of Oogenesis and Menstrual Cycle are given in table 3.5 and Fig. 3.37.

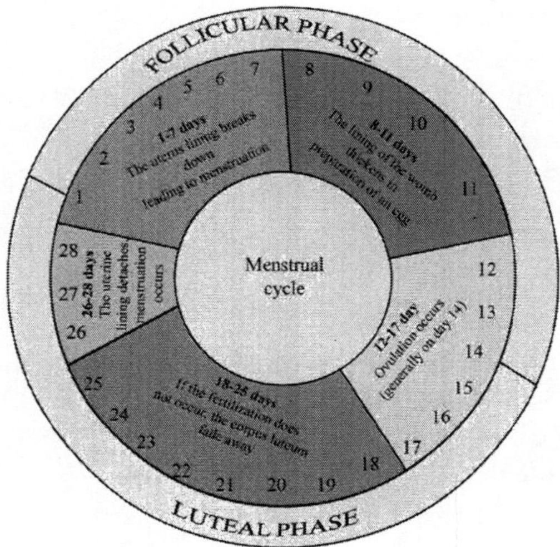

Fig. 3.37 Menstruation cycle

Table 3.5 Events of Oogenesis and Menstrual Cycle

Example of dates (given for understanding	Events of Oogenesis and Menstrual Cycle
1st to 5th (Menstruation Period)	• In case the fertilization does not occur, the thickened inner lining of the uterus breakdown along with blood vessels. The degenerated part of uterus along with the blood moves out of the vagina in the form of bleeding, called menstruation (the 'period'). • Follicle stimulating hormone (FSH), secreted from pituitary gland, stimulates the development of primary oocyte in primary ovarian follicle.
1st to 14th (Proliferative Phase)	• The developing follicle releases the hormone – estrogen. This hormone stimulates repair and development of the lining of the uterus and fallopian tube. It also regulates maturation of only one follicle so that only one ovum is released at a time. • The hormone – estrogen builds up conditions so that a mature ovum is released from the ovary into the body which passes into the fallopian tube. • Release of ovum from the ovary is called ovulation. • At the same time, another hormone-progesterone is released from the ovary. This hormone stimulates the uterus to maintain its thickening, as well as, glandular activity in uterus and vagina.

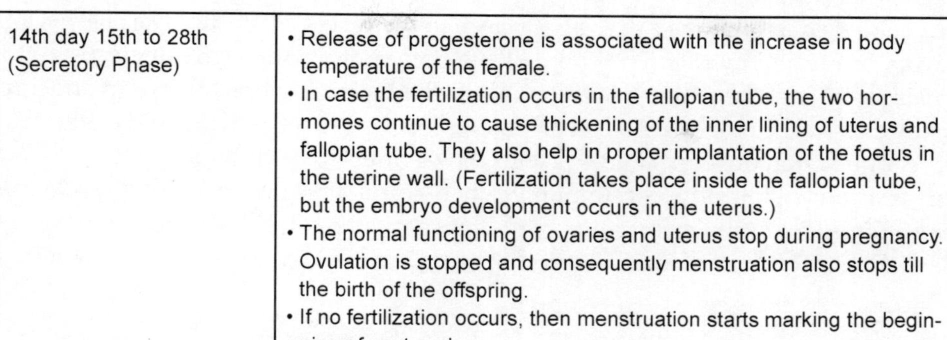

14th day 15th to 28th (Secretory Phase)	• Release of progesterone is associated with the increase in body temperature of the female. • In case the fertilization occurs in the fallopian tube, the two hormones continue to cause thickening of the inner lining of uterus and fallopian tube. They also help in proper implantation of the foetus in the uterine wall. (Fertilization takes place inside the fallopian tube, but the embryo development occurs in the uterus.) • The normal functioning of ovaries and uterus stop during pregnancy. Ovulation is stopped and consequently menstruation also stops till the birth of the offspring. • If no fertilization occurs, then menstruation starts marking the beginning of next cycle.

Table 3.6 Differences between Menarche and Menopause

Sr. No.	Menarche	Menopause
1.	It is the stage of initiation of menstruation in the life of female (girl).	It is the stage at which menstruation stops in the life of women.
2.	It generally occurs at an age of 11-14 years.	It occurs in women at an age between 45-50 years.
3.	It refers to the beginning of the reproductive phase in girls.	It refers to the end of the reproductive phase in the women.

In human females, onset of puberty (menarche) is marked by production of sex hormones, development of secondary sexual characters, maturation of ovum (ovulation) and menstrual flow (menstruation) every 28 to 30 years.

Menarche is the beginning of the reproductive life of a girl (or woman). It generally occurs at an age of 11-14 years. It refers to the beginning of the reproductive phase in girls. The cycle of events taking place in female reproductive organs in every 28 days and marked by bleeding or menstrual flow is called menstrual cycle. It ceases at the age of 45 to 50 years. This is called menopause.

In human females, fertilization takes place inside the fallopian tube, but the embryo development occurs in the uterus. It refers to the end of the reproductive phase in the women. Stoppage of menstruation (period) is the earliest sign of pregnancy. Pregnancy is known as gestation and completed in about 240 days or 40 weeks from the first day of the last menstruation.

3.6.4 Fertilization:

Internal fertilization is an essential part of reproduction in human beings (Fig. 3.34). This is achieved by copulation (or mating). During copulation, the human male inserts his penis into the vagina of the female and releases gametes (sperms) inside the genital tract. This is called ejaculation. At a time, approximately 3cm^3 of semen (sperm-bearing fluid produced by testes and accessory glands) is discharged during ejaculation, of which about 10% are sperms. They are millions in number. The sperms are deposited at the top of the vagina close to the cervix of uterus. The sperms are highly active and mobile. They travel from here upward through the uterus to the top of fallopian tube within 5 minutes. There they spend several hours and undergo many chemical changes. Finally, only one sperm fertilizes the ovum in the fallopian tube. The fusion of sperm nucleus with the egg nucleus a diploid zygote is called fertilization. Fertilization occurs only when copulation occurs during ovulation period (middle of the menstrual cycle). It is marked by the stoppage of menstrual flow. Thus, stoppage of menstruation or period is the earliest sign of pregnancy.

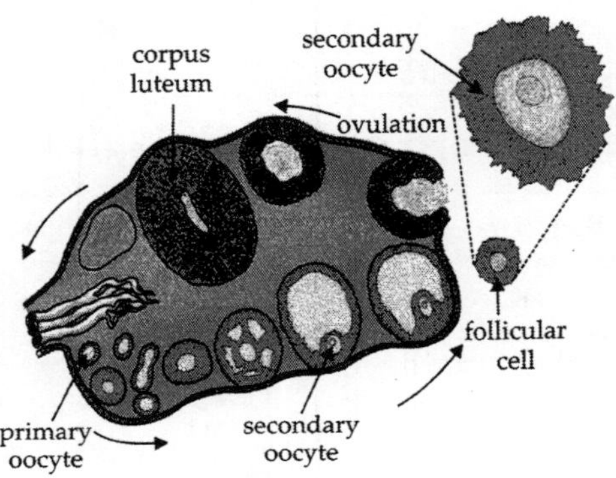

Fig. 3.38 T.S. of Ovary showing egg cell at different stages of development

3.6.5 Post Fertilization Changes:

Immediately after conception (i.e., fusion of male and female gametes to form a zygote), the embryonic development begins in the fallopian tube. This stage marks the beginning of pregnancy. As the zygote moves from the fallopian tube down towards the uterus, It starts dividing by successive nuclear and cell divisions resulting in the formation of a small multicellular ball of cells by the process, called as cleavage. The young multicellular embryo, formed as a result of cleavage, is now called blastocyst. The blastocyst gets embedded into the thickened inner wall of the uterus. The close attachment of the blastocyst to the uterine wall is called implantation. It takes place about seven days after fertilization.

Implantation is followed by a number of development changes in the blastula, as well as, in the wall of uterus. These changes include the formation of different embryonic layers (ectoderm, mesoderm and endoderm) and successive development of young foetus (embryo). The first organs, formed in developing foetus, are neural tube and spinal cord. These organs begin to develop in the third week. Subsequently, a special umbilical cord develops in the fourth week which establishes an intimate connection between the foetal membrane and the uterine wall, called placenta. In the placenta, the foetal blood comes very close to the maternal blood. Here the exchange of materials between the mother's blood and the blood of foetus take place. The foetus gets nutrients, water, oxygen, minerals, vitamins, etc. from the maternal blood and gives off wastes carbon dioxide, urea, etc. Thus, the placenta serves as the nutritive, respiratory and excretory organ of the foetus. The complete development of foetus, from the initial stage of conception till the birth of the young one, is called gestation/ pregnancy. Gestation is completed in about 280 days or 40 weeks from the first day of the last menstruation. Gestation is followed by parturition. Parturition is the act of expelling the full-term young one from the mother's uterus at the end of gestation. The new born child, after 280 days of gestation, weighs about 3.5 kg and measures about 240 mm from head to bottom.

3.7 BIRTH CONTROL

The population of India is increasing rapidly day by day. It is creating the problem to provide food, shelter and nutrition and health. Though our country has sufficient food resources but still many people do not get sufficient food for their large families (having many children) due to poverty. So, every year it is becoming very difficult for Indian Government to provide clothing, good housing and proper education to every citizen of the country. It is, therefore, very important for the couples (husbands and wives) who are in the reproductive stage of their lives to control the size of their families by having fewer children by practicing family planning through birth control measures. Family planning enables a couple to decide on the number of children they wants to have and when to have them. If a couple has less number of children, it can provide good food, good clothes, and good education to each child. This will make the parents as well as the children happy. So, a small family is a happy family.

Family planning can be done by practicing birth control measures. Birth control can be adopted by preventing pregnancy in females or women. Pregnancy can be prevented by adopting a method or procedure by which sperms produced during copulation between man and woman can be prevented from meeting the ovum or egg and fertilizing it. It is possible to prevent fertilization (and hence prevent pregnancy) because the ovum is available for fertilization only for a short period. Since frequent pregnancies have a very bad effect on the mother's health, and also add to our already exploding population. A number of techniques or methods have been developed to prevent pregnancies in women are given here as (Fig. 3.39):

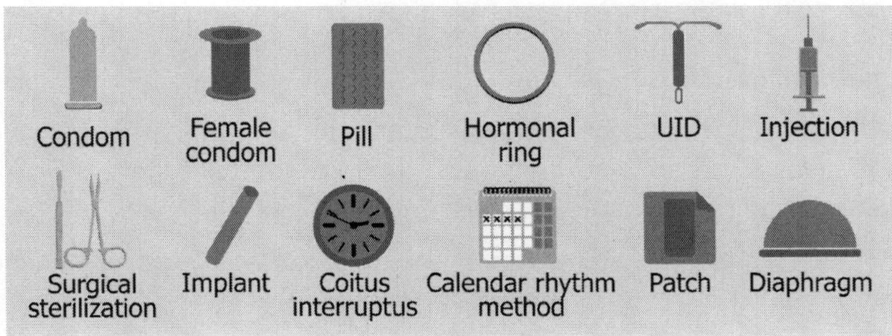

Fig. 3.39 Birth Control Methods

1. Barrier Methods 2. Chemical Methods 3. Intrauterine Contraceptive Device (IUCD).

4. Natural Methods and 5. Surgical Methods

3.7.1. Barrier Methods :

These are physical devices to prevent the entry of sperm in the female genital tract during copulation. They also protect against sexually transmitted diseases. Some common barrier methods are such as:

(i) **Condoms:** These are thin, strong rubber sheaths used by man to cover the erect penis. It is simple but effective and widely used contraceptive that has no side effects. It checks pregnancy by preventing deposition of semen in the vagina. This method gives best results when used regularly and put on before starting coital activities

(ii) **Femidom:** It is not a common contraceptive method. A femidon is a thin rubber or polyurethane tube which fits inside vagina. It is used by female just before starting coital activities.

(iii) **Diaphragm (Cervical Cap):** It is a flexible rubber/plastic cover that is fitted over the cervix in the female's vagina. It checks the entry of sperms into the uterus. These are smeared with a spermicidal jelly or cream and must be kept fitted for at least six hours after sexual intercourse (Fig 3.40).

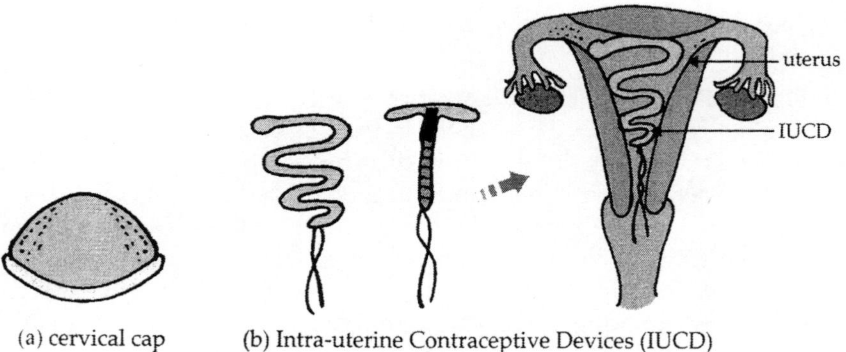

(a) cervical cap (b) Intra-uterine Contraceptive Devices (IUCD)

Fig. 3.40 Mechanical Methods of Contraception

3.7.2. Chemical Methods:

Foam tablets, jellies, pastes, creams and spermicides are some common chemicals used by females. These are placed in vagina. These chemicals adhere to the mucous membrane and immobilize and kill the sperms.

- Vaginal pills are drug preparations which contain spermicides. They are used before copulation by women.
- Oral pills are also used to check ovulation. These are mainly hormonal preparation and contain estrogen and progesterone. These prevent development of egg and ovulation by inhibiting secretion of FSH. The oral pills act on hypothalamus, pituitary and the ovaries. They are called oral contraceptives (OCs).

3.7.3. Intrauterine Contraceptive Device (IUCD) :

These are contraceptive devices made of copper, plastic or stainless steel. A copper-T is inserted into uterus by a practicing doctor or a skilled nurse and left in place (fig 3.39B). It prevents implantation in the uterus. The other intrauterine contraceptive devices are loop, spiral ring, bow and shield. These can be placed for long periods (upto 5 years). The drawbacks with these devices are bleeding and discomfort. IUCD may come out even without the women's knowledge.

3.7.4. Natural Methods

These methods include self-restraint (avoid copulation), rhythm method (avoid copulation around the time of ovulation and coitus interrupts (withdrawal of penis before ejaculation).

3.7.5. Surgical Methods

Surgical methods are safe in the long run but they may cause infections and other problems if not done properly. These methods include – (i) Vasectomy and (ii) Tubectomy.

(i) Vasectomy: This is a small surgical operation performed in males. It involves removal of a small portion of the sperm duct (or vas deferens) by surgical operation. The two cut ends are then ligated (tied) with threads. This prevents the sperms from coming out (Fig 3.41 A).

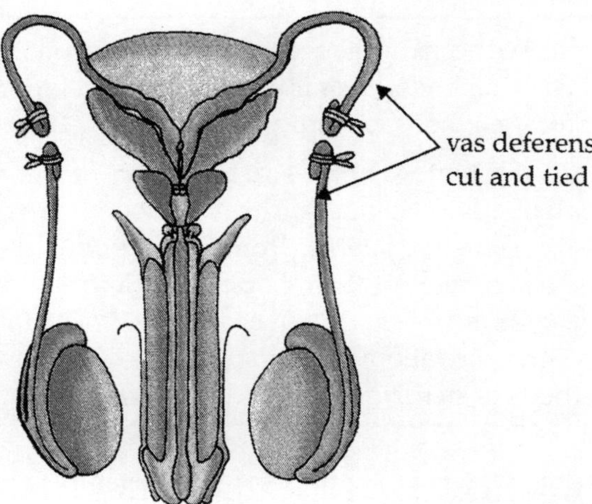

Fig. 3.41A Tubectomy in Males

(iii) Tubectomy: This is a surgical operation performed in females. It involves removal of a small portion of the fallopian tubes (or oviducts) by surgical operation. The cut ends are then ligated (tied) with threads. It prevents the egg (ovum) to enter the fallopian tube (Fig 3.41 B).

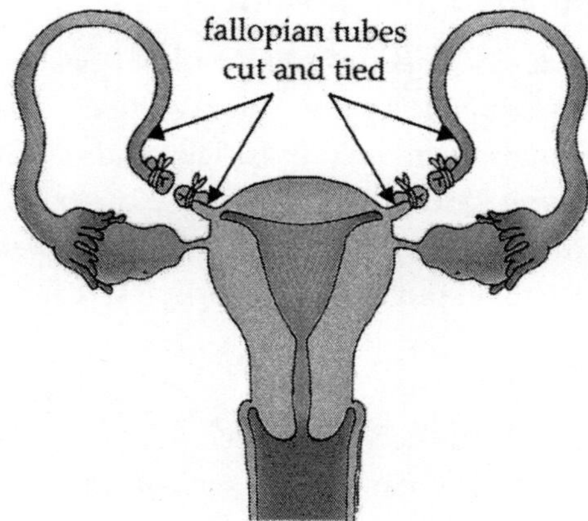

Fig. 3.41B Vasectomy in Females

Both vasectomy and Tubectomy are also known as methods of sterilization. These do not affect the normal sexual life.

Surgery can also be used for abortion. Abortion is the medical termination of pregnancy (MTP) before the foetus becomes viable. However, this method is widely misused by people who do not want a female child. It is illegal (the prenatal sex determination has been prohibited by law). Moreover, the female-male sex ratio must be maintained to have a healthy society. Before using contraceptives, particularly oral pills, vaginal pills and copper-T etc., one must be aware of their side effects and take regular medical advice.

A number of devices have been developed to prevent conception and control child birth. These are barrier methods, chemical methods, Intrauterine Contraceptive Device (IUCD), natural methods and surgical Methods.

Vasectomy involves the removal of a small portion of the sperm duct (vas deferens) by surgical operation in males. The two cut ends are then ligated (tied) with threads. This prevents the sperms from coming out. Tubectomy involves the removal of a small portion of the fallopian tubes (or oviducts) by surgical operation in females. The cut ends are then ligated (tied) with threads. It prevents the egg (ovum) to enter the fallopian tube. Both vasectomy and tubectomy are also known as methods of sterilization, and do not affect the normal sexual life.

3.8 REPRODUCTION HEALTH AND SEXUALLY TRANSMITTED DISEASES (STDS):

Reproduction is an important biological process like nutrition, respiration, transportation, excretion, etc. The reproductive life of human begins begin in puberty and continues throughout life in males and upto menopause in females. Like physical, mental and social fitness, the human beings need fitness of reproductive life. This is called reproductive health. It includes such aspects that ensure a responsible, safe and satisfying reproductive life. It is our responsibility and the responsibility of the Nation to provide necessary information and general awareness regarding reproductive health to each and every individual. Everyone must know that **(i)** Marriage and child bearing during more mature stages of life, are important for reproductive health of a society, **(ii)** Secondary school education, which is lagging behind in many parts, must be enhanced, **(iii)** Complications during pregnancy and child birth and unsafe abortion are the causes of death of women, and **(iv)** Sexually transmitted diseases are spread by unsafe sex. All these aspects of reproductive health have been considered by World Health Organization (WHO) and National Population Policy and National Health Policy of India.

As we know, the diseases are caused by bacteria, viruses, protozoans, fungi, mycoplasmas and many other kinds of pathogens. These are called infectious diseases. Some other kinds of diseases are caused by physiological disorders. They are not infectious. The infectious diseases are spread by means of air, water and food. Similarly, some infectious diseases concerned with the reproductive parts are transmitted through sex. These are sexually transmitted diseases (STDs). These diseases are also known as venereal diseases (VD). Human beings suffer from over 30 different kinds of sexually transmitted diseases (STDs), which are usually caused by bacteria, protozoa and viruses. The common among them are:

Syphilis:

Syphilis is caused by *Spirochaete bacterium, Treponema pallidum.* The disease is a chronic illness which affects the mucous membrane in genital, rectal and oral regions, and causes lesions (Fig. 3.41). Infection occurs by sexual intercourse; and occasionally by kissing close body contact. Syphilis is easily cured with antibiotics.

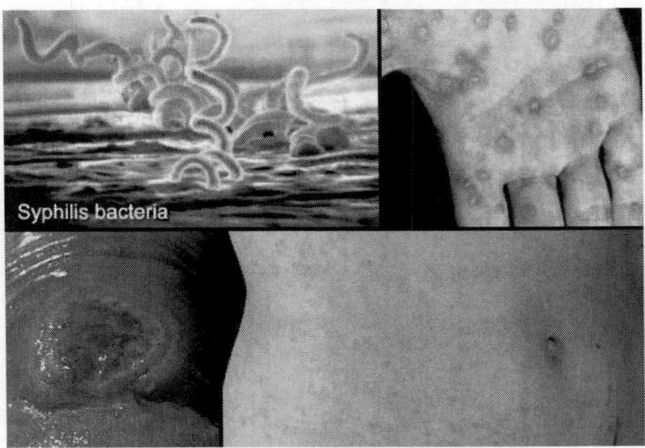

Fig. 3.41 *Syphills* caused by *Spirochaete* bacterium

Gonorrhea:

Gonorrhea is caused by a bacterium Nisseria gonorrhoeae. The victim feels burning sensation and pain during urination (Fig. 3.42). The disease causes inflammation of the mucous membrane of the urino-genital tract, rectum, throat and eye. There may be pus from the penis and excessive secretion from vagina. It spreads by sexual contact and through infested clothes. Gonorrhea is easily curable with antibiotics.

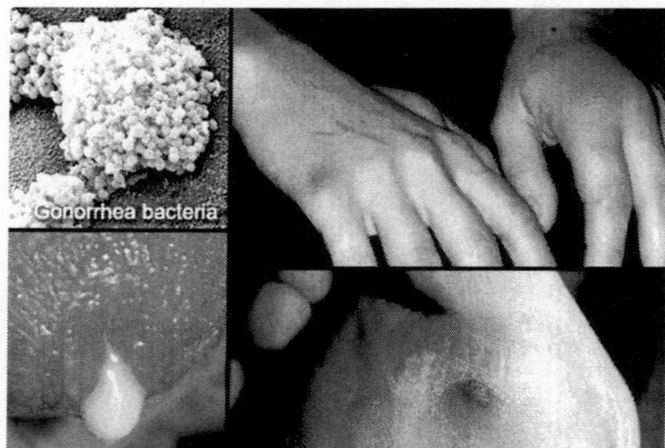

Fig. 3.42 Gonorrhea, caused by *Nisseeria gonorrhoae*

The disease AIDS is caused by a virus, called HIV. It is fast spreading incurable diseases which weaken the body's immune system (Fig. 3.43). The virus (HIV) was discovered in 1984 by an American Scientist, named Robert Gallo and a French Scientist–Luc Montagnier, independently. It now seems that the disease AIDS may have passed into humans from chimpanzees by eating butchered chimps. The first case of AIDS was noticed in U.S.A. in 1981. In India, the first case of AIDS was reported in Chennai prostitute in 1986. December 1st is celebrated every year as the World AIDS Day to spread information about AIDS among the public.

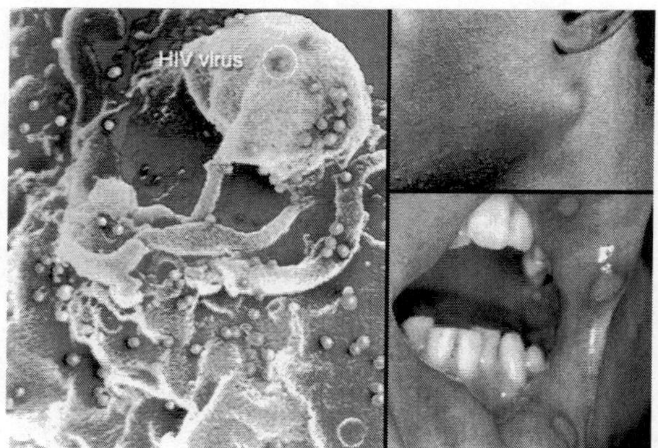

Fig. 3.43 HIV (Human Immunodeficiency Virus / AIDS (Acquired Immunodeficiency Syndrome)

The disease can manifest in two major ways such as: (i) Malignant tumours in connective tissue, and (ii) viral, bacterial, protozoan and fungal infections of any system of the body. The main symptoms of the disease are damage to brain, unexplained fever, unexplained loss of appetite, unexplained loss of weight over a short time, chronic diarrhea, cough, night sweats, shortness of breath and severe weakness. In people with AIDS, infectious diseases neurological damage and physiological wasting lead to death.

Spread of Disease:

AIDS is transmitted only by a contact of infected cells-containing blood of a patient with the blood of a healthy person as in (a) Transfusion of infected blood or blood products. (b) Use of contaminated razors for shaving (c) Use of contaminated needles and syringes to inject drugs or vaccines, and (d) Unprotected (without condom) sex with an infected partner if there is tissue injury to permit blood contact.

Prevention:

People, particularly those in high-risk group, should be educated about AIDS transmission, advantage of using condom, danger of sharing needles and virtue of monogamy.

The infective diseases occurring due to sexual contact are called sexually transmitted diseases (STDs). These are gonorrhea, syphilis and AIDS. Gonorrhea is caused by a bacterium Nisseria gonorrhoeae, Syphilis by Spirochaete bacterium, and HIV by Acquired Immunodeficiency Syndrome).

3.9 WOMEN'S HEALTH DURING PREGNANCY

Malnutrition (lack of proper nutrition) and stress during pregnancy are the major causes of ill health of women. She is the most important member of a family and all care must be taken to provide her highly nutritious and easily digestible food. She must be kept free from any stress on account of family disputes or financial problems. It is, therefore, advisable that:

- Sufficient and balanced nutrition must be provided to pregnant and lactating women.
- The number of births should be limited and proper interval between subsequent pregnancies should be maintained.
- Would-be-mothers should not take heavy doses of such medicines which are probably not allowed during pregnancy.
- Cigarette smoking, LSD and marijuana should not be used by would-be-mothers as they cause harmful effects on the foetus.
- Very high level of drinking alcohol during pregnancy should be avoided as it can cause foetal alcohol syndrome including mental retardation.

3.10. STOP FEMALE FOETICIDE

Fig. 3.44 Stop Female Foeticide

Surgical operations can also be used for the termination of pregnancies in women particularly after eight weeks of conception. Such surgical termination of pregnancy (or abortion) is allowed by law only in certain circumstances where the doctors decide that continuance of pregnancy can affect the health of mother or of unborn child gravely. Some people, however, misuse the technique of surgical termination of pregnancy for ulterior motive to get rid of female foetus or unborn girl child. In order to have a male child (son), some misguided people in our country get the sex of their unborn child determined by ultrasound technique (though it is illegal to do so), and if it is a female foetus, they get it removed by surgery. This killing of the unborn girl child is called female foeticide. Female foeticide is reducing the number of girls drastically in some societies of our country.

For a healthy society, the male-female sex ratio must be maintained at almost the same level. Due to reckless female foeticide, the male-female child sex ratio is declining at an alarming rate in some sections of our society. This must be stopped immediately. Female foeticide is a matter of shame for the couples who for it and also for the doctors who perform such surgical abortions (or rather murder the unborn girl child) for earning some quick money!

Let us not forget that our mother is a female, our sister is a female, and our wife is also a female. Then why not have a female as our daughter? Remember, daughters are no less than sons. Our Government has already enacted laws to ban prenatal (before-birth) determination of sex of foetus. These laws must be enforced strictly to save our society from the grave dangers of female foeticide in the long run.

Malnutrition (lack of proper nutrition) and stress during pregnancy are the major causes of ill health of women. She must be kept free from any stress on account of family disputes, financial problems and from other causes. We should avoid prenatal (before-birth) sex determination, and save to girl child.

REMEMBER SOCKETS

- Reproduction is the production of new generation of individuals of the same species that are physically independent of their parents.

- Reproduction ensures the continuation of life, perpetuation of species, replacement of dead individuals and introduction and transfer of variation.
- Reproduction is an energy spending process which is not at all essential to maintain the life of an individual organism. Even then, the organisms reproduce.
- Reproduction involves a mechanism, called DNA (Deoxyribo-nucleic Acid) copying or DNA replication which is capable of producing similar multiple of blue prints.
- Reproduction is an essential biological process for multiplication, perpetuation and continuity of the species.
- Each species of organisms reproduces in a different way. All the different ways of reproduction are grouped under two broad categories such as (a) Asexual reproduction and (b) Sexual reproduction.
- Asexual reproduction is the production of offsprings by a single parent without meiosis, formation of gametes, fertilization and transfer of genetic materials between individuals.
- In asexual reproduction, a part of the parent organism separates off and grows into a new organism. Only one parent is needed to produce a new organism in asexual reproduction. But no sex cells are involved in asexual reproduction.
- Asexual reproduction takes place in unicellular animals and plants, micro-organisms (like bacteria) and simple multicellular animals (like Hydra and Planaria) and some multicellular plants (like Bryophyllum and Rose plants etc.).
- The common methods of asexual reproduction are :
- (i) Fission, (ii) Budding, (iii) Spore formation, (iv) Regeneration, (v) Vegetative propagation (vi) Micro propagation *in vitro* and (vii) Parthenogenesis.
- Binary fission is an asexual method of reproduction of organism. In binary fission, the parent organism splits (or divides) to form two new organisms.
- Budding is an asexual method of reproduction. In budding, a small part of the body of the parent organism grows out as a 'bud' which then detaches and becomes a new organism. The asexual reproduction by budding is observed in hydra and yeast.
- Spores are asexual reproductive bodies which are covered by a hard protective coat. This coat enables them to survive under unfavourable conditions like lack of food, lack of water and extreme temperatures. But when the conditions are favourable, then the spores grow to produce new plants. Thus, spores are a kind of seeds of plants.
- The process of growing many plants from one plant by man-made methods is called artificial propagation of plants.

- The three common methods for the artificial propagation of plants are such as (a) Cuttings, (b) Layering, and (c) Grafting.
- The production of new plants from a small piece of plant tissue (or cells) removed from the growing tips of a plant in a suitable growth medium (culture solution) is called tissue culture.
- The tissue culture technique is being used increasingly for the production of ornamental plants like Orchids, Dahlia, Carnation, Chrysanthemum, etc.
- Sexual reproduction is the production of new individuals by the fusion of two gametes (usually one male and the other female) following the act of fusion and formation of zygote. Thus, two parents are needed to produce a new organism in sexual reproduction.
- The humans, fish, frogs, hens, cats, dogs, cows, horses, deer, rabbits, lions and tigers all reproduce by the method of sexual reproduction.
- The basic difference between asexual and sexual reproduction is that only one parent is needed in asexual reproduction whereas two parents are needed in sexual reproduction. Another difference is that no sex cells (or gametes) are involved in asexual reproduction but sex cells take part in sexual reproduction.
- Fertilization is the fusion of two gametes to form a single cell (zygote).
- Reproductive parts of the angiospermous plants are found in the flowers.
- A flower is a specialized condensed reproductive shoot of flowering plants on which are inserted the essential reproductive parts.
- A typical angiospermous flower consists of (a) Calyx (whorl of sepals), (b) Corolla (whorl of petals), (c) Androecium (whorl of stamens), and (d) Gynoecium (whorl of carpels).
- Anthers produce pollen grains containing male gamete.
- The transfer of pollen grains from the opened anther of the stamen to the receptive stigma of the carpel is called pollination.
- The common agents responsible for successful pollination are such as wind, water, insects, birds, bats and many other living organisms.
- Pollination is of two types such as Self-pollination and Cross pollination
- Self-Pollination involves the transfer of pollen grains from the anther of a flower to the stigma of the same flower or to the stigma of another flower borne on the same plant.
- Cross Pollination involves the transfer of pollen grains from the flower of one plant to the stigma of the flower of another plant of the same species.
- Pollen grain germinates to produce pollen tube which acts as carrier of male gamete upto the egg.

- Finally male gamete fuses with the egg to form zygote. This process occurs inside the embryosac and called syngamy. The second gamete fuses with the polar nuclei. This is called double fertilization.
- After fertilization, zygote develops embryo, ovules develop seeds and ovary grows to develop fruit.
- Human beings are unisexual and the human reproduction is highly evolved. The primary sex organs in males are testes and in females are ovaries.
- Testes produce male gametes (sperms) and male hormone (testosterone).
- Ovaries produce female gametes (ova) and female hormones (estrogen and progesterone).
- Puberty is the age of human males and females at which the reproductive organs become functional, gonads start producing gametes and sex hormones, and the boys and the girls become sexually mature. It is attained between the ages of 13 to 14 years in males and between 10 to 12 years in females.
- In human males, puberty is triggered by the secretion of the hormone testosterone from the testes. The hormone brings about the development of secondary sexual characters during puberty and maintains these throughout adult life.
- In human females, onset of puberty (menarche) is marked by production of sex hormones, development of secondary sexual characters, maturation of ovum (ovulation) and menstrual flow (menstruation) every 28 to 30 days.
- Male reproductive system consists of testes, scrotum, vas deferens, urethra and penis.
- Female reproductive system consists of ovaries, fallopian tubes (oviducts) uterus and vagina.
- A fallopian tube is about 10-12 cm long muscular tube which carries egg from the ovary to the uterus and also provides the appropriate environment for its fertilization.
- The uterus is a large, inverted pear-shaped, muscular structure that lies behind the bladder. If fertilization takes place, the embryo gets attached to the wall of uterus and grows there until birth.
- A zygote is formed by the fusion of male and female gametes (sperm and egg). It is the beginning of the formation of a baby.
- An embryo is formed by the repeated cell division of a zygote. An embryo is an unborn baby in the uterus in the early stages of development (up to 8 weeks after fertilization).
- A foetus is formed by the growth and development of an embryo. A foetus is an unborn baby in the uterus in the later stages of development (after 8 weeks till birth).

- Release of ovum from the ovary is called ovulation.
- Menarche is the beginning of the reproductive life of a girl (or woman).
- The cycle of events taking place in female reproductive organs in every 28 days and marked by bleeding or menstrual flow is called menstrual cycle. It ceases at the age of 45 to 50 years. This is called menopause.
- In human females, fertilization takes place inside the fallopian tube, but the embryo development occurs in the uterus.
- Stoppage of menstruation (period) is the earliest sign of pregnancy.
- Pregnancy is known as gestation and completed in about 240 days or 40 weeks from the first day of the last menstruation.
- A number of devices have been developed to prevent conception and control child birth. These are Barrier Methods, Chemical Methods, Intrauterine Contraceptive Device (IUCD), Natural Methods and Surgical Methods.
- Vasectomy involves the removal of a small portion of the sperm duct (or vas deferens) by surgical operation in males. The two cut ends are then ligated (tied) with threads. This prevents the sperms from coming out.
- Tubectomy involves the removal of a small portion of the fallopian tubes (or oviducts) by surgical operation in females. The cut ends are then ligated (tied) with threads. It prevents the egg (ovum) to enter the fallopian tube.
- The infective diseases occurring due to sexual contact are called sexually transmitted diseases (STDs). These are gonorrhea, syphilis and AIDS.
- The first case of AIDS was noticed in U.S.A. in 1981. In India, the first case of AIDS was reported in Chennai prostitute in 1986.
- December 1st is celebrated every year as the World AIDS Day to spread information about AIDS among the public.

VERY SHORT ANSWER TYPE QUESTIONS:

1. Define the term reproduction.
2. Name the two types of reproduction.
3. Define the term asexual reproduction.
4. Define the term sexual reproduction.
5. Give one example of animal which reproduces asexually by budding
6. Define the term fission.
7. Give one example of plant which reproduces asexually by spore.
8. Full form of DNA.
9. Define the term spore.
10. What is bread mould?

11. Give one example of vegetative reproduction.
12. Define the term tissue culture.
13. What is the use of Agar?
14. Agar is obtained by which algae.
15. What is external fertilization?
16. What is internal fertilization?
17. Name the various parts of a flower.
18. Write names of one male and one female sex hormone.
19. What do we call the undifferentiated mass of cells formed during tissue culture?
20. Mention any one limitation of vegetative propagation.
21. Define the term parthenogenesis.
22. Give two examples in which asexual reproduction occurs by spore formation.
23. What is triple fusion?
24. What is pollination?
25. What is zygote?
26. What is embryo?
27. What is foetus?
28. Give the name of one agent responsible for pollination.
29. What is the duration of reproductive life of man?
30. Name the male and female gametes in animals.
31. Where does fertilization take place in human females?
32. What is ovulation?
33. What is menarche?
34. What is menopause?
35. Define the term menstrual cycle.
36. Give the name of one method of birth control.
37. What is Vasectomy?
38. What is Tubectomy?
39. Full form of IUCD.
40. Full form of HIV.
41. Full form of AIDS.
42. Full form of STDs.
43. Give the name of one sexually transmitted disease.

SHORT ANSWER TYPE QUESTIONS:

1. Define the terms unisexual and bisexual giving one example of each.
2. Mention the reproductive parts of a flower.
3. Fertilization is possible if copulation has taken place during the middle of the menstrual cycle.
4. Give the limitations of vegetative propagation.
5. Give the illustration on study of *Spirogyra* in laboratory.
6. Give the advantages of artificial vegetative propagation.
7. What is artificial propagation of plants?
8. Give the advantages of tissue culture.
9. Methods of artificial propagation of plants.
10. List two advantages of vegetative propagation.
11. Give the primary and secondary sex organs.
12. Give the primary and secondary sexual characters.
13. Give the events of menstrual cycle.
14. Mention the parts of male reproductive system:
15. Mention the parts female reproductive system:
16. Give reasons for the following (a) Scrotum remains outside the body of human males (b) Petals of flowers are variously coloured (c) Some plants are propagated only by vegetative methods (d) Regeneration cannot be regarded as reproduction'. Why?
17. What is the function of placenta?
18. Give one word for the following (a) Birth canal (b) Small sac-like muscular structure that encloses testes (c)The organ where foetus develops during gestation
19. Name the mode of reproduction that ensures variation in the offspring. What mechanisms, do you thing, is actually responsible for variation leading to the survival of the species?
20. Define the term Sexually Transmitted Diseases (STDs) and give one example of STD.
21. Write a short notes on (a) Syphilis (b) HIV (Human Immunodeficiency Virus/ AIDS (c) Women's Health during Pregnancy (d) Stop Female Foeticide **(e)** Bryophyllum (f) Post Fertilization Changes (g) Barrier Methods (h) Surgical Methods (i) Intrauterine Contraceptive Device (j) Vas differentia (k) Oviducts (l) Grafting (m) Tissue Culture (n) Agar

22. Differentiate between (a) Binary Fission and Multiple Fission (b) External Fertilization and Internal Fertilization (c) Androecium and Gynoecium (d) Self and Cross Pollination (e) Zygote, Embryo and Foetus (f) Menarche and Menopause

LONG ANSWER TYPE QUESTIONS:

1. Describe the asexual reproduction in *Amoeba*.
2. Explain the vegetative propagation with examples, and give the advantages also.
3. Asexually reproducing organism can never produce an offspring with its own exact set of genetic material. Justify the statement.
4. Explain the process of fertilization in flowering plant with the help of diagram.
5. Explain human male reproductive system with the help of a diagram
6. Explain human female reproductive system with the help of a diagram
7. Explain the Sexually Transmitted Diseases (STDs) with examples
8. Explain the following methods of contraception giving one example of each (a) Barrier Method (b) Chemical method (c) Surgical method

MULTIPLE CHOICE QUESTIONS:

1. Reproduction is essential for living organisms in order to
   (a) Fulfill their energy requirement
   (b) Keep the individual organism alive
   (c) Maintain growth
   (d) Continue the species generation after generation
2. Offspring formed as a result of sexual reproduction exhibit more variation because
   (a) Sexual reproduction is a lengthy process
   (b) Genetic material comes from two parents of the same species
   (b) Genetic material comes from many parents
   (d) Genetic material comes from two parents of different species
3. Characters that are transmitted from parents to offspring during reproduction show
   (a) Only similarities with parents
   (b) Only variations with parents
   (c) Neither similarities nor variations
   (d) Both similarities and variations with parents

4. The number of chromosomes in parents and offspring's of a particular species remains constant due to
   (a) Halving of chromosomes after gamete formation.
   (b) Doubling of chromosomes during gamete formation
   (c) Halving of chromosomes during gamete formation
   (d) Doubling of chromosomes before zygote formation
5. Offspring formed by asexual method of reproduction have greater similarity among themselves because
   (i) Asexual reproduction does not involve gametes
   (ii) Asexual reproduction involves only one parent
   (iii) Asexual reproduction occurs after sexual reproduction
   (iv) Asexual reproduction occurs before sexual reproduction
   (a) (i) and (ii) (b) (i) and (iii) (c) (ii) and (iv) (d) (iii) and (iv)
6. Out of two daughter cells produced as a result of binary fission in Amoeba
   (a) One is large and other is small
   (b) One is uninucleate and other is without nucleus
   (c) One is spherical and other is irregular
   (d) The two are nearly equal sized and identical
7. During the binary fission, the DNA molecule of parent cell
   (a) Coil (b) Replicate (c) Elongate (d) Disintegrate
8. Name of the most common type of reproduction in Amoeba
   (a) Budding (b) Binary Fission
   (c) Vegetative (d) Spore formation
9. Amoeba and Yeast are
   (a) Unicellular organism (b) Multicellular Organism
   (c) Protozoa (d) Prokaryotic Organism
10. Binary fission and budding are the example of
    (a) Copulation (b) Heterothallism
    (c) Sexual reproduction (d) Asexual reproduction
11. Characters transmitted from parents to offspring are present in
    (a) Ribosome (b) Genes (c) Golgi bodies (d) Cytoplasm
12. A feature of reproduction that is common to Amoeba, Spirogyra and Yeast is that
    (a) They are all unicellular (b) They reproduce only sexually
    (c) They are all multicellular (d) They reproduce asexually

13. In *Rhizopus*, tubular thread-like structure bearing sporangia at their tips are called
    (a) Roots (b) Rhizoids (c) Hyphae (d) Filaments
14. Vegetative propagation refers to formation of new plants from
    (a) Stem, leaves and flowers (b) Stem, flowers and fruits
    (c) Stem, roots and leaves (d) Stem, roots and flowers
15. In *Spirogyra*, asexual reproduction takes place by
    (a) Division of a cell into two cells
    (b) Division of a cell into many cells
    (c) Formation of young cells from older cells
    (d) Breaking up of filaments into smaller bits
16. The ability of a cell to divide into several cells during reproduction in Plasmodium is called
    (a) Budding (b) Binary fission
    (c) Reduction division (d) Multiple fission
17. The process where the unfertilized egg is released out of the body with blood is known as
    (a) Germination (b) Menstruation (c) Pollination (d) Fertilization
18. Asexual reproduction produces offspring that are
    (a) Genetically identical to their parents
    (b) Genetically identical to their siblings
    (c) Both a and b
    (d) None of the above
19. Which of the following is not a vegetative propagule?
    (a) Seed (b) Leaf (c) Root (d) Stem
20. Multiple fission occurs in
    (a) *Spirogyra* (b) *Plasmodium* (c) *Euglena* (d) *Hydra*
21. The multicellular organism that reproduces by budding is
    (a) *Agaricus* (b) Yeast (c) *Rhizopus* (d) *Hydra*
22. Nucleus of the parent cell divides only once to form two daughter nuclei.
    (a) Binary Fission (b) Multiple fission
    (c) Budding (d) Fragmentation
23. Bread mould is
    (a) *Mucor* (b) *Rhizopus* (c) *Amoeba* (d) *Paramecium*

24. *Planaria* reproduces by
    (a) Regeneration (b) Fragmentation
    (c) Fission (d) Budding
25. *Spirogyra* is reproduced by
    (a) Regeneration (b) Fragmentation
    (c) Fission (d) Budding
26. The production of new plants from a small piece of plant tissue (or cells) removed from the growing tips of a plant in a suitable growth medium (culture solution) is called as
    (a) Fragmentation (b) Tissue Culture
    (c) Fission (d) Callus
27. Both male and the female discharge their gametes outside their body.
    (a) External Fertilization (b) Tissue Culture
    (c) Internal Fertilization (d) None of these
28. Male discharges the gametes into the reproductive tract of the female. The female releases the gametes inside its reproductive tract.
    (a) External Fertilization (b) Tissue Culture
    (c) Internal Fertilization (d) None of these
29. Unisexual flowers contain
    (a) Only carpel (b) Both stamen and carpel
    (c) Either stamen or carpel (d) Only stamen
30. Fruit is formed from
    (a) Stigma (b) Stamen (c) Ovary (d) Ovule
31. Which of the following is the correct sequence regarding sexual reproduction in a flowering plant?
    (a) Embryo, seedling, pollination, fertilization
    (b) Seedling, embryo, fertilization, pollination
    (b) Pollination, fertilization, embryo, seedling
    (d) Pollination, fertilization, seedling, embryo
32. Which among the following statements are true for unisexual flowers?
    (i) They possess both stamen and pistil.
    (ii) They possess either stamen or pistil
    (iii) They exhibit both self and cross pollination
    (iv) Unisexual flowers possessing only stamens cannot produce fruits.
    (a) (i) and (iv) (b) (ii) and (iii) (c) (iii) and (iv) (d) (i) and (iv)

33. Which among the following statements are true for sexual reproduction in flowering plants?
    (i) It requires two types of gametes
    (ii) Fertilization is a compulsory event
    (iii) It always results in formation of zygote
    (iv) Offspring formed are clones
    (a) (i) and (iv)
    (b) (i), (ii) and (iv)
    (c) (i), (ii) and (iii)
    (d) (i), (iii) and (iv)
34. Examples of vegetative reproducing plants are
    (a) Lady's-finger, onion, potato, tomato
    (b) Tomato, lady's-finger, onion, cauliflower
    (c) Cauliflower, onion, potato, tomato
    (d) Potato, ginger, onion, cauliflower
35. In the list of organisms given below, those that reproduce by the asexual method are
    (i) Banana (ii) Dog (iii) Yeast (iv) *Amoeba*
    (a) (ii) and (iv) (b) (i), (iii) and (iv) (c) (i) and (iv) (d) (ii),(iii)and(iv)
36. Which of the following is not an advantage of vegetative propagation?
    (a) It is easier method than sowing seeds.
    (b) Such plants produce seeds and fruits much earlier than plants produced from other methods
    (c) It is the best method to introduce new genetic traits in the species
    (d) Plants which produce non-viable seeds can be grown.
37. Which of the following statements are true for flowers?
    (i) Flowers are always bisexual
    (ii) They are the sexual reproductive organs
    (iii) They are produced in all groups of plants
    (iv) After fertilization, they give rise to fruits
    (a) (i) and (iv) (b) (ii) and (iii) (c) (i) and (iii) (d) (ii) and (iv)
38. In a flower, the parts that produce male and female gametes (germ cells) are
    (a) Filament and stigma
    (b) Anther and ovary
    (c) Stamen and anther
    (d) Stamen and style
39. The opening at the base of the ovule is called
    (a) Style (b) Micropyle (c) Stigma (d) Layering

40. Length of pollen tube depends on the distance between
    (a) Pollen grain on upper surface of stigma and ovule
    (b) Pollen grain in anther and upper surface of stigma
    (c) Upper surface of stigma and lower part of style
    (d) Pollen grain and upper surface of stigma
41. After fertilization, which of the following parts develops into seed?
    (a) Stigma (b) Pollen grain (c) Ovary (d) Ovule
42. Factor responsible for the rapid spread of bread mould on slices of bread are
    (i) Large number of spores
    (ii) Availability of moisture and nutrients in bread
    (iii) Presence of tubular branched hyphae
    (iv) Formation of round shaped sporangia
    (a) (i) and (iii) (b) (ii) and (iv) (c) (i) and (ii) (d) (iii) and (iv)
43. In human females, an event that reflects onset of reproductive phase is
    (a) Changes in hair pattern (b) Menstruation
    (c) Growth of body (d) Change in voice
44. In human males, the testes in the scrotum, because it helps in the
    (a) Formation of sperm (b) Process of mating
    (c) Easy transfer of gametes (d) All the above
45. The correct sequence of organs in the male reproductive system for transport of sperms is
    (a) Testes → ureter → urethra
    (b) Testes → vasdeferens → ureter
    (c) Testes → vasdeferens → urethra
    (d) Testes → urethra → ureter
46. The number of chromosomes in human ovum is
    (a) 22 (b) 21 (c) 24 (d) 23
47. Which of the following is not the function of the seminal vesicles present in human males?
    (a) To store the sperms in a fluid medium
    (b) To make transport of sperms easier.
    (c) To provide nutrition to developing spermatozoa.
    (d) None of the above.

48. The common passage meant for transporting urine and sperms in males is
(a) Anus (b) Ureter (c) Urethra (d) Vas deferens

49. Oral contraceptives prevent
(a) Fertilization (b) Entrance of sperms in vagina
(c) Ovulation (d) All of these

50. Menstrual cycle is generally of
(a) 28 days (b) 38 days (c) 40 days (d) 21 days

51. Growing Foetus derives nutrition from mother's blood through.
(a) Placenta (b) Cervix
(c) Uterus (d) Fallopian tube

52. What is the puberty age in human males?
(a) 8-10 years (b) 14-16 years (c) 13-14 years (d) 10-12 years

53. The process of release of an ovum from the ovary is called
(a) Ovulation (b) Oogenesis (c) Menstruation (d) None of these

54. The condition when the menstrual cycle stops is known as
(a) Gestation (b) Puberty (c) Menstruation (d) Menopause

55. Why are the testes located outside the abdominal cavity in scrotum?
(a) Because sperm formation requires a higher temperature.
(b) Because sperm formation requires more space.
(c) Because sperm formation requires a lower temperature
(d) None of the above

56. Which of the following is not a sexually transmitted disease?
(a) Hepatitis (b) Syphilis (c) AIDS (d) Gonorrhea

57. Which of the following diseases is transmitted sexually?
(a) Elephantiasis (b) Syphilis
(c) Jaundice (d) Sleeping sickness

58. Which of the following is a contraceptive device?
(a) Condom (b) Diaphragm (c) Copper-T (d) All of these

59. What is the surgical method of contraception used in human males?
(a) Condoms (b) Tubectomy
(c) Vasectomy (d) Contraceptive pills

60. IUCD is for
(a) Increasing fertility (b) Vegetative propagation
(c) Avoiding miscarriage (d) Contraception

61. Full Form of AIDS
   (a) Acquired Immune Deficiency Syndrome
   (b) Acquired Immune Deficiency System
   (c) Acquired Immortal Deficiency Syndrome
   (d) Acquired Immortal Deficiency System
63. Which of the following is celebrated as World AIDS Day?
   (a) 1st September (b) 1st October
   (c) 1st November (d) 1st December
64. Full Form of HIV
   (a) Hetero Immunodeficiency Virus
   (b) Homo Immunodeficiency Virus
   (c) Human Immunodeficiency Virus
   (d) Human Immunodeficiency Venous
65. HIV is
   (a) Virus (b) Fungi (c) Algae (d) Bacteria

ANSWER KEY

1. (d)	2. (b)	3. (d)	4. (b)	5. (a)	6. (d)	7. (b)	8. (b)
9. (a)	10. (d)	11. (b)	12. (d)	13. (c)	14. (c)	15. (d)	16. (d)
17. (b)	18. (a)	19. (a)	20. (b)	21. (b)	22. (c)	23. (b)	24. (a)
25. (b)	26. (b)	27. (a)	28. (c)	29. (c)	30. (c)	31. (c)	32. (b)
33. (c)	34. (b)	35. (b)	36. (b)	37. (b)	38. (b)	39. (b)	40. (a)
41. (d)	42. (c)	43. (b)	44. (a)	45. (a)	46. (a)	47. (c)	48. (c)
49. (d)	50. (a)	51. (a)	52. (c)	53. (a)	54. (a)	55. (c)	56. (a)
57. (b)	58. (d)	59. (b)	60. (d)	61. (a)	62. (d)	63. (d)	64. (c)
65. (a)							

4

CHAPTER

HEREDITY AND EVOLUTION

Heredity; Mendel's Contribution- Laws for Inheritance of Traits: Sex Determination: Brief Introduction; Basic Concepts of Evolution. Origin of Life

A recognizable feature of a human being or any other organism like height, complexion, shape of hair, colour of eyes, and shape of nose and chin, etc. are called 'characters' or 'traits'. The transmission of characters or traits from the parents to their offsprings (next generation) is called heredity. Two parents, a male and a female, are involved in sexual reproduction. The sexually reproducing organisms produce sex cells (gametes). The male gamete called sperm fuses with a female gamete called ovum or egg to form a zygote which gradually develops into a young one (or offspring), showing some similarities with the parents. Actually, the hereditary information is present in the sex cells of the parents. Thus, gametes constitute the link between one generation and the next, and pass on the paternal (father's) and maternal (mother's) characters or traits to the offspring. This relation that continues to exist between successive generations is referred to as heredity. In brief, genetics is the branch of biology that deals with the study of heredity and variation. The term genetics was coined by William **Bateson in 1906**.

A recognizable feature of a human being or any other organism like height, complexion, shape of hair, colour of eyes, and shape of nose and chin, etc. are called characters (traits). The transmission of characters or traits from the parents to their offsprings (next generation) is called heredity. The term genetics was coined by William Bateson in 1906. Any differences in traits between individual of species, caused either through genetic or environmental factor is called as variation. These variations could be observed in physical appearance, metabolism, behaviour etc. The number of variations produced during sexual reproduction is, however, very large In the case of human beings, skin colour, heights, hair, color of eyes, shape of nose and ears, chin are the characters (traits). Variations are the occurrence of differences among individuals of a species

4.1 VARIATIONS

The offspring are never a true copy of the parents. Although the offsprings inherit the characters of the parents and resemble them very closely, but their resemblance is not complete in all respects In fact, no two individuals are exactly alike and the members of any one species differ from one another in some characters or the other. Thus any differences in traits between individual of species, caused either through genetic or environmental factor is called as variation. These variations could be observed in physical appearance, metabolism, behaviour etc. The recognizable features of an organism are called character or traits. Some amount of variations is produced even during asexual reproduction put it is very small. The number of variations produced during sexual reproduction is, however, very large In the case of human beings, skin colour, heights, hair, color of eyes, shape of nose and ears, chin are the characters (traits). Variations are the occurrence of differences among individuals of a species (Fig. 4.1). Different peoples have different skin colour, heights, hair, color of eyes, shape of nose and ears, etc.

(A) Human Female **(B) Human Male**

(C) Monkey **(D) Chimpanzee**

Fig. 4.1 Diagram shows the Variations

(i) Less variations in member of species (A and B)
(ii) More variations in member of different species (C and D)

4.1.1 Types of Variations:

Variations are of two types such as (a) Germinal and (b) Somatic variations.

Germinal Variations: Variations caused by differences in number or structure of chromosome or gene are called as germinal variations, and these are heritable i.e. Colour of eyes, height etc. Germinal variations play an important role in organic evolution.

Somatic Variations: Variations caused by environmental factors like climate, food supply and interaction with other organisms are called as somatic variations. These variations are not transferred from one generation to next.

Depending on the utility, these variations are of three types as (a) Useful (b) Harmful and (c) Neutral.

- **Useful variations** provide some advantages to the individuals to survive in nature and also help in evolution and artificial selection.
- **Harmful variations** are eliminated as the individuals possessing them are unable to survive unlike individuals having useful variations.
- **Neutral variations** may or may not take part in survival of organism.

For example, the sugar cane plants reproduce by the process of asexual reproduction, so if we observe a field of sugar cane, we will find very little variations in various sugar cane plants. All the sugar cane plants almost look alike. But in animals (including human beings) which reproduce by the process of sexual reproduction, a large number of variations are produced. It is due to these variations that no two human beings look alike except identical twins. From this discussion we conclude that the number of successful variations is maximized by the process of sexual reproduction. The variation is a necessity for organic evolution.

A recognizable feature of a human being or any other organism like height, complexion, shape of hair, colour of eyes, and shape of nose and chin, etc. are called characters (traits). Any differences in traits between individual of species, caused either through genetic or environmental factor is called as variation. These variations could be observed in physical appearance, metabolism, behaviour etc. The number of variations produced during sexual reproduction is, however, very large In the case of human beings, skin colour, heights, hair, color of eyes, shape of nose and ears, chin are the characters (traits). Variations are the occurrence of differences among individuals of a species

Variations are of two types such as (a) Germinal and (b) Somatic variations. Germinal caused by differences in number or structure of chromosome or gene are called as germinal variations, and these are heritable. Somatic Variations caused by environmental factors like climate, food supply and interaction with other organisms are called as somatic variations. These variations are not transferred from one generation to next. Depending on the utility, these variations are of three types as (a) Useful (b) Harmful and (c) Neutral. of characters or traits from the parents to their offsprings (next generation) is called heredity. The term genetics was coined by William Bateson in 1906. Any differences in traits between individual of species, caused either through genetic or environmental factor is called as variation. These variations could be observed in physical appearance, metabolism, behaviour etc. The number of variations produced during sexual reproduction is, however, very large In the case of human beings,

skin colour, heights, hair, color of eyes, shape of nose and ears, chin are the characters (traits). Variations are the occurrence of differences among individuals of a species

4.1.2 Accumulation of Variations:

The reproduction of organism produces variations. The variations produced in organisms during successive generations get accumulated in the organisms. The significance of a variation can be shows up with the example of bacterium reproduction. Suppose a bacterium produces two bacteria by asexual reproduction. Again suppose that one of the offspring bacterium has a variation due to which it can tolerate a little higher temperature (or little more heat) than the other one. Now, this variation of little more heat resistance will go on accumulating in the offsprings of successive generations of this bacterium, and this will ultimately give rise to a variant of bacteria which will be highly heat resistant and able to survive even at very high temperatures. The great advantage of variation to a species is that it increases the chances of its survival in a changing environment.

4.2 SOME IMPORTANT TERMS OF MENDEL'S EXPERIMENTS:

Before we describe for explaining the transmission of characteristics (or traits) from parents to their offsprings or progeny, we should know the meaning of some terms such as chromosome, gene, dominant gene, recessive gene, genotype, phenotype, F_1 generation of F_2 generation.

4.2.1 DNA (Deoxyribonucleic Acid)

DNA is a macromolecule or polymer. It is made of very large number of 'nucleotide' units and hence is termed polynucleotide. Each nucleotide unit in a DNA molecule is made up of three components such as Deoxyribose Sugar (Pentose sugar), Phosphate Group and Nitrogenous bases- (i) Purines bases : Adenine (A) and Guanine (G), and (ii) Pyrimidines bases : Cytosine (C) and Thymine (T) (Fig. 4.2)

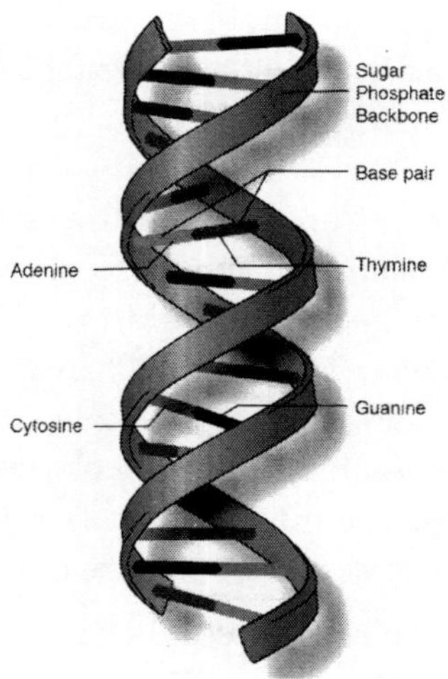

Fig. 4.2 Deoxyribonucleic Acid (DNA)

4.2.2 Chromosome

Chromosome is a thread-like structure in the nucleus of a cell formed of DNA which carries the genes (Fig. 4.3). Different organisms have different number of chromosomes in their nuclei. A normal cell of human body contains 23 pairs of chromosomes match in male and females but 23rd pair is different X and Y are the sex chromosomes in a male but a female has two X chromosomes. A gene is a unit of DNA on a chromosome which governs the synthesis of one protein that controls a specific characteristic (or trait) of an organism. There are thousands of genes on a chromosome which control various characteristics of an organism.

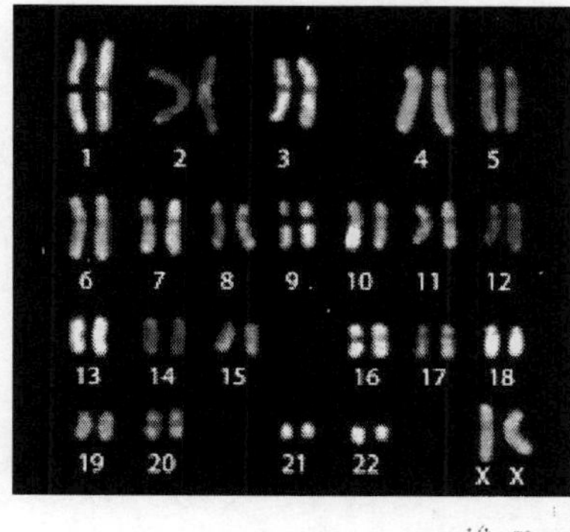

Fig. 4.3 Full set of Chromosome of Human

4.2.3 Genes

Genes are actually units of heredity which transfer characteristics or traits from parents to their offsprings during reproduction (Fig. 4.4). Genes work in pairs. Genes controlling the same characteristics are given the same letters. For example, the gene for tallness is represented by the letter 'T' whereas the gene for dwarfness is represented by the letter't;. The letters 'T' and't' actually represent two forms of the same gene (which controls the length of an organism, say the length of stem of a plant). The term 'factors' which were used by Mendel as carriers of heredity information are now known as 'genes'.

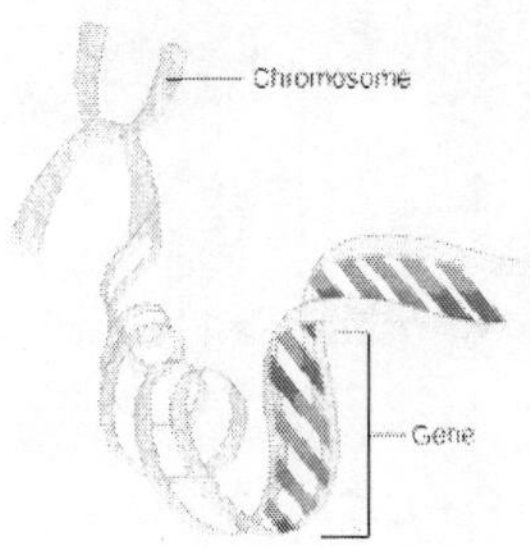

Fig. 4.4 Genes of Chromosome

4.2.4 Dominant and Recessive Genes: Genes for controlling the same characteristic of an organism can be of two types such as (i) dominant genes or (ii) recessive genes. The gene which decides the appearance of an organism even in the presence of an alternative gene is known as a **dominant gene**. It dominates the recessive gene for the same characteristic on the other chromosome of the pair. The gene which can decide the appearance of organisms only in the presence of another identical gene is called a **recessive gene**. A single recessive gene cannot decide the appearance of an organism. The dominant gene is represented by a capital letter and the corresponding recessive gene is represented by the corresponding small letter. For example, in pea plants, the dominant gene for tallness is T and the recessive gene for dwarfness is t. Thus, when we write the genetic cross for pea plant, then the capital 'T' represents 'tall' and small 't' represent dwarf.

4.2.5 Genotype shows the genetic constitution of an organism. In simple words, genotype is the description of genes present in an organism. Genotype is always a pair of letters such as TT, Tt or tt (where T and t are the different forms of the same gene). Thus, the genotype of a tall plant could be 'TT 'or 'Tt' whereas that of a dwarf plant is 'tt' (Fig. 4.5). The characteristic (or trait) which is visible in an organism is called its phenotype. For example, begin 'tall' or 'dwarf' are phenotypes of a plant because these traits can be seen by us or they are visible to us. The phenotype of an organism is actually its physical characteristic which is determined by its genotype. For example, genotype 'TT' or 'Tt' results in a tall phenotype and the genotype tt results in a dwarf phenotype.

Fig. 4.5 Genotypes or traits (Tall plant TT or small plant tt)

4.2.6 When two parents cross (breed) to produce progeny or offsprings, then their progeny is called **first filial (family) generation or F_1 generation** (where F progeny cross among themselves to produce second generation progeny, then this progeny is called **second filial (family) generation of F_2 progeny is called F_2 generation**. An example will make it clearer such as mother and father are parental generation. Their children are F_1 generation.

DNA is a macromolecule or polymer. It is made of very large number of 'nucleotide' units and hence is termed polynucleotide. Each nucleotide unit in a DNA molecule is made up of three components such as Deoxyribose Sugar (Pentose sugar), Phosphate Group and Nitrogenous bases- (a) Purines bases - Adenine (A) and Guanine (G), and (b) Pyrimidines bases – Cytosine (C) and Thymine (T).

Chromosome is a thread-like structure in the nucleus of a cell formed of DNA which carries the genes. Different organisms have different number of chromosomes in their nuclei.

A gene is a unit of DNA on a chromosome which governs the synthesis of one protein that controls a specific characteristic (or trait) of an organism. Genes are actually units of heredity which transfer characteristics or traits from parents to their offsprings during reproduction.

The gene which decides the appearance of an organism even in the presence of an alternative gene is known as a dominant gene. It dominates the recessive gene for the same characteristic on the other chromosome of the pair. The gene which can decide the appearance of organisms only in the presence of another identical gene is called a recessive gene. A single recessive gene cannot decide the appearance of an organism.

Genotype shows the genetic constitution of an organism. In simple words, genotype is the description of genes present in an organism. A new form of plant resulting from a cross or breeding of different varieties of a plant is known as a hybrid. When we breed two pea plants having one contrasting characteristic each (or one trait each) to obtain new plants, then it is called a monohybrid cross. On the other hand, if we breed two pea plants having two contrasting characteristics each (or traits each) to obtain new plants, then it is called a dihybrid cross.

A new form of plant resulting from a cross or breeding of different varieties of a plant is known as a hybrid. When we breed two pea plants having one contrasting characteristic each (or one trait each) to obtain new plants, then it is called a **monohybrid cross**. In monohybrid cross we will study the inheritance of one pair of contrasting characteristics 'tallness' and 'dwarfness' of the pea plants by their first generation and second generation progeny. On the other hand, if we breed two pea plants having two contrasting characteristics each (or traits each) to obtain new plants, then it is called a **dihybrid cross**. Inheritance of two pairs of contrasting characteristics of pea plants such as round-yellow seeds and wrinkled-green seeds will be explained in the dihybrid cross.

Table 4.1: Seven contrasting characteristics noted by Mendel in Garden Pea Plant

Sr No.	Characters	Contrasting Traits	
		Dominant	**Recessive**
1.	Plant Size or Height	Tall	Dwarf
2.	Position of Flower on the Stem	Axial	Terminal
3.	Colour of Unripe Pod	Green	Yellow
4.	Shape of Pod	Inflated	Constricted
5.	Shape of Seed	Round (smooth)	Wrinkled
6.	Colour of Seed	Yellow	Green
7.	Colour of Flower	Violet	White

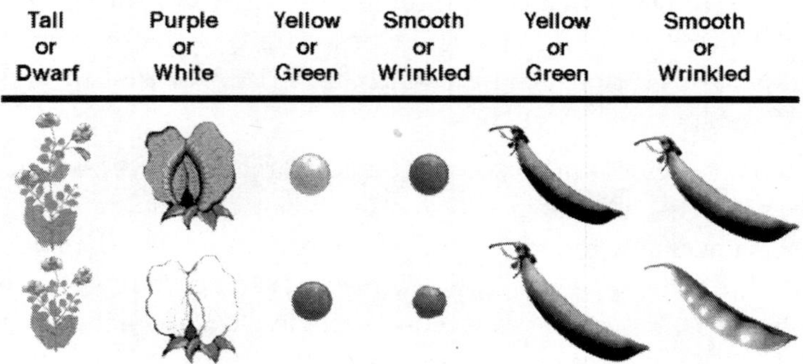

Fig. 4.6 Some of the characters of pea plants studied by Mendel

4.3 MENDEL'S CONTRIBUTION OR RULES FOR THE INHERITANCE OF TRAITS

Inheritance is the transmission of genetically characteristics (traits) from one generation to the next. Mendel studied the inheritance of characters or traits in various generations of a single characteristic or single trait such as plant height.

4.3.1 Monohybrid Inheritance and the Law of Segregation:

In order to trace the inheritance of a single pair of contrasting characteristics among the pea plants (like tall stem and short stem), Mendel crossed (cross-bred) the pure-bred pea plants differing in these traits and noted their occurrence in the progeny of succeeding generations.

- Mendel first crossed pure-bred tall pea plants with pure-bred dwarf pea plants and found that only tall pea plants were produced in the first generation of F_1 generation (Fig. 4.7). No dwarf pea plants were obtained in the first generation of progeny. From this Mendel concluded that the first generation (F_1 Cross) showed the traits of only one of the parent plants: tallness. The trait of other parent plant, dwarfness, did not show up in the progeny of first generation.

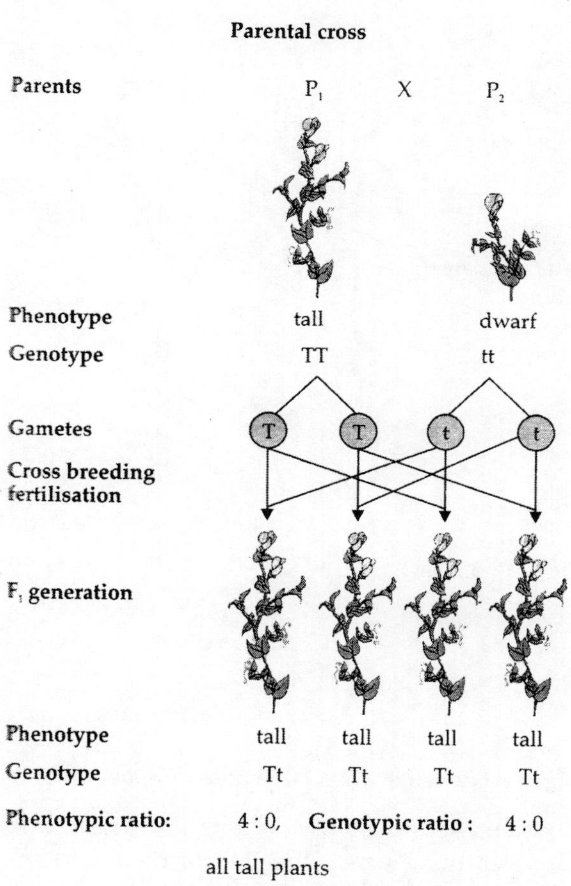

Fig. 4.7 Monohybrid Cross

- Mendel then crossed the tall pea plants of the first generation (F_1 generation) and found that tall plants and dwarf plants were obtained in the second generation (F_2 generation) in the ratio of 3: 1. In other words in the F_1 generations, three-fourth plants were tall and one-fourth were dwarf (Fig. 4.8). Mendel noted that the dwarf trait of the parent pea plant which had seemingly disappeared in the first generation progeny reappeared in the second generation. Mendel said that the trait of dwarfness of one of the parent pea plant had not been lost; it was simply suppressed in the first generation to re-emerge in the second generation. Mendel called the repressed trait of 'dwarfness' as 'recessive trait' and dwarf pea plants showed that the traits may be dominant or recessive.

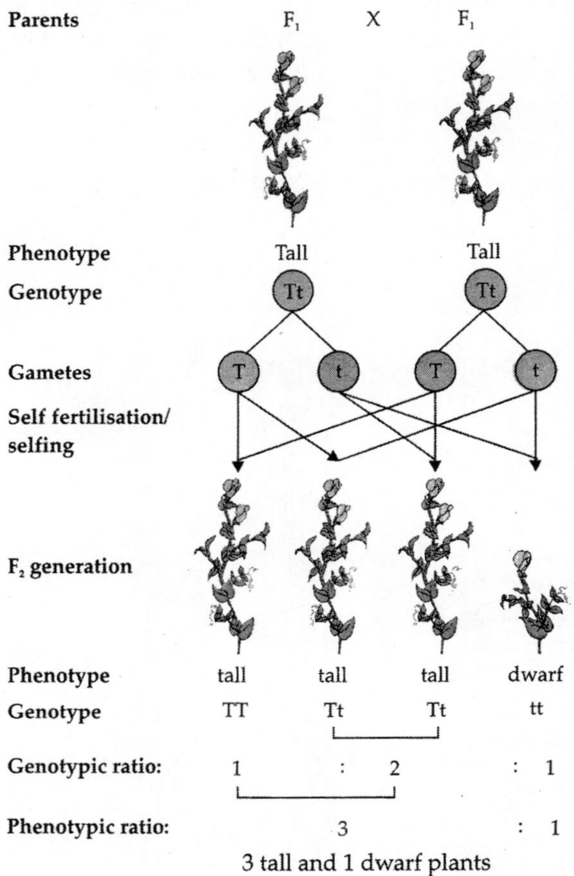

Fig. 4.8 Cross between tall plants of F_1 generation

- Mendel also noted that all the pea plants produced form the hybrid tall parents of F_1 generation, were either tall or dwarf. There were no plants with intermediate height (or medium height) in-between the tall and dwarf plants. In this way, Mendel's experiment showed that the traits (like tallness and dwarfness) are inherited independently. This is because if the traits of tallness or dwarfness had blended (or mixed up), then medium sized pea plants would have been produced. Out of total 1064 pea plants of F_2 generation, Mendel found that there were 787 tall pea plants and 277 dwarf pea plants. The ratio of tall plants to dwarf comes to be 787: 277 = 2.84: 1, which is approximately equal to 3: 1. Thus, a yet another result obtained from Mendel's monohybrid inheritance experiment is that the ratio of tall plants to dwarf plants in the F_1 generation is 3: 1. Since tallness is a dominant trait and dwarfness is a recessive trait, so we can also say that the contrasting progeny in the F_1 generation occur in the ratio of 3 dominant to 1 recessive. The ratio 3: 1 is known as the monohybrid ratio. The results of monohybrid cross enabled Mendel to formulate his first law of inheritance which is called as law of Segregation. According to Mendel's first law of inheritance, the characteristics (or traits) of an organism are determined by internal 'factors' which occur in pairs. Only one of a pair of such factors can be present in a single gamete.

Explanation for Monohybrid Inheritance:

(a) Mendel said that each trait is determined by a pair of 'factors'. This means that the pure-bred tall pea plant has two factors TT for the trait of tallness, and the pure-bred dwarf pea plant also has two factors for the trait of dwarfness.

(b) The factors of inheritance of tallness TT separate into two gametes T and T, and the factors for inheritance of dwarfness it separate into two other gametes t and t (The traits are transmitted to progeny through these gametes).

(c) The gametes of tall pea plant then cross with the gametes of the dwarf pea plant by the process of fertilization to form zygotes which then produce various progeny in the F_1 generation which consists of all tall plants. Thus, the F_1 generation possesses one factor of inheritance from each parent plant which was carried in gametes. The parental cross is shown clearly in the Fig 4.9.

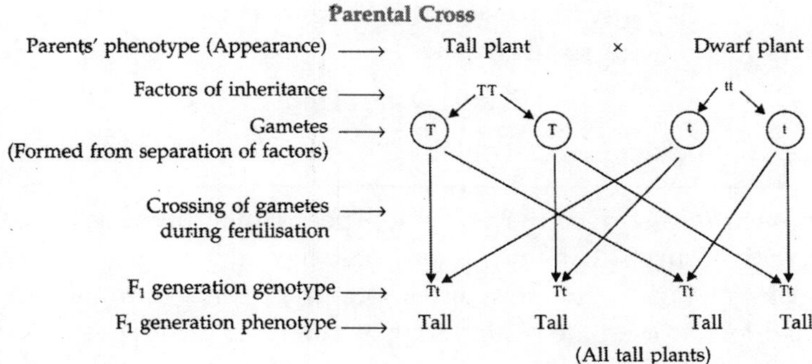

Fig. 4.9 Parental cross : (All Tall Plants)

In the F_1 generation shown above, all the progeny plants have factors Tt in which T is the factor for tallness which is a dominant trait. Since, all the plants in the F_1 generation have the factors Tt, so all of them are tall. The small letter t represents recessive trait to dwarfness, which does not show up in first generation in the presence of dominant trait T.

(d) When two hybrids, tall pea plants (Tt) produced in the first generation (F_1) are now cross-bred with each other, then they will produce second generation (F_2) pea plants. It again happens by the separation of factors of inheritance of these tall plants into individual gametes and then crossing of the gametes during fertilization as shown in the Fig. 4.10:

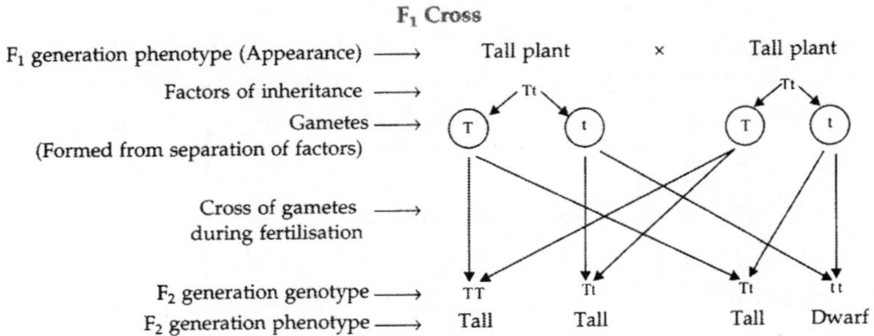

Fig. 4.10 F_1 Cross : Three tall and one dwarf plant

Figure 4.10 shows that in the second generation (F_2), the pea plants produced has genotype or inheritance factors TT, Tt, Tt and tt. Now, the plants having genotype TT, Tt and Tt all contain the factor T for dominant trait 'tallness', so all the three plants (TT, Tt and Tt) are tall. The plant having the genotype it has both factors t for the recessive trait 'dwarfness', so it is a dwarf plant. Please note that though a single copy of factor T is enough to make a plant tall but both copies of factor t (that is tt) are necessary to make a plant dwarf. In the F_2 generation, we get 1 plant having genotype TT, 2 plants having genotype Tt and 1 plant having genotype tt. So, the genotype ratio in monohybrid cross will be:

$$TT : Tt : tt = 1 : 2 : 1$$

Again, in the F_2 generation, we get 3 tall plants and 1 dwarf plant, so the phenotypic ratio in monohybrid cross will be:

Tall Plants: Dwarf Plants = 3: 1

This result is the same as that obtained by Mendel through experiments.

> Mendel crossed (cross-bred) the pure-bred pea plants differing in these traits and noted their occurrence in the progeny of succeeding generations, and found ration 3: 1, known as the monohybrid ratio. The results of monohybrid cross enabled Mendel to formulate his first law of inheritance which is called as Law of Segregation. According to Mendel's first law of inheritance, the characteristics (or traits) of an organism are determined by internal 'factors' which occur in pairs. Only one of a pair of such factors can be present in a single gamete.

4.3.2 Dihybrid Inheritance or Law of Independent Assortment:

Dihybrid inheritance involves the inheritance of two pairs of contrasting characteristics or traits at the same time. The two pairs of contrasting characteristics chosen by Mendel were shape and colour of seeds: round-yellow seeds and wrinkled-green seeds. In order to trace the inheritance of two pairs of contrasting traits, Mendel crossed pea plants having wrinkled-green seeds and noted their occurrence in the succeeding generations of pea plants. Mendel made the following observations:

(i) Mendel first crossed pure-bred pea plants having round-yellow seeds with pure-bred plants having wrinkled-green seeds and found that only round-yellow seeds were produced in the first generation. No wrinkled-green seeds were obtained in the F_1 generation. From this it was concluded that round shape and yellow colour of the seeds were dominant traits over the wrinkled shape and green colour of the seeds.

(ii) When the F_1 generation pea plants having round-yellow seeds were cross-bred by self-pollination, then four types of seeds having different combinations of shape and colour were obtained in second generation of F_2 generation. These were round-yellow, round-green, wrinkle-yellow and wrinkle-green seeds. Mendel collected a total of 556 F_2 seeds and counted them shape wise and colour wise. The phenotypic ratio of different type of seeds can be written as:

Round	Round	Wrinkled	Wrinkled
Yellow (315)	Green (108)	Yellow (101)	Green (32)
Seeds (9)	Seeds (3)	Seeds (3)	Seeds (1)

Thus, the ratio of each phenotype (or appearance) of the seeds in the F_2 generation is 9: 3: 3: 1. This is known as the dihybrid ratio.

Mendel observed that he had started with two combinations of characteristics in seeds such as round-yellow and wrinkled-green, and two new combinations of characteristics had appeared in the F_2 generation : round-green and wrinkled-yellow (Figure 4.11A, B, C, D). On the basis of this observation, Mendel concluded that though the two pairs of original characteristics (seed shape and colour) combine in the F_1 generation but they separate and behave independently in subsequent generations. The results of dihybrid cross enabled Mendel to formulate his second law of inheritance which is called the Law of Independent Assortment. According to Mendel's second law of inheritance, in the inheritance of more than one pair of traits in a cross simultaneously, the factors responsible for each pair of traits are distributed independently to the gametes.

(A) Round yellow

(B) Round green

(C) Wrinkled yellow

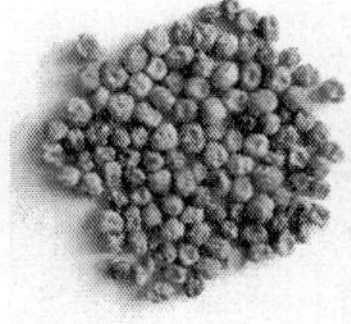

(D) Wrinkled green

Fig. 4.11 Diagram showing the various peas

Explanation for Dihybrid Inheritance:

In the dihybrid cross, the parent plants having the phenotype round-yellow seeds have the factors of inheritance of gene combination RRYY (in which RR are the dominant genes for round shape whereas YY are the dominant genes for yellow colour). On the other hand, the parent plants having the phenotype wrinkled-green seeds have the factors of inheritance or gene combination rryy (in which rr are the recessive genes for wrinkled shape and yy are the recessive genes for green for green colour). The chart showing the dihybrid cross between pea plants having round yellow seeds and wrinkled green seeds is given in Fig. 4.12.

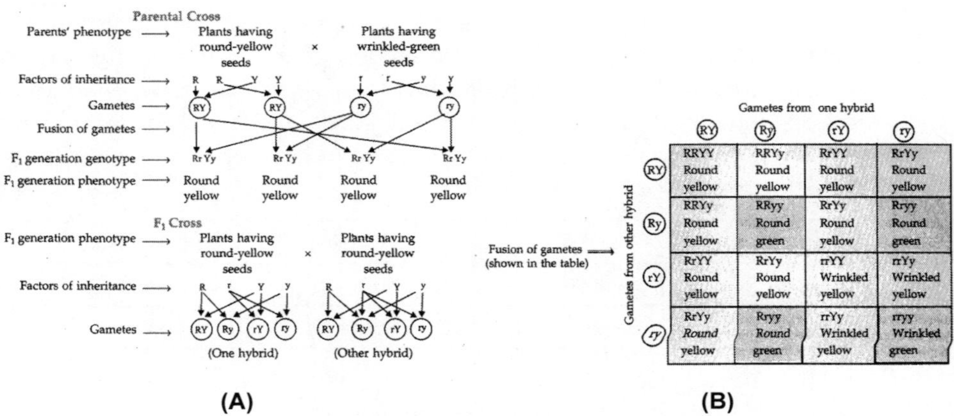

Fig. 4.12 Explanation of Dihybrid Inheritance

The result is the same as that obtained by Mendel through experiments

An amazing thing about Mendel's work is that he worked out the underlying rules of inheritance before any knowledge of DNA, chromosomes or genes became available.

Sample Problems:

(a) What do the progeny of a tall plant with round seeds and a short plant with wrinkled seeds look like? Why?

(b) What happens when F_1 progeny obtained above are used to produce F_2 progeny by self-pollination?

Answer:

(a) The progeny of a tall plant having round seeds crossed with short plants having wrinkled seeds are all tall plants having round seeds. This is because tallness and round shape of seeds are dominant traits. On the other hand shortness and wrinkled shape of seeds are recessive traits.

(b) When F_1 progeny are cross bred by self-pollination, then we will get four types of progeny in the F_2 generation. Of these four types of progeny, two types will have traits like parents and the other two will have new combinations of traits. Thus,

- Some F_2 progeny will be tall plants with round seeds (9).

- Some F_2 progeny will be tall plants with wrinkled seeds (3).
- Some F_2 progeny will be short plants with round seeds (3).
- Some F_2 progeny will be short plants with wrinkled seeds (1).

Please note that though Mendel studied the inheritance of characteristics by using plants (or rather pea plants) but the rules for the inheritance of traits given by Mendel are also applicable to the inheritance of traits in animals (including human beings).

Dihybrid inheritance involves the inheritance of two pairs of contrasting characteristics or traits at the same time. The two pairs of contrasting characteristics chosen by Mendel were shape and colour of seeds: round-yellow seeds and wrinkled-green seeds. In order to trace the inheritance of two pairs of contrasting traits, Mendel crossed pea plants having wrinkled-green seeds and noted their occurrence in the succeeding generations of pea plants. The ratio of each phenotype (or appearance) of the seeds in the F2 generation is 9: 3: 3: 1. This is known as the dihybrid ratio.

4.4 TRANSMISSION OF CHARACTERISTICS (TRAITS) TO PROGENY (OFFSPRING)

Genes are responsible for the characteristic features (traits) of an organism such as plant or animal. The characteristics or traits of parents are transmitted to their progeny (offsprings) through genes present on their chromosomes during the process of sexual reproduction. Genes work in pairs. There is a pair of genes for each characteristic of an organism (one is dominant gene and the other is recessive gene). Each parent possesses a pair of genes for each characteristic on a pair of chromosomes. However, each parent passes only one of the two genes of the pair for each characteristic to its progeny through gametes. Thus, the male gamete and female gamete carry one gene for each characteristic from the gene pairs of parents (which are located on the pair of chromosomes). But when a male gamete fuses with a female gamete during fertilization, they make a new cell called zygote with a full set of genes (on a full set of chromosomes). This zygote grows and develops to form a new organism having characteristics or traits from both the parents which it has inherited through genes.

The two genes or pair of genes responsible for a particular characteristic is always present on the corresponding positions of the pair of chromosomes. For example, two genes for the same characteristic (length of plant stem), are present on the corresponding positions of the pair of chromosomes (Fig. 4.13). One gene of the pair is for 'tallness' and the other is for 'dwarfness'. Please note that though the progeny inherits two genes (or a pair of genes) for each trait from its parents but the trait shown by the progeny depends on which inherited gene is dominant of the two. For example, if a pea plant progeny (or hybrid) inherits the gene for tallness (T) from one parent and the gene for dwarfness (t) from the other parent, then it will show the trait of 'tallness' and become a tall plant because the gene for tallness is dominant over the gene for dwarfness. So, although the gene for dwarfness (t) is present in all the cells of the hybrid plants, it does not show its effect (because it is a recessive gene). If, however, both the parent plants pass on one copy each of the recessive gene for dwarfness (t) making the genotype (tt), then the traits of dwarfness will appear in the progeny plant.

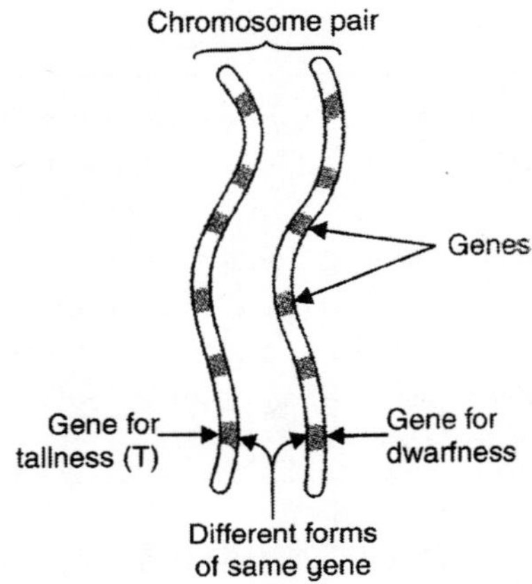

Fig. 4.13 Gene works in pairs

Please note that the gene for 'tallness' and 'dwarfness' are not to be considered two different genes. They are just the two forms of the same gene which controls only one characteristic feature of a plant: length of its stem. But there can be increase in length of stem making the plant tall or decrease in the length of stem, making the plant dwarf.

4.5 MECHANISM OF GENES CONTROLS THE CHARACTERISTICS (TRAITS)

A gene is the section of DNA on a chromosome which codes for the formation of a protein controlling a specific characteristic or trait of the organism. Suppose a plant progeny has gene for the characteristic called 'tallness'. Now, the gene for tallness will give instructions to the plant cells to make a lot of plant growth hormones, and due to the formation of excess of plant growth hormones, the plant will grow too much and hence become tall. One the other hand, if the plant has the gene set for dwarfness, then less plant growth hormones will be produced due to which the plant will grow less, remain short and hence become a dwarf plant. Just like plants, the characteristics or traits in animals including human beings are also transmitted from the parents through genes by the process of sexual reproduction. We will now give an example of the transmission of colour of hair from the parents (father and mother) to the child. Before we do that please keep in mind that black hair is a phenotype produced by the genotype HH or Hh. On the other hand, blonde hair (pale yellow hair) is a phenotype produced by the genotype hh.

Example:

A mother has black hair, the father has blonde hair (pale yellow hair), and the child has black hair. It can be explained on the basis of transmission of genes for 'hair colour' from the mother and father to the child as follows: Mother's cell contains two genes HH for black hair. Both the genes HH are dominant genes, so the mother has black hair (Fig. 4.14A). Father's cells contain two genes (hh) for blonde hair. The two genes hh are recessive genes, so the father has blonde hair (or pale yellow hair) (Fig. 4.14B). Now, during the process of reproduction, the mother transmits one of the dominant genes H for black hair to the child and the father transmits one of his recessive genes h for blonde hair to the child. Due to this, the child has the genes Hh for her hair. Now, the gene G for black hair is the dominant gene but the gene h for blonde hair is the recessive gene. The dominant gene H for black hair shows its effect due to which the child has black hair (Fig. 4.14C). The recessive gene h for blonde hair cannot show its effect in the presence of dominant gene H for black hair. Please note that the genes which dominate other genes are called dominant genes, and the genes which get dominated are called recessive genes.

(A) Mothers cells contain two dominant genes HH for black hair, so she has black hair

(B) Father cell contains two recessive genes hh for blonde hair, so he has blonde hair

(C) Child has one domaint gene H for black hair (from mother) and one recessive gene h for blonde hair (from father, so her genetypes is Hh and ptenotype is black hair.

Fig. 4.14 Mechanism of Genes Control

4.6 INHERITANCE OF BLOOD GROUPS

A person has one of four blood groups such as A, B, Ab and O (Fig 4.15). This blood group system is controlled by a gene (represented by I) which has three different forms denoted by the symbols I^A, I^B and I^O. The genes I^A and I^B show no dominance over each other, that is, they are codominant. However, genes I^A and I^B both are dominant over the gene I^O. In other words, the blood gene I^O is recessive in relation to genes I^A and I^B. Although there are three gene forms (called alleles) for blood, but any one person can have only two of two forms of the genes he/she possesses.

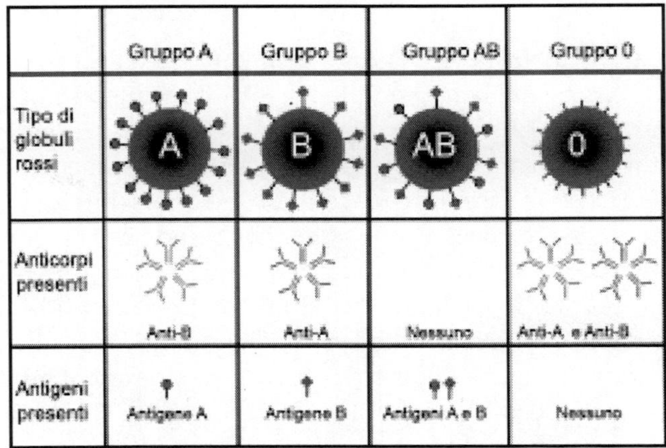

Fig. 4.15 Blood Groups : A, B, AB and O

- If the genotype (gene combination) is $I^A\,I^A$, then the blood group of the person is A. And if the genotype is $I^A\,I^O$ even then the blood group is A (because I^O is a recessive gene).
- If the genotype is $I^B\,I^B$, then the blood group of the person is B. And if the genotype is $I^B\,I^O$ even then the blood group is B (because I^O is a recessive gene).
- If the genotype is $I^A\,I^B$, then the blood group of the person is AB.
- If the genotype is $I^O\,I^O$, then the blood group of the person is O.

Sample Problem.

A man having blood group A marries a woman having blood group O and they have a child. What will be the blood group of the child?

Answer. The answer to this question depends on whether the blood group A of the man has gene combination $I^A\,I^A$ or $I^A\,I^O$.

(a) When the blood group A has genotype $I^A\,I^A$. In this case the genotype of man's blood is $I^A\,I^A$ and that of woman's blood is I^OI^O. So, the child will have blood group A (because the gene I^A is dominant over gene I^O).

(b) When the blood group A has genotype I^AI^O. Here the genotype of man's blood is I^AI^O and that of woman's blood is I^OI^O. So, in this case there is an equal chance that the genotype of child's blood can be either I^AI^O. Due to this, there is an equal chance of the child acquiring blood group A or blood group O.

Just as the blood group in inherited by a child from its parents, in the same way the sex of a child (boy or girl) is also inherited from the parents: mother and father. We will now describe the inheritance of sex by a child from the parents. Inheritance of sex is also known as sex determination. Please note that while discussing the determination of sex of a child, we use letter symbols to describe whole sex chromosomes rather individual genes. The sex chromosomes are: XX for a Female (Girl) and XY for a Male (Boy).

4.7 SEX DETERMINATION

A person can have a male sex or a female sex. The process by which the sex of a person is determined is called sex determination. Genetics is involved in the determination of the sex of a person. The chromosomes which determine the sex of a person are called sex chromosomes. There are two types of sex chromosomes, one is called X chromosome and the other is called Y chromosome.

4.7.1 Genetic Chromosomal Determination of Sex:

It can be understood by two mechanisms as:

4.7.1.1 XX-XY Mechanism:

The mechanism of sex determination in human beings and fruit fly (Drosophila melanogaster) is called XX-XY mechanism. Both have two types of chromosomes, autosomes and allosomes. Autosomes are those chromosomes which control normal body functions, other than sex determination. Allosomes or Sex chromosomes are those chromosomes which determine the sex of the individual. Each cell in human beings consists of 46 chromosomes (23 Pairs). 22 pairs of chromosomes are found common in both males and females. These are autosomes. The 23rd pair of chromosomes in human beings takes part in sex determination and thus is termed as sex chromosomes.

(a) A male has one X chromosome and one Y chromosome i.e. XY (heteromorphic) (Fig. 4.16). It means that half the male gametes or half the sperms will have X chromosomes and the other half will have Y chromosomes. Therefore, male is called as heterogamic.

(b) A female has two X chromosomes (but no Y chromosomes) i.e. XX (homomorphic) (Fig. 4.16). It means that all the female gametes called ova (or eggs) will have only X chromosomes. Therefore, female is called as homogametic.

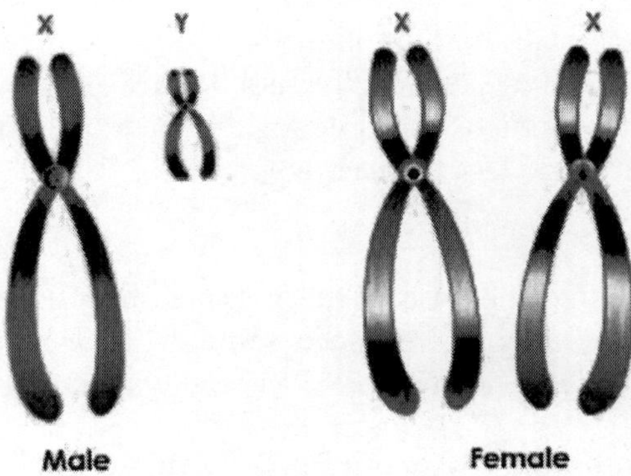

Fig. 4.16 Human Chromosomes : Male (X and Y) and Female (X and XX) chromosome

The sex of a child depends on what happens at fertilization **(a)** If a sperm carrying X chromosomes fertilizes and ovum (or egg) which carries X chromosome, then the child born will be a girl (or female). This is because the child will have XX combination of sec chromosomes (Fig. 4.17). **(b)** If a sperm carrying Y chromosome fertilizes an ovum (or egg) which carries X chromosome, then the child born will be a boy (or male). This is because the child will have XY combination of sec chromosomes (Fig. 4.17).

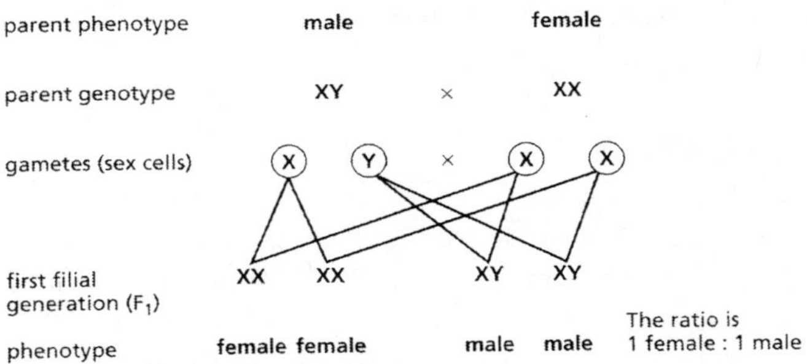

Fig. 4.17 Inheritance of Sex in Humans

It is noted here that it is the sperm which determines the sex of the child. This is because half of the sperms have X chromosomes and the other half have Y chromosomes. Thus, there is a 50% chance of a boy and a 50% chance of a girl being born to the parents. This is why the human population is roughly half males and half females.

It is concluded that if the father (man or husband) contributes X sex chromosome at fertilization through his sperm, the baby born will be a girl. On the other hand, if the father contributes a Y sex chromosome at fertilization through his sperm, then the baby born will be a boy. It means that it is the sex chromosome contributed by father which decides the sex of the baby which the mother will give birth to. Thus, father is responsible for the sex of the baby (boy or girl) which is born. The belief that mother (woman or wife) is responsible for the sex of her baby is absolutely wrong. In many ignorant Indian families, the mother is held responsible for the birth of a girl child and unnecessarily harassed. Such people should understand that it is the husband who is responsible for the birth of a girl child (and not his wife). Moreover, a girl is no less than a boy.

4.7.1.2 XX-XO Mechanism

In grasshoppers and some other insects, the male has only one sex chromosome (XO) whereas the female has two homotrophic (of same size and shape) sex chromosomes (XX). This type of sex mechanism is called as XX-XO mechanism.

4.7.2 Role of Environmental Factors in Sex Determination:

In some of the animals, sex determination is also controlled by the environmental factors.

- In some reptiles, the temperature at which the fertilized egg is incubated before hatching plays a role in determining the sex of the offspring.
- In turtle (*Chrysema picta*), high incubation temperature (above 33^{0}C) leads to the development of female offsprings (or female progeny) but temperature below 28^{0}C produces male offspring.
- In lizard (*Agama agama*), high incubation temperature results in male offsprings (or male progeny).
- In snails, individuals can change sex, indicating that sex is not determined genetically in this case.
- In marine worm (*Bonelia viridis*) and marine mollusk (*Crepidula*), the larvae develop into female if they live alone. However, if they come in company of a female the larvae develops in males.
- In a marine vertebrate (*Echiuorids*), the CO_2 concentration of water is responsible for the sex determination
- In plants, genetic basis of sex determination exists but is different from that of animals.

The process by which the sex of a person is determined is called sex determination. Genetics is involved in the determination of the sex of a person. The chromosomes which determine the sex of a person are called sex chromosomes. There are two types of sex chromosomes, one is called X chromosome and the other is called Y chromosome. Genetic Chromosomal Determination of Sex can be understood by two mechanisms as XX-XY Mechanism and XX-XO Mechanism.

The mechanism of sex determination in human beings and fruit fly (*Drosophila melanogaster*) is called XX-XY Mechanism. Both have two types of chromosomes, autosomes and allosomes.

In grasshoppers and some other insects, the male has only one sex chromosome (XO) whereas the female has two homotrophic (of same size and shape) sex chromosomes (XX). This type of sex mechanism is called as XX-XO mechanism.

In some of the animals, sex determination is also controlled by the environmental factors such as temperature and CO_2 concentration of water etc.

Table 4.2 Differences between Autosomes and Allosomes

Sr No.	Parameter	Autosomes	Allosomes
1	Similarity	These are similar in both males and females	These are different in both males (XY)- nd females (XX)
2.	Number in humans	22 pairs or 44	1 pair or 2
3.	Functions	Chromosomes which are not involved in sex determination	Chromosomes which are involved in sex determination

4.8 ACQUIRED AND INHERITED TRAITS

A trait (or characteristic) of an organism which is not inherited but develops in response to the environment is called an acquired trait. For example, if a beetle does not get sufficient food for a considerable time, its weight will be reduced due to starvation. The low weight of this beetle is an acquired trait of the beetle which has been acquired in response to the environment which contained insufficient food. Again, suppose the tail of a mouse gets cut. The cut tail of this mouse is also an acquired trait which has been brought about by some agent in its environment. A human know how to do exercise or build the body swimming and learning the music (Fig. 4.18A, B, C). All these are acquired traits which the man has picked up (or acquired) himself as he goes through life. The man is not born with these traits and he cannot pass on these traits to his children. The acquired traits of organisms cannot be passed on to their future generations. In other words, only those traits can be transmitted to future generations in which changes have occurred in the genes (DNA) present in the reproduction to future generations in which changes have occurred in the genes (DNA) present in the reproductive cells (or gametes) of parent organisms. The changes in the non-reproductive body cells of an organism cannot be inherited by its offsprings.

(A) Muscles Building

(B) Swimming

(C) Music learning

Fig. 4.18 Acquired traits

Example:

If we breed some mice, all the progeny of mice will have tails, just like their parents. Now, if we cut the tails of these first generation mice surgically and breed them, we will get new mice, all with full fails. It has been observed that even after cutting the tails of mice for a number of generations, a tail-less mouse is never born. Actually, the cut tail of mice is an acquired trait which is never passed on to their progeny. This is because cutting the tails of mice does not change the genes of their reproductive cells (or gametes). And since the acquired trait of 'cut tails' does not bring about a change in the genes of mice, this trait cannot be passed on to their next generations (Fig. 4.19). From this discussion we conclude that the experiences of an individual during its life time (called acquired traits) cannot be passed on to its progeny, and hence cannot lead to evolution (because they are not caused by the change in genes).

(a) These are first generation mice. All these mice have full tails

(b) The tails of all the mice are cut surgically so that they become tail-less mice.

(c) All the second generation mice produced by breeding cut-tail mice have full tails. No tail-less mouse is born

Fig. 4.19 Diagram shows acquired traits in Mice

A trait (or characteristic) of an organism which is caused by a change in its genes (or DNA) is called an inherited trait. Inherited traits can be passed on to the progeny of the organism. Because they have produced changes in the genes or DNA of the organism. Suppose there is a population of red beetles in the green bushes. Again suppose that a colour variation arises during reproduction so that there is one beetle which is green in colour (instead of red). This change of green colour in the beetle has been brought about by a change in the genes of the reproductive cells (Fig. 4.20A,B). The green colour of this beetle is an inherited trait which can be passed on to the next generations. The change from red beetle to green beetle can be considered to be an example of evolution because it helps in its survival by mixing with green bushes.

(A) Red beetle

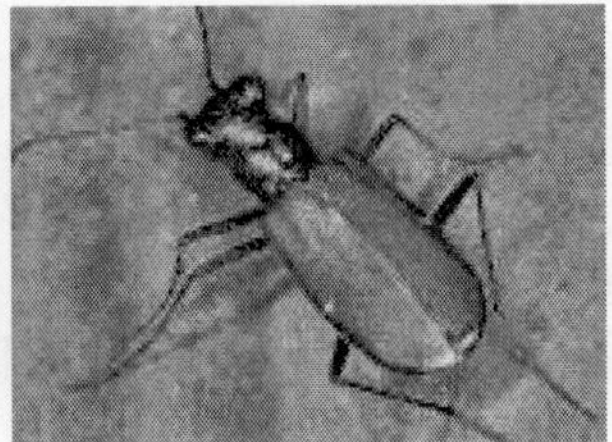
(B) Green beetle

Fig. 4.20 Change of colour due to change in the genes of DNA

A trait (or characteristic) of an organism which is not inherited but develops in response to the environment is called an acquired trait. For example, if a beetle does not get sufficient food for a considerable time, its weight will be reduced due to starvation. The low weight of this beetle is an acquired trait of the beetle which has been acquired in response to the environment which contained insufficient food. The acquired traits of organisms cannot be passed on to their future generations.

Table 4.3 Differences between Acquired and Inherited Traits

Sr No.	Parameter	Acquired Traits	Inherited Traits
1	Origin	These are obtained from parents	These develop in an organism during its life span
2	Type of variations	Genetic variations	Somatic Variations

3	Development	These develop due to crossing over during meiosis and mutations	These develop due to use and disuse of the organs and are affected by the environment
4	Heredity	These are transmitted to successive generations	These are transmitted to other generations, and die with the death of the individuals

4.9 EVOLUTION

There is a massive number and types of living organisms (plants and animals) on the earth. In addition to this wide variety of living organisms, the remains of the dead organisms which lived in the remote past (called fossils) are also known. The word 'evolution' has been derived from the Latin word 'evolvere' which means to 'unroll' or 'unfold'. Evolution is a kind of gradual unfolding (or formation) of the new organisms from the pre-existing primitive organisms through slow and steady changes. Thus, evolution is the sequence of gradual changes which take place in the primitive organisms over millions of years in which new species are produced. Since the evolution is on the living organisms, so it is also called 'organic evolution'. It is through the constant process of evolution taking place in the organisms since the origin of life that such an enormous variety of plants and animals have come to exist on this earth at present. All the plants and animals (or organisms) which we see today around us have evolved from some or the other ancestors that lived on this earth long, long ago.

The process of evolution will become clear from the following example of 'pterosaur'. Pterosaur is an ancient flying reptile which lived on the earth about 150 million years ago. The development of pterosaur is an example of evolution. It began life as a big lizard which could just crawl on land (Fig. 4.21A). Over millions of years, small folds of skin developed between its feet which enabled it to glide from tree to tree (Fig. 4.21B). Over many, many generations, spread over millions of year, the folds of skin, and the bones and muscles supporting them grew to form wings which could make it fly (Fig. 4.21C). In this way, an animal which crawled on ground evolved into a flying animal. This evolution led to the formation of a new species of a flying reptile.

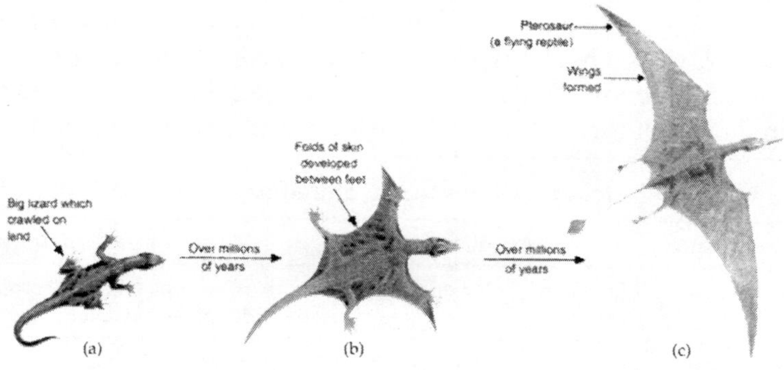

Fig. 4.21 Development of 'Pterosaur'

4.9.1 Evidences for Evolution:

Various biological studies tell us that since their origin, living organisms have been undergoing changes in their organization to evolve into new forms. A number of common features of different types of organisms provide evidence in favour of evolution because they can be considered to have evolved from the common ancestor. The more characteristics (or features) two species have in common, the more closely they will be related. And the more closely they are related, the more recently they will have had a common ancestor. There are several evidences which show the incidence of evolution. Some of the important sources which provide evidences for evolution are (a) Homologous Organs (b) Analogous Organs, and (c) Fossils.

4.9.1.1 Homologous Organs:

Organs which have the same basic structure (or same basic design) but different functions are called homologous organs. The homologous organs of different animals provide evidence for evolution. There are many organs in different groups of animals (or plants) which all seem to be built from the same basic design but are used for many different purposes. These are called homologous organs. For example, the forelimbs of a man, a lizard (reptile), a frog (amphibian), a bird and a bat (mammal) seen to be built from the same basic design of bones (Fig. 4.22), but they perform different functions. The forelimbs of a man are used for grasping, the forelimbs of a lizard are used for running, the forelimbs of a frog are used to prop up the front end of its body when at rest, and also act as shock absorbers when the frog lands back on the ground after a leap; whereas the forelimbs of bird and bat are modified for flying, Since the forelimbs of a human, a lizard, a frog, a bird and a bat have similar structures (or design) but perform different functions, they are the homologous organs. The presence of homologous forelimbs in humans (man), a lizard, a frog, a bird and a bat indicate that all these forelimbs have evolved from a common ancestral animal which has a 'basic design' limb. In other words, we can say that a human, a lizard, a frog, a bird and a man, all have evolved from a common ancestor. Thus, the presence of homologous organs in different animals provides evidence for evolution as they are derived from the same ancestor who had the 'basic design' of the organ on which all the homologous organs are based. It should be noted that the wings of a butterfly (an insect) and the wings of a bat cannot be considered to be homologous organs because they have different basic designs (though they are used for the same purpose of flying). Plants also posses homologous organs *i.e.* a thorn of *Bougainvillea* (family Nyctaginaceae) plant and a tendril of *Passiflora* (family Passionflower) or *Cucurbita* plant are homolgons organs. Both perform different function but have some best structural design. Other examples are potato tubers and runners of thorn of *Bougainvillea* and *Tendrils* of *Cucurbita* grass (both are modified stem), and human teeth and tusks of elephant. Thus, presence of homologous organs show divergent evolution i.e. some structure modified differently in a different organisms due to different adaptations.

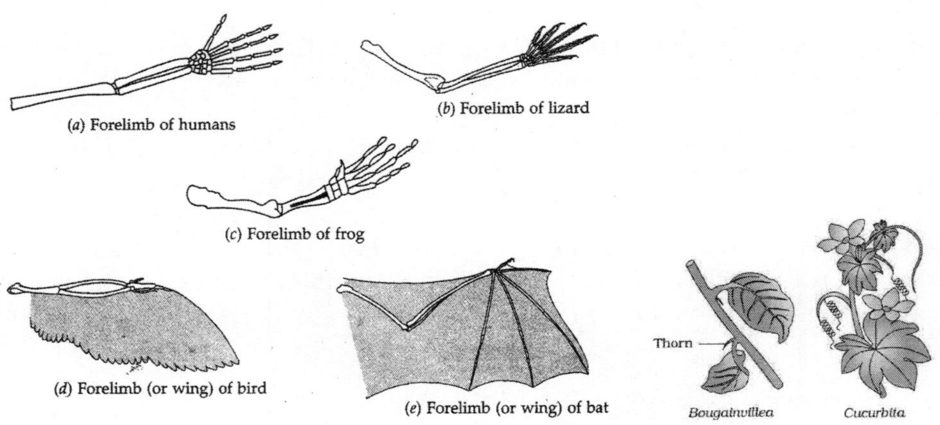

Fig. 4.22 Homologous Organs (Forelimbs of human, lizard, frog, bird and bat; Thorn of *Bougainvillea* and Tendrils of *Cucurbita*)

4.9.1.2 Analogous Organs:

Organs which have different basic structure (or different basic design) but have similar appearance and perform similar functions are called analogous organs. The analogous organs provide the evidence for evolution. There are many organs in different groups of animals which seem to be built from different basic structure but appear to be similar in shape and perform similar functions. These are called analogous organs. For example, the wings of an insect and a bird have different structures (the insects have a fold of membranes as wings which are associated with a few muscles whereas a skeleton, flesh and feathers support bird's wings) but they perform the same function of flying (Fig. 4.23). Since the wings of insects and birds have different structures but perform similar functions, they are analogous organs. Now, since the analogous organs have different basic design, so they do not show a common ancestor for the organism. The presence of analogous organs indicates that even the organisms having organs with different structures can adapt to perform similar functions for their survival under hostile environmental conditions. Thus, the presence of analogous organs in different animals provide evidence for evolution by telling us that though they are not derived from common ancestors, They can still evolve to perform similar functions to survive, flourish and keep on evolving in the prevailing environment. The analogous organs actually provide a mechanism for evolution.

The winds of bat are skin folds stretched mainly between elongated fingers. However, the wings of a bird are feathery covering all along the arm. But all of these are adopted for flying. Plants also have analogons organs *i.e.* sweet potato and potato, stem tendrils of *parriflora* and leaf tendrils of *Pisum satiyum*.

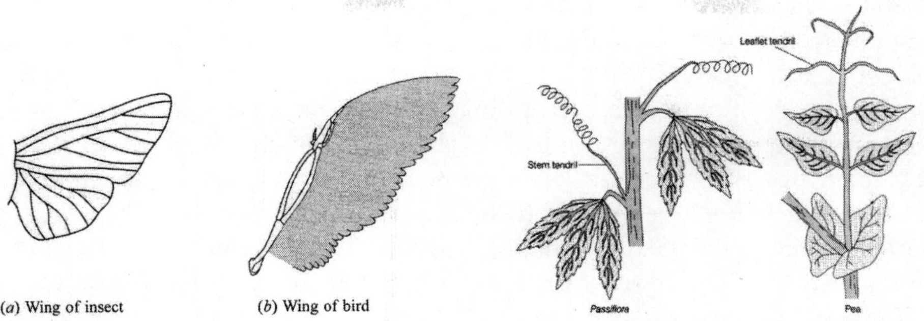

(*a*) Wing of insect (*b*) Wing of bird

Fig. 4.23 Analogons Organs (Wings of Insect and Bird); Stem tendrils of *Passiflora* and leaf tendrils of *Pissum sativum*

Evolution is a kind of gradual unfolding (or formation) of the new organisms from the pre-existing primitive organisms through slow and steady changes. Thus, evolution is the sequence of gradual changes which take place in the primitive organisms over millions of years in which new species are produced. Since the evolution is of the living organisms, so it is also called 'organic evolution'.

The process of evolution will become clear from the following example of 'pterosaur'.

Various biological studies tell us that since their origin, living organisms have been undergoing changes in their organization to evolve into new forms. Some of the important sources which provide evidences for evolution are (a) Homologous Organs (b) Analogous Organs, and (c) Fossils.

Organs which have the same basic structure (or same basic design) but different functions are called homologous organs. The homologous organs of different animals provide evidence for evolution. There are many organs in different groups of animals (or plants) which all seem to be built from the same basic design but are used for many different purposes. These are called homologous organs.

Organs which have different basic structure (or different basic design) but have similar appearance and perform similar functions are called analogous organs. The analogous organs provide the evidence for evolution. There are many organs in different groups of animals which seem to be built from different basic structure but appear to be similar in shape and perform similar functions. These are called analogous organs.

4.9.1.3 Vestigial Organs

The organs which occur in reduced form and are useless to the process or, but are homologous to the fully developed, functional organs in the ancestors or related forms are called vestigial organs i.e. vermiform appendix of the large intestive and initiating membrane in the eye of human beings are vestigial organs. Vermiform appendix, blind tube at the end of large intestine in certain mammals including human beings. In herbivorous mammals, it is a large structure and played role for digestion & cellulose of the food by the action of bacteria, while in human beings, the habbit of taking soft and cooked food has reduced the need for bacterial digestion. The appendix has reduced consequently and has no useful role now, its presence in human beings shows ancestral forms. Nichtating membrane (third eyelid) is a rarely, small, pinkish, triangular flap in the inner angle of each eye in human beings. It has become non functional and reduced because the cornea of eye is being boned & ported by the upper eyelid in humans. However, it is well developed & functional in birds for protection of eyes. Hence, no vestigial organs in birds. Wisdom tooth (third molar), auditing muscles of ear, tail bone, excessive hairs on body are the examples of vestigial organs in human beings.

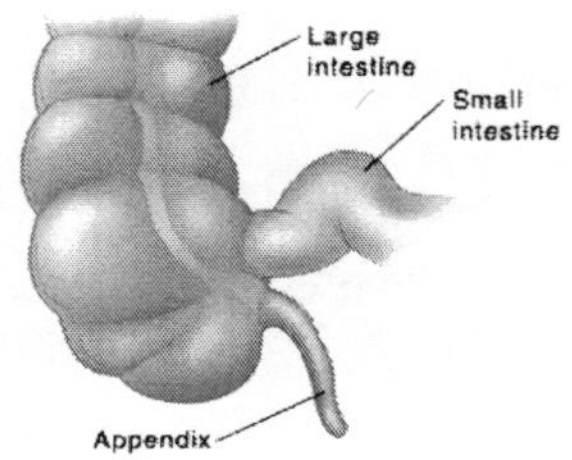

(A) Vermiform appendix

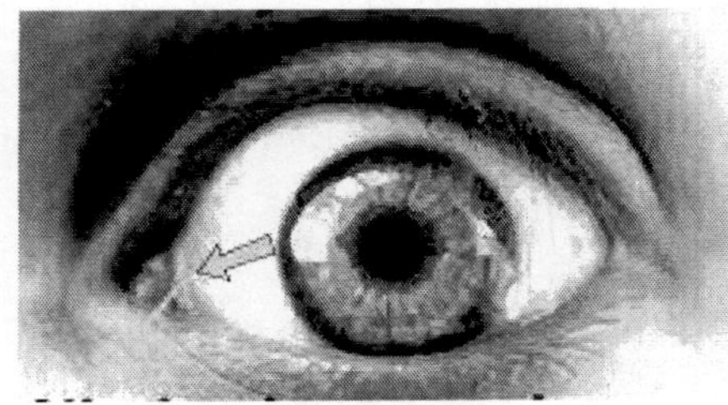

(B) Nictitating membrane

Fig. 4.24 Vestigial organs

4.9.1.4 Fossils:

The remains or impressions of dead plants or animals that lived in the remote past are known as fossils. Palaeontology is the study of past life based on the fossil record. Normally, when the organisms die, their bodies are decomposed and lost. However, if the body or some of its parts remain in an environment/medium that does not let it decompose completely, the body or its parts will eventually be preserved or alternately leave the impressions. For instance, if a dead insect gets caught in mud, it will not decompose quickly; the mud will eventually harden and retain the impressions of the body parts of the insect. Thus, the fossils are the remains or impressions of the dead animals and plants that lived in the remote past, and provide evidence for evolution. Fossils occur in sedimentary rocks. They may also be found in peat, amber, asphalt, lava and snow

The fossil bird, Archaeopteryx lived in the Jurassic period about 180 million years ago. It had feathers which are exclusively bird structures. It also had fused bones, beak and some other bird characters. But, it also had number of features which are found in reptiles, e.g., teeth in the jaws, claws on free fingers, a long tail, etc. Thus, Archaeopteryx represents a stage midway between the reptiles and birds, and is often called the lizard bird. This example provides a clue that the birds have evolved from reptiles. Thus, fossils provide the evidence that the present animals and plants have originated from the previously existing ones through the process of continuous evolution.

4.9.1.4.1 Formation of Fossils:

Usually, when organisms (plants or animals) die, their bodies will decompose by the action of micro-organisms in the presence of oxygen, moisture, etc. Sometimes, however, the conditions in the environment are such (like absence of oxygen or moisture etc.), which do not let the body of the organism to decompose completely. It is such body or part of body of an organisms which we get as fossil on digging the earth (Fig. 4.25). In many cases, the soft parts of the organisms get decomposed and what we get as a fossil is a skeleton of hard parts (like bones etc.). Even the soft parts of the plants and animals (which usually decompose quickly) are sometimes preserves as fossils in the form of their impressions inside the rocks.

Fig. 4.25 *Archaeopteryx*

Some 100 million year ago, for instance, some invertebrates died on the sea bed and got buried in the soil/sand. More soil/sand was carried by rain into the streams and rivers and ultimately into the oceans, where it settled at the bottom layer after layer. In the course of time, the layers so deposited changed into rocks (sand stone) due to the presence of the water above and also due to chemical reactions. Various kinds of fossils have been unearthed from sedimentary rocks. These include many fossil invertebrates (e.g., Ammonites, Trilobites), dinosaur's skull of Rajasaurus (from Narmada valley) and fossil fish (Knightia), etc. the fossils provide direct evidence of evolution.

Ammonites were the invertebrate animals (molluscs) with a flat, coiled, spiral shell which lived in the sea (Fig. 4.26A). The estimation of the age of ammonite fossils has told us that they are about 180 million years old. This means that ammonites lived in the sea about 180 million years ago. Another invertebrate animal fossil which has been studied is that of trilobite (Fig. 4.26B). Trilobites were marine arthropods which were common between 400 to 600 million years ago, fossil fish (Knightia), Knightia is an extinct genus of clupeid clupeiform bony fish that lived in the fresh water lakes (Fig 4.26C). Dinosaurs (Fig. 4.26D) are extinct carnivorous or herbivorous reptiles (The word 'dinosaur' means 'terrible lizard'). The estimation of the age of dinosaur fossils have told us that they first appeared on earth about 250 million years ago and became extinct about 65 million years ago. It is clear from the above discussion that we can even study about those species which are extinct (no longer exist), by studying their fossils which are found during the digging of earth.

(A) Fossil of Ammonite

(B) Fossil of Trilobite

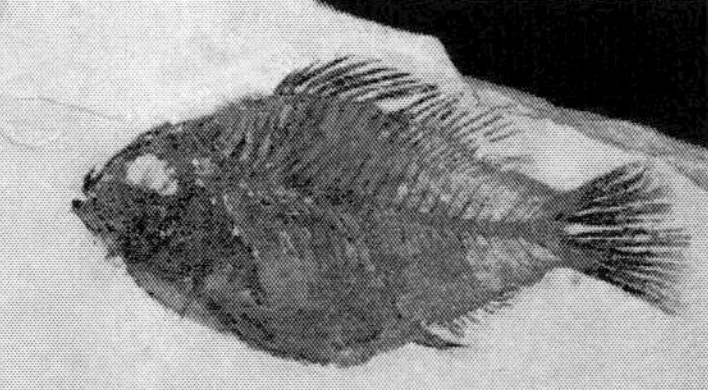

(C) Fossil of Fish-Knightia

(D) Fossil of Dinosaur

(E) Fern Fossil

Fig. 4.26 Various kinds of Fossils

For example, if a dead leaf gets jammed in mud, it will not decompose quickly. The mud around the leaf will set around it as a mould, gradually harden to form a rock and retain the impression of the whole leaf. This forms a leaf fossil which can be dug out from the earth a long time later (Fig. 4.26 E). The fossil of a dead insect caught in mud is also formed in a similar way to leaf fossil. All such preserved impressions of the body parts of the once living organisms are also called fossils.

4.9.1.4.2 Determination of Age of Fossils:

Fossils are obtained by digging into the earth. The age of fossils can be estimated by two methods such as (a) Relative method, and (b) Carbon dating method.

(a) **Relative Method:** When we dig into the earth, we find fossils at different depths. The fossils which we find in layers closer to the surface of the earth are more recent; whereas the fossils found in the deepest layers of earth are the oldest ones. Fossils which we find today were once living objects. All the living objects contain some carbon-14 atoms which are radioactive.

(b) **Carbon dating Method:** Carbon (^{14}C) dating is also done for calculating the age of recent fossils. Similarly, dating the rocks or fossils is done by working out the ratio of uranium to lead in a rock or fossil. It is so because radioactive uranium decays spontaneously to lead at a constant and known rate. This property is used to determine the age of rocks or fossils. Radioactive Potassium and Argon dating is also used where the age of fossils is several millions of years.

The remains or impressions of dead plants or animals that lived in the remote past are known as fossils. Palaeontology is the study of past life based on the fossil record. Archaeopteryx represents a stage midway between the reptiles and birds, and is often called the lizard bird. This example provides a clue that the birds have evolved from reptiles.

Fossils are obtained by digging into the earth. The age of fossils can be estimated in two methods such as (a) Relative method, and (b) Carbon dating method.

4.10 DARWIN'S THEORY OF EVOLUTION

Charles Robert Darwin gave the theory of evolution in his famous book 'The Origin of Species'. The theory of evolution proposed by Darwin is known as 'The Theory of Natural Selection'. This theory is called the theory of natural selection because it suggests that the best adapted organisms are selected by nature to pass on their characteristics or traits to the next generation. Darwin's theory of evolution applies to plants as well as animals. Darwin's Theory of Evolution can be described as follows:

- Within any population, there is natural variation. Some individuals have more favourable variations than others.
- Even though all species produce a large number of offsprings, populations remain fairly constant naturally.
- This is due to the struggle between members of the same species and different species for food, space and mate.
- The struggle for survival within populations eliminates the unfit individuals. The fit individuals possessing favourable variations survive and reproduce. This is called natural selection (or survival of the fittest).
- The individuals having favourable variations pass on these variations to their progeny from generation to generation.
- These variations when accumulated over a long period of time, lead to the origin of a new species.

We will now understand *Darwin's theory of evolution by 'natural selection'* by taking an example. No two animals are ever exactly alike. So some changes always appear when animals produce their progeny by sexual reproduction. For example, one of the progeny may be tall (having long legs) than the other progeny. Thus, there may be a variation of height in the progeny (Fig. 4.27A). Now, the advantage of long legs to the progeny is that when no food (grass, etc.) is available on the ground, then this progeny having long legs can reach the leaves on tall trees, eat them as food and survive (Fig. 4.27B). On the other hand, the progeny which have short height (due to short legs) cannot reach the leaves on tall trees, they will not get any food, and they will starve and hence die. Thus, nature has selected the animal with long legs to survive (because it is the fittest animal under these circumstances). Now, since long legs help in survival, the long-legged animals will live long enough to produce their offsprings. The offspring will inherit long legs. So, all the future generations will have long-legged animals (Fig. 4.27C). In this way, the animals having short legs have evolved into animals having long legs due to variation. This is an example of evolution.

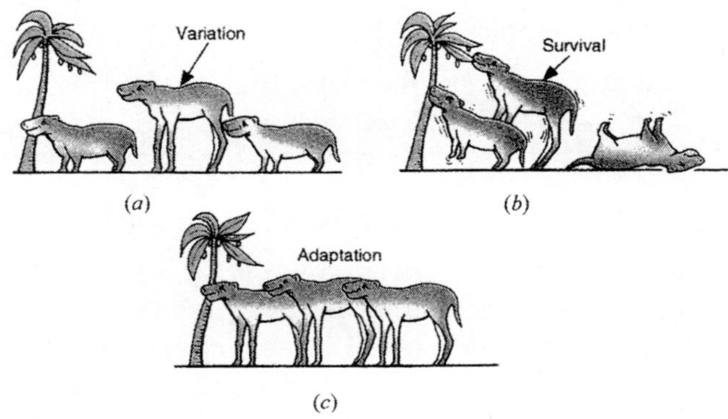

Fig. 4.27 Diagram shows Darwin's Theory of Evolution by Natural Selection

We can now define natural selection as follows: Natural selection is the process of evolution of a species whereby characteristics which help individual organisms to survive and reproduce are passed on to their offspring, and those characteristics which do not help are not passed on. Though Darwin's theory was widely accepted, but it was criticized on the ground that it could not explain 'how the variations (which lead to origin of new species) arise'. With the progress in genetics, the source of variations was explained to be the 'genes'. Genes vary in natural population. Genetic variation is the raw material of evolution. So, the Darwin's theory was modified accordingly. These days, the most accepted theory of evolution is the Synthetic Theory of Evolution in which the origin of species is based on the interaction of 'genetic variation' and 'natural selection'.

Sometimes, a species (a type of animal or plant) may completely die out. It may become extinct. Dodo was a large flightless bird which has become extinct (Fig. 4.28). Once a species is extinct, its genes are lost forever. It cannot re-emerge at all. The small numbers of surviving tigers are a cause of worry from the point of view of genetics because if they all die out and become extinct, their genes will be lost forever (Fig. 4.29). Our coming generations will not be able to see tigers at all. We should, therefore, make all out efforts to protect tigers (and other endangered species) to prevent them from extinction.

Fig 4.28A Dodo (flight less bird) found in Mauritius is no longer alive, become extinct

Fig 4.28B Very few Tigers are alive today. They are threatened with extinction in the near future.

Charles *Robert Darwin* gave the theory of evolution in his famous book 'The Origin of Species'. The theory of evolution proposed by Darwin is known as 'The Theory of Natural Selection'. Darwin's theory of evolution applies to plants as well as animals.

4.11 SPECIATION

A species is a population of organisms consisting of similar individuals which can breed together and produce fertile offspring. Species can be of plants or of animals. Wheat, paddy, sunflower, lotus, mango, humans, tiger, dog and cat, etc. are all examples of various types of species. The human beings who look so different from each other in terms of size, colour and looks are said to belong to the same species (*Homo sapiens*) because they can interbreed to produce fertile offsprings (sons and daughters). The process by which new species develop from the existing species is known as speciation. In simple, words, the formation of new species is called speciation. In most of the cases, new species are formed when the population of same species splits into two separate groups which then get isolated from each other geographically by the barriers such as mountain ranges, rivers or the sea.

The geographical isolation of the two groups of population leads to their reproductive isolation due to which no genes are exchanged between them. However, breeding continues within the isolated populations producing more and more generations. Over the generations, the processes of genetic drift (random change in gene frequency), and natural selection operate in different ways in the two isolated groups of population and make them more and more different from each other. After thousands of years, the individuals of these isolated groups of population become so different that they will be incapable of reproducing with each other even if they happen to meet again. We say that two new species have been formed.

4.11.1 Mechanism of Speciation

To explain the mechanism of speciation, let us consider a situation that the bushes on which the beetles live are widely spread over a mountain range and over a period of time beetle population becomes very large. In the large beetle population, individual beetles normally do not travel far. Instead, they continue to feed on few nearby bushes only throughout their life time. Thus, the large population of beetles, in fact, is divided into sub-populations called demes in neighbourhood. Most of the male and female beetles reproduce sexually only within the respective demes. Rarely an adventurous beetle might go from one sub-population to another or alternatively it is picked up by predator crow from one site and dropped in the other site without being eaten. In either case, the migrant beetle gets an opportunity to reproduce with local inhabitant of this sub-population. As a result, genes of migrant beetle enter into the gene pool of this sub-population. Such kind of gene flow happens between sub-populations that are only partly separated. However, if some large river somehow comes into existence in between these two sub-populations, then these two sub-populations will further be isolated resulting in minimum level of gene flow between them.

After few generations, genetic drift will accumulate different variations in each of the two geographically separated sub-populations. Natural selection may also operate simultaneously in a different way in these geographically isolated sub-populations. For instance, in one geographical region, crows are systematically killed by eagles whereas in the other geographical region no such change occurs and number of crows is very high. Consequently, green body colour variation will not be selected at the first geographical region while it will be strongly selected at the other geographical site. It this way, both genetic drift and natural selection make these geographically isolated sub-populations to become more and more different from each other. Ultimately, reproductive isolation occurs between individuals of these two groups. As a result, they are unable to reproduce even if they happen to meet somehow. These two groups then transform into new species. Reproductive isolation refers to the mechanisms which check the population of two different groups from interbreeding. Reproductive isolation may happen in a number of ways:

- **Temporal Isolation:** It is the failure of two groups of individuals to interbreed because they breed at different times (seasons) of the year.
- **Behavioural Isolation:** It prevents interbreeding of two groups due to differences in courtship and mating behavior.
- **Physiological Isolation:** It is the functional incompatibility in mating, or in production and survival of gametes.
- **Mechanical Isolation:** It is the difference in size or structures of genitalia (i.e. mismatched genitalia) between two group that make interbreeding difficult or impossible.

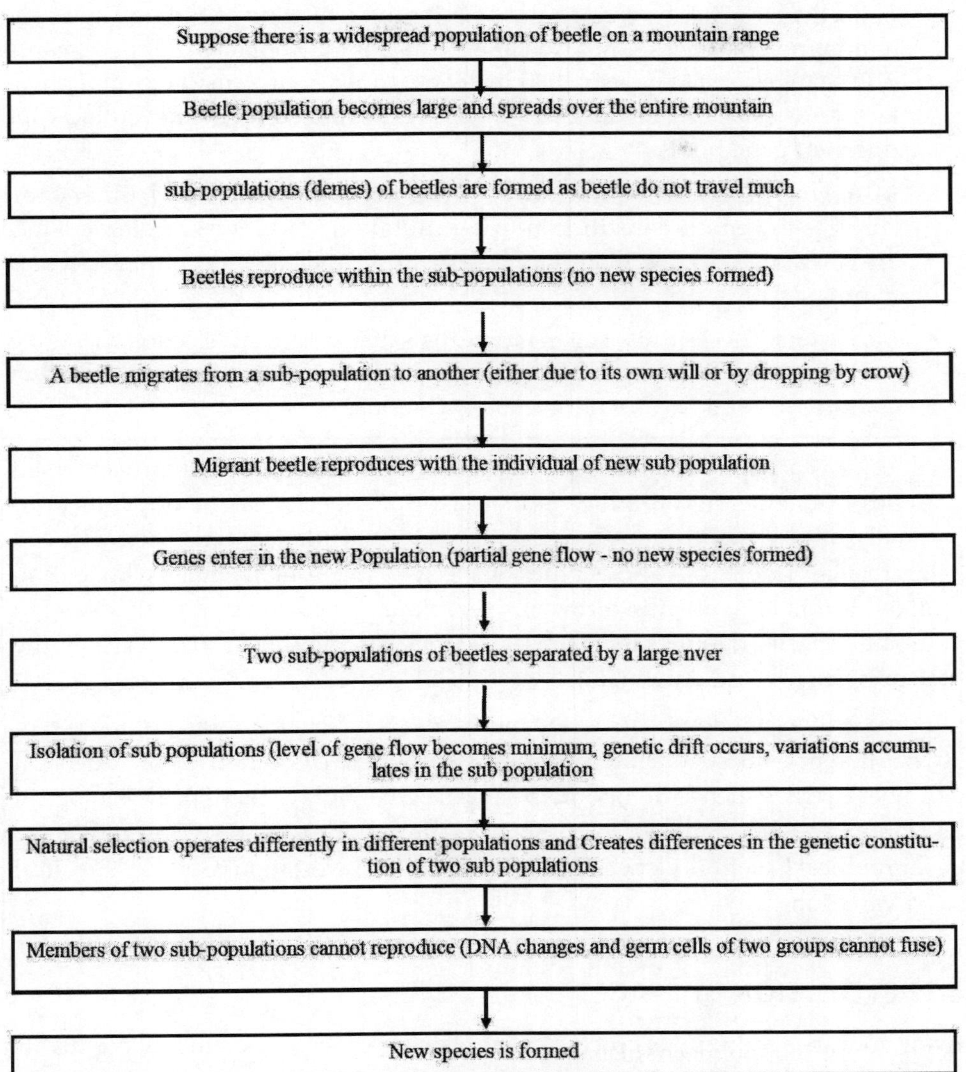

Fig. 4.29 Mechanism of Speciation

4.11.2 Causes of Speciation

- **Geographical Isolation:** It is the most common reason of reproductive isolation and speciation. It prevents inbreeding among the individuals of a population caused by various types of barriers (such as mountain ranges, rivers and sea). The geographical isolation leads to reproductive isolation due to which there is no flow of genes between separated groups of population.
- **Genetic Drift:** It is also known as Wright effect. It always happens in small randomly isolated populations. It causes drastic changes in the frequencies

of particular genes by chance alone i.e. it causes fixation of certain alleles and elimination of others. The fixed alleles may or may not have adaptive value. The changes due to genetic drift in gene frequency of a small population are known as sampling error. A new species is produced due to new alleles and different gene pools.

- **Mutation (Genetic Isolation):** Mutations are sudden and inheritable changes in genetic constitution. After mutation, variations originate which are considered as raw materials for evolution and/or speciation to occur in nature.
- **Variations:** These are caused in individuals due to natural selection. Generally variations are the result of mutation. These are of two types such as (a) Continuous and (b) Discontinuous variations.

A species is a population of organisms consisting of similar individuals which can breed together and produce fertile offspring. Species can be of plants or of animals. Wheat, paddy, sunflower, lotus, mango, humans, tiger, dog and cat, etc. are all examples of various types of species. The human beings who look so different from each other in terms of size, colour and looks are said to belong to the same species (Homo sapiens) because they can interbreed to produce fertile offsprings (sons and daughters).

Reproductive isolation refers to the mechanisms which check the population of two different groups from interbreeding. Reproductive isolation may happen in a number of ways such as Temporal Isolation, Behavioural Isolation, Physiological Isolation and Mechanical Isolation. Most important causes of Speciation are geographical Isolation, Genetic Drift (Wright effect), Mutation (Genetic Isolation) and Variations

4.12 EVOLUTION BY STAGES

Evolution is the continuous process which changes the characteristics of population with the passage of time. This is clear from that we need new antibiotics because widespread use of antibiotics such as penicillin has led to the development of resistant strains of bacteria. This justifies that evolution is still in action. The evolution of new organs, due to adaptation, occurs in stages through natural selection. During the course of evolution, intermediate forms must have evolved before the evolution of full ledged forms. This evidence shows that those intermediate stages were actually useful for the organism possessing them even if not in an obvious way.

4.12.1 Evolution of Eyes:

The eye is a very important organ for animals. The eye is a complicated organ which cannot be generated by a single DNA change. The complex body organs of animals such as eyes have been created in 'stages' over many generations. First of all, the rudimentary eye (basic eye) like that of a flatworm (*Planaria*) was formed (Fig. 4.30A). The eyes of flatworm are very simple that are actually just 'eye-spots' which can detect light. Even these rudimentary eyes provide a survival advantage to flatworm. Starting from this basic design, more and more complex eyes were then evolved in various organisms. Most of the animals have eyes. For example, the insects, octopus and invertebrates, all have eyes. The structure of eyes in each of these organisms is, however, different which suggests their separate evolutionary origins. The evolution of eye is an example of evolution by stages. In insects and crustacean (both are arthopods), both simple and compound eye evolved. Perfect eyes developed in Mollusca, for example, octopus, but their eyes were different from vertebrate eyes in origin.

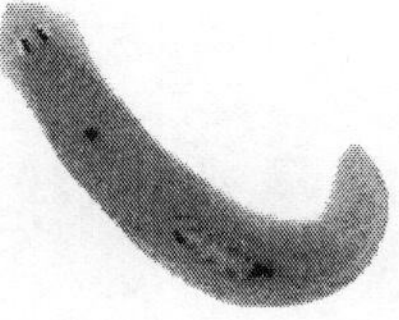

(A) Rudimentary Eye in Flatworm (*Planaria*)

(B) Compound Eye in Arthropods

Fig. 4.30 Examples of Evolution of Eyes

4.12.2 Evolution of Feathers:

Sometimes an evolutionary change produced in an organism for one purpose later on becomes more useful for an entirely different function. For example, birds evolved feathers as a means of providing insulation to their bodies in cold weather but later on these feathers became more useful for the purpose of flying. Even some dinosaurs had feathers though they could not fly by using these feathers. Birds, however, adapted feathers for flying. The presence of feathers on birds (Fig. 4.31A, B) tells us that the birds are very closely related to reptiles because dinosaurs (which had feathers) were reptiles.

(A) Use of feathers for Insulation of body in cold atmosphere

(B) Use of feathers for purpose of flying

Fig. 4.31 Evolution of Feathers

4.12.3 Evolution by Artificial Selection:

In the evidence for evolution we have studied that very dissimilar looking structures can evolve from a common ancestral body design. But that was all guesswork about what happened in history long time ago. We will now give an example from the present time which will show that different looking organisms can in fact be created from the same basic design of the ancestor. The wild cabbage plant is a good example to prove that entirely different looking organisms can evolve from the same organism by the process of evolution. The only difference is that here we are using artificial selection for evolution in place of natural selection. The farmers have been cultivating wild cabbage as a food plant for over two thousand years and have produced (or evolved) entirely different looking vegetables like cabbage, broccoli, cauliflower, kohlrabi and kale from it by artificial selection (Fig. 4.32A).

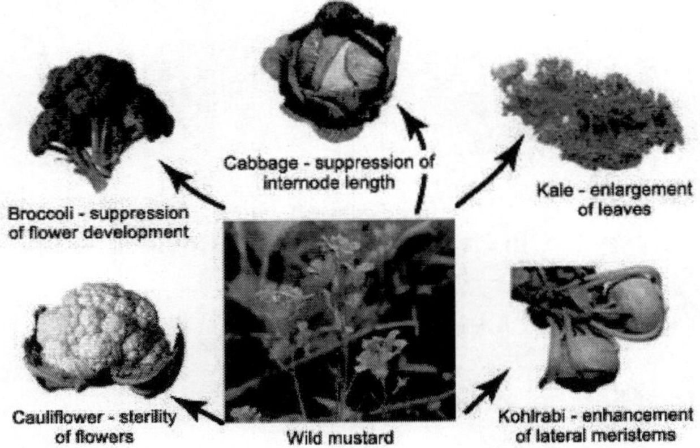

Fig. 4.32A Production of various vegetables form via mustard by the farmers by Artificial Selection Method

(a) Some farmers wanted to have very short distances between the leaves of wild cabbage and produced the common variety of 'cabbage'.

(b) When farmers opted for the arrested flower development of wild cabbage plant, it led to the production of another variety of cabbage called 'broccoli'.

(c) Some farmers went in for sterile flowers of wild cabbage and developed another variety of cabbage called 'cauliflower'.

(d) When farmers opted for the swollen parts of wild cabbage, it led to the evolution of a yet another variety of cabbage called 'kohlrabi'.

(e) Finally, the farmers wanted to grow large leaves of wild cabbage and ended up producing a leafy vegetable called 'kale' which is also a variety of wild cabbage.

Now, wild cabbage is the ancestor and cabbage, broccoli, cauliflower, kohlrabi and kale are all its varieties which have been obtained by evolution 'induced artificially' by the farmers. The ordinary cabbage, broccoli, cauliflower, kohlrabi and kale look so different from their ancestor wild cabbage that if people had not seen it being done with their own eyes, they would never have believed that vegetables having such different structures can be evolved from the same ancestral vegetable plant.

Table 4.4 Differences between Natural selection and Artificial Selection

Sr No.	Parameter	Natural Selection	Artificial Selection
1	Type	This is an natural process	This is an artificial process
2	Control	It is conducted by nature on a vast	It is controlled by human beings in a laboratory at a limited scale
3	Duration	Results are found after a very long period of time. It is a very slow process	Results are found after a very short period of time. It is a rapid process
4	Characters	Characters are useful for concerned species	Characters selected during this process are useful for humans
5	Organism	It operates on all organisms	It operates on some selected organisms
6	Adaptive value	Selective characters have adaptive value	It has no adaptive value
7	Care	Selective character does not require human care	Character targeted by artificial selection requires human protection

4.13 PROGRESS VS. EVOLUTION

Evolution should not be equated with 'progress'. Evolution, in fact, results in the generation of diversity and the shaping of the diversity by environmental selection. The only progressive trend that appears in evolution is the development of more and more complex designs over a period of time. However, many of the older and simpler designs still survive. For instance, bacteria (one of the simplest life forms) still inhabit the most inhospitable habitats such as hot springs, deep sea thermal vents and the ice of Antarctica. Thus, human beings are not the pinnacle of evolution, but simply yet another species in the spectrum of evolving life. In tracing the family tree of species, we must remember that:

There exists the possibility of multiple branching at each and every stage of this process.

Species does not necessarily get eliminated to give rise to a new one (e.g., example of beetles). This, of course, will depend on the prevailing environment.

The newly formed species is not, in any way, better than the older one. It is, in fact, the natural selection and genetic drift which have together led to the formation of a population that cannot reproduce with the original one.

Let us take the example of evolution of human beings and chimpanzees. Human beings are closely related to chimpanzees but have not evolved from them.

Both these species had a common ancestor a long time ago. This common ancestor diverged into several forms and each form probably evolved in its own separate way to give rise to the present forms of human beings, chimpanzees and other closely related great apes (Gorilla, Orangutan etc.). Bacteria are considered to be the most primitive life form present on the earth. Still, they are capable of inhabiting most unfavourable habitats on the planet like hot springs, deep sea, thermal vents and ice in Antarctica etc., where no other complex form can survive. Evolution does not proceed continuously in a particular form or direction. Therefore, there is no species at the topmost level. Rather, it is like a tree with several branches with some branches terminating in present day living species. In other words, evolution does not proceed as 'ladder of progress' but occurs as branches of trees (Fig. 4.32 B).

(A) A ladder of progress (B) Tree branching type

Fig. 4.32B Diagram shows the Evolution Concept

4.14 HUMAN EVOLUTION

Human beings and great apes (e.g., gorilla, chimpanzee etc.) are closely related species. These had a common ancestor a long time ago. This common ancestor diverged into several forms and each form probably evolved in its own separate way to give rise to the present forms of human beings and great apes.

Even in human beings, there occurs a great diversity in form and features across the globe. For a long time, we used to consider these human forms as 'human races'. Our this belief was based on skin colour and other morphological features and accordingly we designated them as yellow, black, white or brown skin coloured forms. For a considerable period of time, biologists debated on the possibility of evolution of these apparent 'human forms' in different ways. However, on the basis of time dating, study of fossils and molecular phylogeny involving DNA sequences, it is now clear that all humans belong to a single species, Homo sapiens. Two other subspecies namely, Homo sapiens neanderthalensis and Homo sapiens fossils also evolved but disappeared shortly due to geological and climatic changes. The earliest members of Homo sapiens have been traced back to Africa. Some of our ancestors stayed back in Africa and spread across this continent; others left Africa and slowly spread across the planet over a period of time (Fig. 4.33). They moved forward and backward in groups, revealing separation or mixing of groups at times. These groups of ancestors, like other species, lived their lives in the then prevailing environment and developed genetic variations to become different coloured with specific features in different geographical regions in modern times. Mendel's experiments tell us the mode of inheritance of traits from one generation to the next and Darwin's theory of evolution tells us how organisms develop from simple to more complex forms. But neither tells us anything about how life originated on earth (or began on earth).

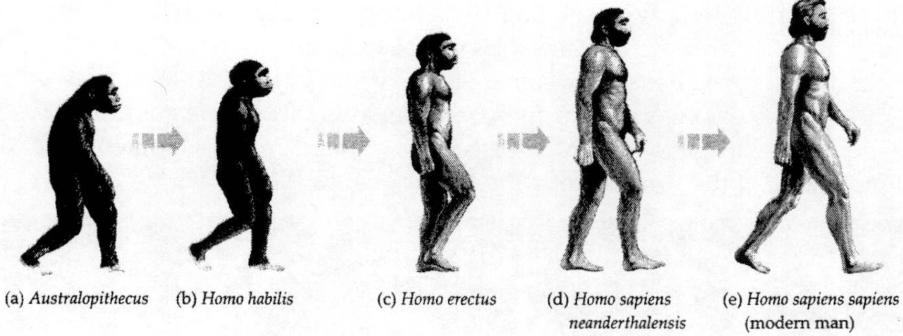

(a) *Australopithecus* (b) *Homo habilis* (c) *Homo erectus* (d) *Homo sapiens neanderthalensis* (e) *Homo sapiens sapiens* (modern man)

Fig. 4.33 Human Evolution

> Evolution is the continuous process which changes the characteristics of population with the passage of time. This is clear from that we need new antibiotics because widespread use of antibiotics such as penicillin has led to the development of resistant strains of bacteria.
>
> The eye is a very important organ for animals. First of all, the rudimentary eye (basic eye) like that of a flatworm (Planaria) was formed.
>
> Human beings and great apes (e.g., gorilla, chimpanzee etc.) are closely related species. These had a common ancestor a long time ago. it is now clear that all humans belong to a single species, Homo sapiens. Two other subspecies namely, *Homo sapiens neanderthalensis* and *Homo sapiens* fossils are also found.

4.15 ORIGIN OF LIFE

According to the dictionary meaning, life is the property that distinguishes living beings from the non-living objects. Biologists, however, find it hard to define life, though they have vast knowledge of living organisms such as bacteria, protists, algae, fungi, plants and animals. All living organisms have some fundamental characteristics that make them different from non-living matter such as (a) Presence of protoplasm, (b) Cellular organization (c) Nutrition (d) Metabolism (e) Growth (f) Reproduction, and (g) Responses to environment.

It is in this context that biologist now visualize. Life as the inherent capacity of living organisms to utilize the outside materials (light, water, gases or food) for energy, growth and reproduction through chemical reactions (metabolism) in a controlled manner. After defining 'life', let us attempt to know about following two events such as **(a) How, when and where the life originated, and (b) How the original simple forms gave rise to the present day wide variety of complex organisms.**

4.4.1 Formation of Solar System

Origin of life is inseparably linked with the formation of our solar system, particularly with the origin of earth. According to Big Bang Theory, some 15 billion years ago, the universe originated by a big bang (thermonuclear explosion) of a dense entity. This big explosion threw out numerous pieces of cosmic material (nebulae) which slowly condensed and became organized slowly into groups of stars called galaxies (Fig. 4.34). These galaxies appear to be moving away from one another at a speed nearly equal to the speed of light. Our galaxy is called Akash Ganga or The Milky Way. Sun is one of the millions of stars of our galaxy.

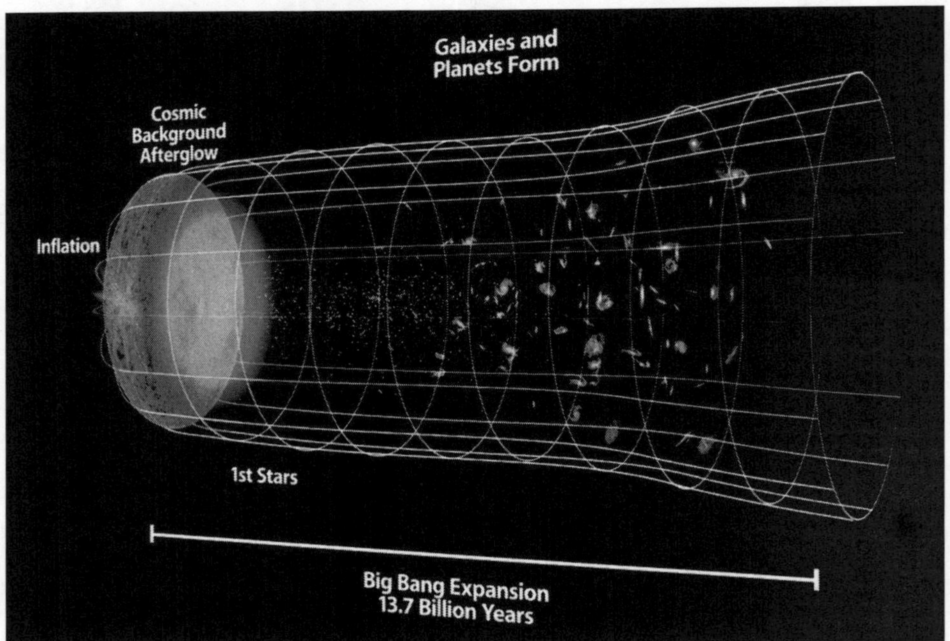

Fig. 4.34 Diagram show the Big Bang Theory

Sun, alongwith its orbiting planets and their respective natural satellites (moons), the comets and asteroids, constitutes our solar system. There are nine planets in our solar system, namely, mercury, venus, earth, mars, Jupiter, Saturn, Uranus, Neptune and Pluto. All these planets are revolving around the sun in elliptical orbits.

Evidences suggest that our solar system was formed about 4.5 billion (4500 million) years ago from a huge spinning cloud of cosmic dust and gas called solar nebula. As this cloud condensed, the rotational movement threw a ring of cloud which kept on rotating around the central mass. The whole matter condensed under the force of its own gravity to become a flattened spinning disc of atom and dust particles. The central mass condensed further and formed the sun (Fig. 4.35). As the sun condensed, it became hotter and brighter due to conversion of gravitational energy into heat. Atoms and dust particles orbiting around the central mass, i.e., sun then aggregated into clumps (about the size of asteroids). These clumps finally swept up more particles and grew into different sized planets. The latter kept revolving round the sun in different elliptical orbits. Earth originally had solid mass called lithosphere surrounded by a gaseous envelope termed atmosphere. The liquid component, known as hydrosphere, appeared later when the earth cooled down to below 100°C.

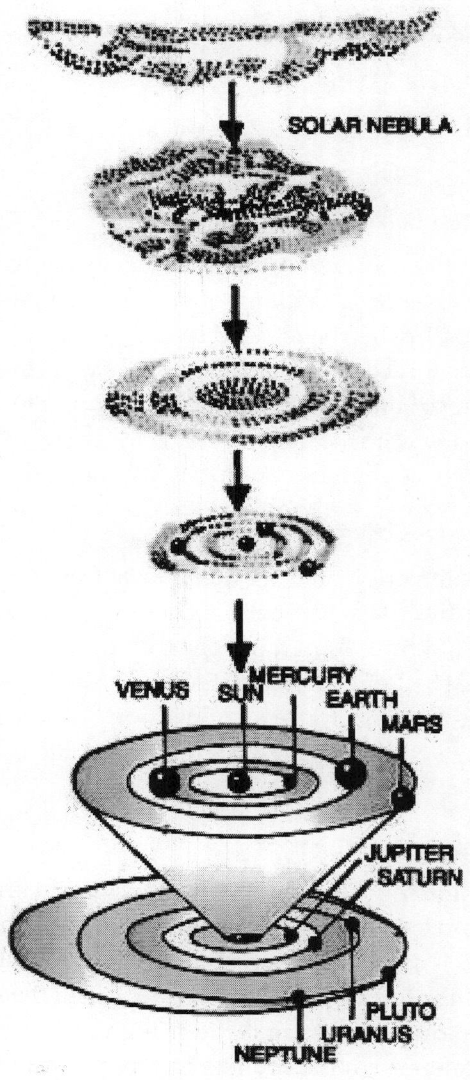

Fig. 4.35 Schematic representation of condensation of Solar Nebula

All living organisms have some fundamental characteristics that make them different from non-living matter such as (a) Presence of protoplasm, (b) Cellular organization (c) Nutrition (d) Metabolism (e) Growth (f) Reproduction, and (g) Responses to environment.

Life as the inherent capacity of living organisms to utilize the outside materials (light, water, gases or food) for energy, growth and reproduction through chemical reactions (metabolism) in a controlled manner. After defining 'life', let us attempt to know about following two events such as (a) How, when and where the life originated, and (b) How the original simple forms gave rise to the present day wide variety of complex organisms.

4.4.2. Theories on the Origin of Life

Many theories have been advanced to explain the origin of life. These theories are briefly described as:

4.4.2.1 Theory of Special Creation

According to this theory, life was created by the wishes of a divine being or some supernatural power, the Creator or God out of nothing at a particular time in the remote past. According to Christian belief, God created the universe, plants, animals and human beings in six natural days. Adam and Eve were the first man and woman created by God. According to Hindu mythology, Lord Brahma created the world with his wish. Manu and Shradha were the first man and woman respectively. The theory of special creation is purely a religious concept. It is not based on scientific truths.

4.4.2.2 Theory of Spontaneous Generation:

According to this theory, life originated spontaneously from non-living materials-abiogenesis. For example, fly larvae were believed to develop from rotten meat; frogs, snakes and crocodiles could form from the moist soil (mud) of the river. However, this idea was placed to rest by Louis Pasteur (1862) who experimentally proved that micro-organisms (protists and bacteria) arise from the pre-existing micro-organisms of their kind. Since then, it is being widely accepted that life originates from pre-existing life-biogenesis.

4.4.2.3 Cosmozoic or Interplanetary Theory:

According to this theory, life could have originated once or several times in various parts of the galaxy in universe and it reached the barren earth from some other heavenly body in the form of resistant spores or seeds. The latter grew and then evolved into various existing forms. This theory is also called theory of panspermia. Because of hazards like low temperatures, very high influx of cosmic and UV radiation, lack of atmosphere, extreme dryness etc. that exist in the interplanetary space and are not conducive to life, this theory of origin of life was rejected.

4.4.2.4 Abiogenic or Naturalistic Theory or Chemical Origin of Life:

According to this theory, life originated from inanimate matter by a series of chemical reactions on the primitive earth. This concept was put forward by Alexander. I. Oparin (1923) and J.B.S. Haldane (1929) independently. It was summarized by Oparin in his book 'origin of life' as 'abiogenesis first, but biogenesis ever since'. This theory is now known as primary abiogenesis. It is now well accepted theory.

A British Scientist **J.B.S. Haldane suggested in 1929** that life must have developed from the formed gases and inorganic compounds, namely, hydrogen (H_2), nitrogen (N_2), ammonia (NH_3), methane (CH_4), carbon dioxide (CO_2) and water vapour (H_2O). The energy for the formation of molecules might have come from reflected solar radiations, lightning or volcanic eruptions. The primitive inorganic molecules later interacted with one another to form simple organic compounds, namely, sugars, fatty acids, glycerol, amino acids and organic bases (purines and pyrimidines). Energy for such interactions was probably provided by solar radiations and electrical discharges. The reactions that led to the formation of organic compounds partly occurred in the primitive atmosphere but mainly in the oceans. Absence of live and reducing atmosphere preserved these compounds long enough.

Major support to this theory was provided by Stanley L. Miller and Harold C. Urey in 1953. They designed a glass apparatus and simulated the conditions that thought to have existed on primitive earth (Fig. 4.36). They circulated a mixture of methane (CH_4), ammonia (NH_3) and hydrogen (H_2) in the ratio of 2:2:1, and water vapour (H_2O) in this apparatus. These gases were believed to prevail in the primitive earth. The mixture was maintained at a temperature just below 100°C. They provided energy for interaction of gases in the form of electric sparks in the gas flask. At the end of a week, 15% of the carbon (from methane) had been converted to simple organic compounds such as amino acids, simple sugars, purines and pyrimidines. These simple organic compounds interacted in oceans to form complex organic polymers such as polysaccharides, polypeptides and proteins, lipids, nucleotides and nucleic acids. These polymers, due to intermolecular interactions, formed large colloidal aggregates called coacervates. These coacervates later organized into the living matter.

Many theories have been advanced to explain the origin of life such as Theory of Special Creation, Theory of Spontaneous Generation, Cosmozoic or Interplanetary Theory and Abiogenic Theory or Chemical Origin of Life. Among all theories, Abiogenic theory (now known as Primary abiogenesis) is well accepted theory.

J.B.S. Haldane suggested in 1929 that life must have developed from the formed gases and inorganic compounds, namely, hydrogen (H_2), nitrogen (N_2), ammonia (NH_3), methane (CH_4), carbon dioxide (CO_2) and water vapour (H_2O). Major support to this theory was provided by Stanley L. Miller and Harold C. Urey in 1953. They designed a glass apparatus and circulated a mixture CH_4, NH_3 and H_2 in the ratio of 2:2:1, and water vapour (H_2O) in this apparatus. At last, he resulted that polymers, due to intermolecular interactions, formed large colloidal aggregates called coacervates. Later, these coacervates are organized into the living matter.

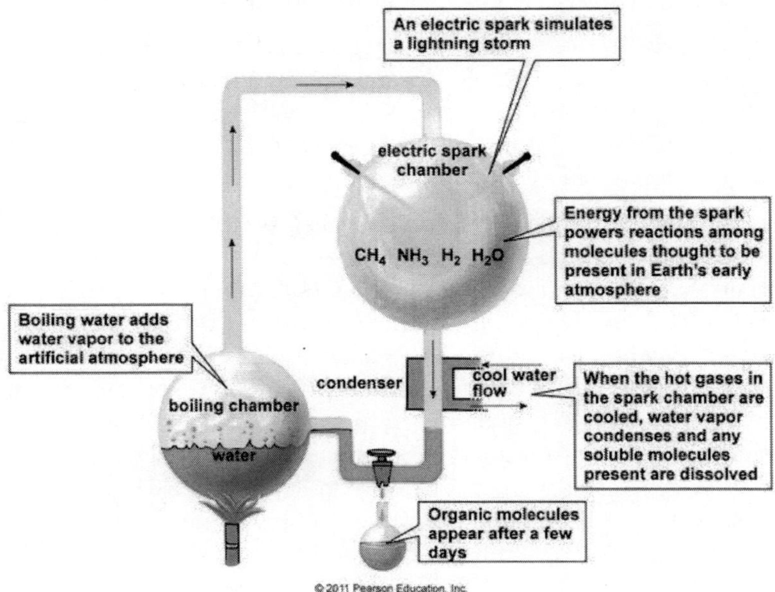

Fig. 4.36 Experiment of Miller and Urey

Short Notes:

Groger Johann Mendel (1822-1884) was born in a peasant family of Moravia. Due to poverty, he became a monk in 1843 and later, in 1847, was made an abbot (head) of Augustinian monastery of St. Thomas at Brunn, Austria (now Bruno in Czech Republic). From here, he went in 1851 to the University of Vienna where he studied natural history and mathematics for two years. He also became interested in the process of hybridization. Mendel returned to the monastery in 1853 and also worked as a teacher in a school. From 1856-1865, Mendel conducted breeding experiments on garden pea plant *(Pisum sativum)* in the garden of his monastery. He published his findings in 1866 in the Annual Proceedings of the Natural History Society of Brunn. His findings remained unknown until these were rediscovered by three biologists namely, **Hugo De Viris (Netherland), Karl Correns (Germany)** and **Erich von Tshermak (Austria)** independently in 1990. The forgotten work of Mendel was republished in 1901. Mendel's findings formed the basis of the science of genetics and he is credited as the Father of Genetics.

Charles Darwin (1809-1882) set on a voyage, at the age of 22, on the ship H.M.S. Beagle. Five-year voyage, from 1831 to 1836, took him to South America and the Galapagoes islands off its coast. During his survey expedition, he made extensive observations of animals and plants. He noted a huge variety of life and also remarkable similarity among organisms and their wonderful adaptation to the environment. After completing his voyage, Darwin returned to England, stayed at home, and analysed the observations of nature during his voyage. He also read population theory given by Malthus. His observations during the voyage and population theory of Malthus led him to formulate Darwin's theory of natural selection. He gave this theory in his book entitled 'The origin of species'. He did not know the mechanism whereby variations arose in the species. He was also not aware of Mendel's work on inheritance. Besides theory of evolution, Darwin also studied the role of earthworms in soil fertility.

DNA (Deoxyribonucleic Acid):

The expanded form of DNA is deoxyribonucleic acid. It was first isolated by the scientist Frederick Meisher from nucleus of the pus cells in 1869. He named it as 'nuclein' or nucleic acid because of its acidic nature. Later, it was experimentally proved by the scientists Griffith (1928), and Avery. McLeod and McCarty (1944), that DNA is the carrier of the genetic information from generation to generation. It transmits the hereditary characters in a coded language from parents to the offspring (i.e., from one generation to another). DNA is a macromolecule or polymer. It is made of very large number of 'nucleotide' units and hence is termed polynucleotide. Each nucleotide unit in a DNA molecule is made up of three components such as:

(a) Deoxyribose Sugar (Pentose sugar).

(b) Nitrogenous bases. Each nucleotide unit has a nitrogen containing base. In a DNA molecule, nitrogenous bases are of two types as (a) Purines bases - Adenine (A) and Guanine (G), and (b) Pyrimidines bases – Cytosine (C) and Thymine (T).

(c) Phosphate Group: The phosphate group contains one phosphorus atom and four specifically linked oxygen atoms.

Thus, there are four types of nucleotides in a DNA molecule depending upon the kind of nitrogenous base present in each nucleotide.

Genes:

Mendel was the first scientist to visualize a gene as a unit of inheritance in 1866. He called it factor. The word 'gene' was, however, introduced by Johannsen in 1909. Before 1950, our knowledge about genes was based on the studies conducted on fruit fly (*Drospohila melanogaster*) and some higher plants. It was then considered as the unit of inheritance. However, nothing was known about its structure. Once the molecular structure of chromosomes and DNA was established, our knowledge about genes also increased. It is now well known that genes are nothing but segments of DNA on a chromosome occupying specific positions. For example, 30,000-40,000 genes are present on 46 human chromosomes. Chemically, each gene has a specific sequence of nucleotides which determines its functional property. It is the genes which determine everything about us, say our appearance, our complexion, the shape and colour of our eyes, the shape of our nose and chin etc. Actually, gene is a segment of DNA molecule which has coded information to form a particular protein in the cell. This protein can function either as a structural protein or as a functional protein (enzyme). On the basis of modern molecular concept, a gene is a hereditary unit, a segment of DNA having specific sequence of nucleotides that determines its specific biological function. It can maintain constancy from generation to generation but at times may also undergo sudden mutations (inheritable changes) to bring variations.

Some important terms used in Genetics

Sr. No.	Terms	Explanation
1.	Genes (Factors)	A unit of DNA on a chromosome which controls the expression of a character in the organism. Mendel Called these units as 'Factors' Later, Johannsen (1909) gave the term genes.
2.	Alleles (Allelomorphs)	Factors which occur on the same loci (site) in the homologous
3.	Homozygous (Pure line or Pure breeding)	When an individual contains alleles for the certain character on homologous chromosomes is called homozygous or pure that particular character. i.e TT and tt alleles are homozygous for height
4.	Heterozygous (hybrid)	When an individual contains two different alleles for the certain character on homologous chromosomes is called heterozygous or hybrid for that particular character. i.e Tt alleles are heterozygous for height
5.	Recessive alleles.	The alleles which does not express itself physically and remains unexpressed due to the presence of dominant allele is called recessive alleles.
6.	Dominant alleles	The alleles which express itself in hybrid is called as dominant alleles
7.	Homologous chromosomes	The pair of exactly similar chromosomes in shape, size and structure which bear same genes at the same loci are called homologous chromosomes
8.	Genotype	Genotype is the genetic constitution of any character of an organism
9.	Phenotype	Phenotype is the external or physical appearance of any character of an individual produced by interaction of genes and environment.
10.	Monohybrid Cross	The cross between two organisms used to study the inheritance of a single pair of alleles or single character is called as monohybrid cross
11.	Dihybrid Cross	The cross between to organisms used to study the inheritance of two pairs of alleles or two characters is called as dihybrid cross
12.	Characters	Distinct and well defined morphological and physiological features of the individual i.e. height
13.	Traits	Distinguishable expression of a character is called trait i.e tallness and dwarfness are the traits of the character height
14.	Test Cross	Cross between an organism of unknown genotype and a homologous recessive organism is called as Test cross
15.	Reciprocal Cross	It is similar as the original cross i.e involves the same trait but is carried by sexes opposite to those in the original cross

16.	First Filial Generation (F_1)	The first generation of offspring or hybrids produced by a cross between genetically different individuals called parents
17.	Second Filial Generation (F_2)	The progeny produced by crossing between the individuals of F_1 generation is called as Second Filial Generation

REMEMBER SOCKETS

- The transmission of characters or traits from the parents to their offsprings (next generation) is called heredity.
- The term genetics was coined by William Bateson in 1906.
- DNA (deoxyribonucleic acid) was first isolated by Frederick Meisher in 1869 who termed it as nuclein or nucleic acid.
- Griffith (1928), and Avery, McLeod and McCarty (1944) experimentally proved that DNA is the genetic material.
- DNA is a macromolecule of very large number of nucleotide units. Each nucleotide unit has three components. (i) deoxyribose (pentose) sugar, (ii) a nitrogenous base, and (iii) a phosphate group.
- The nitrogenous bases are of two types: purines (adenine, guanine) and pyrimidines (cytosine, thymine).
- Gene is a segment of DNA occupying specific position on a chromosome. Chemically, each gene has a specific sequence of nucleotides which determines its specific biological function. Genes are hereditary units which can maintain constancy from generation to generation but at times may undergo mutation to bring variation.
- Since gene is the specific segment of DNA occupying specific position on a chromosome, all the genes present in every cell of an organism control all the characteristics (traits) of the organisms.
- During sexual reproduction, both the parents contribute equally to the DNA of the progeny by contributing one set of genes through germ cells.
- Any differences in traits between individual of species, caused either through genetic or environmental factor is called as variation. These variations could be observed in physical appearance, metabolism, behaviour etc.
- The number of variations produced during sexual reproduction is, however, very large In the case of human beings, skin colour, heights, hair, color of eyes, shape of nose and ears, chin are the characters (traits).
- Variations are the occurrence of differences among individuals of a species
- Variations are of two types such as (a) Germinal and (b) Somatic variations.
- Germinal Variations caused by differences in number or structure of chromosome or gene are called as germinal variations, and these are heritable

i.e. Colour of eyes, height etc. Germinal variations play an important role in organic evolution.

- Somatic Variations caused by environmental factors like climate, food supply and interaction with other organisms are called as somatic variations. These variations are not transferred from one generation to next.
- In the cells, two sets of genes are not present as separate single threads of DNA. Instead, these exist as separate independent pieces, called chromosomes.
- Genes for controlling the same characteristic of an organism can be of two types such as (i) dominant genes or (ii) recessive genes.
- The gene which decides the appearance of an organism even in the presence of an alternative gene is known as a dominant gene. It dominates the recessive gene for the same characteristic on the other chromosome of the pair.
- The gene which can decide the appearance of organisms only in the presence of another identical gene is called a recessive gene. A single recessive gene cannot decide the appearance of an organism.
- Each cell has two copies of each chromosome (homologous chromosomes), one each from male and female parents.
- When the germ cells are formed, every germ cell receives one chromosome from each pair and these may be either of maternal of paternal origin. When these male and female germ cells combine during fertilization, normal number of chromosomes in the progeny is restored ensuring the stability of the DNA of the species.
- The mechanism by which the sex of an individual is determined as it begins life is called sex determination.
- In diploid organisms having separate sexes, a pair of sex chromosomes determines the sex of the individual. All other chromosomes are termed autosomes.
- Human beings and fruit fly (*Drosophila*) have XX-XY type of sex determination mechanism. In these animals, male has XY sex chromosomes (i.e. heteromorphic sex chromosomes having different size and shape). On the other hand, female has XX sex chromosomes, i.e., homomorphic sex chromosomes of same size and shape.
- Each cell in human beings consists of 46 chromosomes (23 Pairs). 22 pairs of chromosomes are found common in both males and females (Autosomes). The 23rd pair of chromosomes in human beings takes part in sex determination (sex chromosomes).
- In some of the animals, sex determination is also controlled by the environmental factors such as temperature and CO_2 concentration of water etc.
- In some reptiles, the temperature at which the fertilized egg is incubated before hatching plays a role in determining the sex of the offspring.

- In lizard (*Agama agama*), high incubation temperature results in male offsprings (or male progeny).
- In a marine vertebrate (Echiuorids), the CO_2 concentration of water is responsible for the sex determination
- Male is heterogametic, i.e., it produce two types of sperms (50% having X chromosome and 50% having Y chromosome). Female is homogametic, i.e., it produces only one type of gametes, each having X chromosome, then the offspring will be a boy (male) having XY combination of sex chromosomes.
- Grasshoppers and some other insects have XX-XO type sex determination mechanism.
- In some animals (e.g., turtle, *Chrysema picta* and lizard (*Agama agama*), sex determination is regulated by environmental factors.
- In plants, genetic basis of sex determination exists but it is different from that of animals.
- The concept of organic evolution was given by Charles Darwin which simply means descent with modification'. Accordingly, the present day complex organisms have originated from the earlier simpler forms of life during the course of ages by the process of gradual change.
- Charles Robert Darwin put forward the theory of natural selection in his famous book 'The Origin of species'. He based the theory on his observation during the voyage, and population theory of Malthus.
- The most accepted theory of evolution now a days is called synthetic theory of evolution. Accordingly, the origin of species is based on the interaction of genetic variation and natural selection.
- Variation arise either because of errors in DNA copying or as a result of sexual reproduction. Individuals with variations (that give them selective advantage) are favoured by nature (natural selection). Alternatively, natural calamity or migration of individuals in a small population changes the frequency of some genes in the population even if these give no survival advantage (genetic drift).
- There are two kinds of traits in every organism: (i) inherited traits, and (ii) acquired traits.
- Inherited traits are controlled by specific genes and are passed on from one generation to another through germ cells.
- Acquired traits involve changes in the non-reproductive tissues during organisms' life time. These are not inherited from one generation to other as they do not bring any change in DNA.
- Origin of new species from the existing one is called speciation.
- Genetic variations, natural selection and reproductive isolation are key factors in the modern concept of evolution.

- Artificial selection is the process by which man selects traits useful to him for improving the qualities of domesticated plants and animals.
- Human beings and chimpanzees are closely related species. Both these had a common ancestor a long time ago. This ancestor diverged into several forms and each form probably evolved in its own separate way to give rise to the present forms of human beings, chimpanzees and other great apes.
- Variation arise either because of errors in DNA copying or as a result of sexual reproduction. Individuals with variations (that give them selective advantage) are favoured by nature (natural selection).
- Origin of new species from the existing one is called speciation.
- Genetic variations, natural selection and reproductive isolation are key factors in the modern concept of evolution.
- There are number of evidences in favour of evolution. These are categorized as (i) morphological and anatomical evidences (e.g. body organization, homologous, homologous organs, analogous organs and vestigial organs), (ii) embryological evidences, and (iii) paleontological evidences i.e., evidences based on fossil records. These help in tracing evolutionary relationship between organisms.
- Geographical Isolation is the most common reason of reproductive isolation and speciation. It leads to reproductive isolation due to which there is no flow of genes between separated groups of population.
- Genetic Drift (Wright effect) always happens in small randomly isolated populations. It causes drastic changes in the frequencies of particular genes by chance alone i.e. it causes fixation of certain alleles and elimination of others.
- Mutations are sudden and inheritable changes in genetic constitution. After mutation, variations originate which are considered as raw materials for evolution and/or speciation to occur in nature.
- Palaeontology is the study of past life based on the fossil record.
- Archaeopteryx represents a stage midway between the reptiles and birds, and is often called the lizard bird. This example provides a clue that the birds have evolved from reptiles.
- The age of fossils can be estimated in two methods such as (a) Relative method, and (b) Carbon dating method.
- The changes due to genetic drift in gene frequency of a small population are known as sampling error. A new species is produced due to new alleles and different gene pools.
- Many theories have been advanced to explain the origin of life such as Theory of Special Creation, Theory of Spontaneous Generation, Cosmozoic or Interplanetary Theory and Abiogenic Theory or Chemical Origin of Life.

- Among all theories, Abiogenic theory is now known as primary abiogenesis. It is now well accepted theory.
- J.B.S. Haldane suggested in 1929 that life must have developed from the formed gases and inorganic compounds, namely, hydrogen (H_2), nitrogen (N_2), ammonia (NH_3), methane (CH_4), carbon dioxide (CO_2) and water vapour (H_2O).
- Major support to J.B.S. Haldane theory was provided by Stanley L. Miller and Harold C. Urey in 1953.
- Stanley L. Miller and Harold C. Urey resulted that polymers, due to intermolecular interactions, formed large colloidal aggregates called coacervates.
- Ernst Haeckel proposed the biogenetic law. This law simply means 'ontogeny recapitulated phylogeny'. It means that an organism repeats or recapitulated its ancestral history during its embryonic development.

VERY SHORT ANSWER TYPE QUESTIONS:

1. Define the term heredity.
2. Define the term genetics.
3. Define the term variations.
4. What is speciation?
5. What is a gene?
6. Who is known as father of genetics?
7. What are acquired traits?
8. What are analogous organs?
9. Define the term genetic drift.
10. What is evolution?
11. Why type of reproduction gives rise to more number of successful variations?
12. Which evidence supports that birds and dinosaurs are closely related?
13. Name the scientist who first isolated DNA from the nucleus of the pus cells.
14. Name two purine nitrogenous bases present in DNA molecule.
15. Name two pyrimidine nitrogenous bases present in DNA molecule.
16. Who wrote the book 'Origin of Species'?
17. Name two human traits that show variations.
18. Which decides the sex of the child, sperm or ova?
19. Why are J.D. Watson and F.H.C Crick famous?
20. Name two common vestigial organs present in human beings.

21. Give example of analogous organs.
22. Give example of homologous organs found in plants.
23. What is the modern name given to Mendel's 'factors' which are said to be the carriers of hereditary information?
24. Name the pentose sugar present in each nucleotide unit of DNA molecule.
25. How many types of nucleotides are present in the DNA molecule?
26. What do you mean by microevolution?
27. Give two examples where sex determination is regulated by environment factors.
28. Name two animals which depict XX-XY type of sex determination mechanism.
29. Which sex chromosomes are heteromorphic?
30. Who gave the biogenetic law?
31. Define the term ontogeny.
32. How would you briefly explain the biogenetic law?
33. Name the fossil bird which had characters of both reptiles and birds.
34. Define the term variation.
35. What is a sex chromosome?
36. What do you understand by evolution?
37. Define the term homologous organs.
38. Name the possible sources of energy in most accepted theory of origin of life.
39. What was sealed in the spark chamber in famous Miller and Urey simulation experiment?
40. What is Archaeopteryx? What is its significance in evolution?
41. How many contrasting traits Mendel noted in garden pea plant?
42. Mention any two contrasting traits noted by Mendel in garden pea plant.
43. Write down Mendel's dihybrid ratio for phenotypes.
44. What do the letters P, F_1 and F_2 represent in heredity.
45. What are inherited traits?
46. What do you mean by reproductive isolation?
47. Give the name the theory of evolution given by Charles Darwin.
48. What is meant by allopatric speciation?
49. State the law of segregation given by Mendel.
50. Give evidence in support of organic evolution.
51. What are vestigial organs?

52. What is meant by evolution?
53. Define the term species.
54. What are the factors which contribute towards speciation?
55. Name the chromosomes responsible for sex determination in humans.
56. What is paleontology?
57. Which combination of sex chromosome produces a male child?
58. Give an example of homologous organs.
59. Give an example of analogous organs.
60. Which is the connecting link between reptiles and birds?
61. On which plant Mendel performed his experiments to study inheritance?
62. How many characters did Mendel study in his experimental plant?
63. What is the scientific name of garden pea?
64. What is the phenotype ratio of dihybrid cross?
65. What was the name of Darwin's book in which he gave the concept of evolution?
66. Is the colour of eyes an acquired trait?
67. What are called as 'hereditary units'?
68. Who gave the theory of natural selection'?
69. Where did the first human evolve?
70. Which is the dominant trait in inflated and constricted pea pods?
71. Give the term used for genetic constitution of an organism.
72. What do we call the filamentous bodies present in the nucleus which are composed of DNA?
73. What is the number of allosomes in humans?
74. What is an offspring?
75. Name the branch of science which deals with the transmission of characters from parents to offspring.
76. Who is known as the father of Genetics?
77. What is artificial selection?
78. What is population?
79. What is a dihybrid inheritance?
80. What is genetic drift?
81. What is sex determination?
82. Name two animals which depict XX-XY type of sex determination mechanism.

83. What are sex chromosomes?
84. How many different nucleotides are present in DNA molecule?
85. Which life process ensures that a plant or animal species will not disappear from this earth?
86. What is the most important requirement for speciation to occur?
87. Who coined the term Genetics and when?
88. Which is the most acceptable theory of origin of life?
89. From which plant broccoli was produced?
90. Which gas was not present in free form in the primitive earth's atmosphere?
91. Which organs of man are homologous to the limbs supporting wings of birds?
92. Name the scientist who experimentally proved that microorganisms originate from the pre-existing microorganisms of their kind.
93. Who gave theory of natural selection?
94. Explain in three words the biogenetic law given by Ernst Haeckel.
95. Give in three words the concept of organic evolution of Charles Darwin.
96. When did the life originate on the primitive earth?
97. Name the scientists who independently put forward the most accepted theory of origin of life.
98. In which rocks are fossils commonly found?
99. Name the scientist who gave the theory of evolution.
100. Name an animal having rudimentary eyes.
101. Where did life originate on the earth?
102. Name the famous book written by Charles Robert Darwin.
103. Name two organisms which are now extinct and studied from their fossils.
104. Write the names of at least three inorganic molecules which helped in the origin of life on the earth.

SHORT ANSWER TYPE QUESTIONS

1. What is meant by chemical basis of heredity?
2. How will you justify selection of pea plant by Mendel for heredity experiments?
3. What were the results of Mendel's monohybrid cross?
4. Write the basic features of mechanism of inheritance.
5. Why variations are more pronounced in sexual reproduction than asexual one?

6. State the most common cause of speciation.
7. What is meant by contrasting characters? Give example.
8. Name five varieties of vegetables which have been produced from 'wild cabbage' by the process of artificial selection.
9. Write down the difference between homologous and analogous organs.
10. What do you understand by genetic drift? In which type of populations, genetic drift helps in evolution?
11. How reproductive isolation is must for speciation?
12. They eyes of planarians, octopus, insects and mammals are different in structure yet they have the same function. How would you explain it?
13. Give the three key factors of the modern concept of evolution.
14. In which year Mendel's findings were rediscovered by three scientists independently?
15. Name the most acceptable theory of origin of life. Who gave this theory?
16. Why DNA is called a polynucleotide?
17. How many different bases are found in a DNA molecule? Which bases pair together?
18. Name one reptile in each case where higher incubation temperature leads to the development of Male progeny and Female progeny
19. Define homologous organs and Give example of homologous organs in animals.
20. What are the units of inheritance in living organisms and where are they located?
21. Why Mendel selected garden pea plant for his experiments?
22. How can you explain the existence of analogous organs.
23. How do natural selection and artificial selection differ?
24. Why are acquired characters not inheritable?
25. Why is artificial selection beneficial?
26. Define the terms homogametic and heterogametic.
27. Define variation in relation to a species. Why is variation beneficial to the species?
28. What are fossils? What do they tell about the process of evolution?
29. What is a test cross? How does it differ from a reciprocal cross?
30. Give one example each of characters that are inherited and the ones that are acquired in humans. Mention the difference between the inherited and the acquired characters.

31. In what way the reducing atmosphere helped in the origin of life?
32. Make a list of theories put forward for the origin of life.
33. Which is the most reliable evidence for evolution and why?
34. How do the homologous and analogous organs differ? Cite one example of each.
35. What is genetic drift? Give the significance in evolution.
36. What is variation? Name the basic processes that cause variation among organisms.
37. Mention different characters of garden pea, observed by Mendel in his hybridization experiments.
38. Write some characteristics on the basis of which forelimbs of frog and that of human beings are considered to be homologous organs.
39. Does genetic combination of mother play a significant role in determining the sex of a new born baby?
40. Give two examples of dominant traits.
41. Give the name of different types of nitrogenous bases found in DNA.
42. Which nucleotide bases pair together in DNA?
43. What is meant by a species? Give two examples of plant species and two of animals.
44. If a normal human cell has 46 chromosomes, what will be the number of chromosomes in (a) sperm cell and (b) zygote?
45. Distinguish between a gamete and zygote.
46. State the law of independent assortment given by Mendel.
47. How does speciation help in survival?
48. What are the differences between acquired traits and inherited traits? Give examples also.
49. Describe the role of analogous organs in organic evolution.
50. Write a short note on environmental determination of sex.
51. Write the role of vestigial organs in the study of evolution.
52. Explain how offspring and parents of organisms reproducing sexually have the same number of chromosomes.
53. Discuss evolution on the basis of analogous and homologous organs.
54. How will you provide evidence for the evolution of birds from reptiles?
55. Define dominant and recessive traits. How do Mendel's experiments show that traits may be dominant or recessive?

56. Describe the laws of inheritance as proposed by Mendel.
57. How do natural selection and artificial selection differ?
58. Define allopatric and sympatric speciation.
59. How can we determine the age of fossils?
60. What is meant by acquired and inherited traits? Explain with one example each.
61. Why are the traits acquired during the lifetime of an individual not inherited?
62. Name two animals having homologous organs and two having analogous organs. Name these organs.
63. Name the various tools of tracing evolutionary relationship which have been used for studying human evolution.

LONG ANSWER TYPE QUESTIONS

1. Mention three characteristics of genes.
2. Define evolution. Why are traits acquired during the lifetime of an individual not inherited?
3. Explain analogous and homologous organs.
4. Give important features of the double helical model of DNA proposed by Watson and Crick.
5. Explain with the help of diagram Mendel's monohybrid cross.
6. Explain diagrammatically a Mendel's monohybrid cross.
7. What is artificial selection? Explain with the help of example. What is its importance?
8. Explain the process of evolution with suitable examples.
9. Explain briefly most acceptable theory of origin of life from inanimate matter.
10. Briefly explain contributions of Mendel in Genetics.
11. State the relationship between classification, speciation and evolution.
12. Explain determination of sex among human beings, with the help of a diagram.
13. What are fossils? How do they tell about the process of evolution?
14. Explain XX-XY type of sex determination in human beings through diagram.
15. Give two examples where environmental factors govern the sex determination.
16. What are heteromorphic chromosomes and homomorphic chromosomes?
17. Why do the animals have certain useless organs (vestigial organs) is their body?

18. How analogous organs provide evidence in favour of evolution?
19. Explain sex determination mechanism in humans and How does environment affect determination of sex?
20. Summarize Miller and Urey simulation experiment for organic synthesis. Comment on its efficiency.

MULTIPLE CHOICE QUESTIONS:

1. The theory of evolution of species by natural selection was given by
   (a) Mendel (b) Darwin (c) Dalton (d) Lamarck
2. The term 'father of genetics' is used for the scientist
   (a) Morgan (b) Mendel (c) Darwin (d) Marie Curie
3. Allele that cannot express itself in presence of another is
   (a) Complementary (b) Recessive Allele
   (c) Codominant Allele (d) Dominant Allele
4. An organism with two unlike genes for a trait is called
   (a) Heterozygous (b) Dominant variety
   (c) Homozygous (d) Wild variety
5. Mendel conducted his famous breeding experiments by working on
   (a) *Pisum sativum* (b) *Escherichia coli*
   (c) *Drosophila* (d) All of these
6. Which of the following is a recessive trait in garden pea plant?
   (a) Wrinkled seeds (b) Tall stem
   (c) Round seeds (d) Yellow coloured seed coat
7. The discipline dealing with the study of inheritance is
   (a) Morphology (b) Genetics
   (c) Evolution (d) Cytology
8. The main reason for Mendel's success in discovering the principles of inheritance was
   (a) He was lucky not to encounter with linkage problem
   (b) He considered each character separately
   (c) The plant was pure breeding
   (d) All the above
9. A segment of DNA providing information for a protein is called
   (a) Chromosomes (b) Gene
   (c) Nucleus (d) Trait

10. Mendel studied seven contrasting characters for his breeding experiment with Pisum sativum. Which of the following character did he not use?
    (a) Pod shape (b) Plant height (c) Pod colour (d) Leaf shape
11. Why were pea plants more suitable than dogs for Mendel's experiments?
    (a) All pea plants have only two chromosomes
    (b) Dogs have many genetic traits
    (c) There were no pedigree records of dogs
    (d) Pea plants can be self-fertilized
12. If a tall plant is crossed with a dwarf plant, this type of cross is called
    (a) Monohybrid (b) Trihybrid (c) Dihybrid (d) Reciprocal
13. Segregation of alleles takes place during
    (a) Cleavage (b) Crossing over (c) Meiosis (d) Fertilization
14. Law of independent assortment can be proved on the basis of which of the following ratios?
    (a) 2: 1: 1 (b) 2: 1 (c) 1: 3: 1 (d) 9: 3: 3: 1
15. Exchange of genetic material takes place in
    (a) Vegetative reproduction (b) Asexual reproduction
    (c) Sexual reproduction (d) Budding
16. Which law was put forward by Ernst Haeckel?
    (a) Law of inheritance (b) Biogenetic law
    (c) Law of segregation (d) Law of dominance
17. The term gene was coined by
    (a) Johannsen (b) Mendel (c) Bateson (d) Darwin
18. The genetic constitution of an organism is called
    (a) Genotype (b) Phenotype (c) Variation (d) Character
19. Which of the following is true about fossils?
    (i) Fossils are the remains or impressions of the dead organisms that lived in the remote past.
    (ii) Fossils cannot be dated.
    (iii) More recent fossils are found in superficial layers of sedimentary rocks.
    (iv) Fossils do not provide evidences of evolution.
    (a) (i) and (ii) (b) (i) and (iii) (c) (i) and (iv) (d) (ii) and (iv)
20. Name the organ which is vestigial in man but functional in birds.
    (a) Vermiform appendix (b) Caecum
    (c) Nictitating membrane (d) Third molar

21. Thorn of *Bougainvillea* and tendril of *Cucurbita* are examples of
    (a) Vestigial organs
    (b) Retrogressive evolution
    (c) Analogous organs
    (d) Homologous organs
22. Which of the following are homologous?
    (a) Sweet potato and ginger
    (b) *Passionflower tendril* and *Bougainvillea thorn*
    (c) Insect wing and bird wing
    (d) Tortoise shell and Mollusc shell
23. Analogous organs have
    (a) Different fundamental structure but same function
    (b) Similar fundamental structure but different function
    (c) Different fundamental structure and different function
    (d) Similar fundamental structure and similar function
24. Homologous organs have
    (a) Similar basic structure and also look similar and perform similar functions.
    (b) Different basic structure and also look different and perform different functions.
    (c) Similar basic structure but looks different and perform different functions.
    (d) Different basic structure but looks similar and perform similar functions.
25. Forelimbs of vertebrates possess
    (a) Different shapes and serve different purposes
    (b) Different shapes but serve similar purposes
    (c) Similar shapes and serve similar purposes
    (d) Similar shapes but serves different purposes
26. Insect wing is
    (a) Homologous to bird wing
    (b) An extension of the internment
    (c) Formed of limb bones covered with skin
    (d) Not used for flying in the air
27. In evolutionary terms, we have more in common with a:
    (a) Japanese Boy
    (b) Bacterium
    (c) Spider
    (d) Chimpanzee

28. A gradual change, over a long period, in a form of life is known as:
    (a) Erosion (b) Evolution (c) Revolution (d) Evaluation
29. According to scientists, aves have evolved from:
    (a) Mammals (b) Amphibians (c) Reptiles (d) Arthropods
30. Only one of the following characteristic of the parents can be inherited by their children. This one is:
    (a) Deep scar on chin (b) Cut nose
    (c) Technique of swimming (d) Snub nose
31. The organs which perform different functions but have the same basic structure are known as:
    (a) Homolytic organs (b) Analogous organs
    (c) Homologous organs (d) Analytic organs
32. The organs which perform similar functions but have different basic structure are called:
    (a) Analogous organs (b) Asymmetric organs
    (c) Homologous organs (d) Homophonic organs
33. According to the evolutionary theory, formation of a new species is generally due to:
    (a) Sudden creation by nature.
    (b) Clones formed during asexual reproduction
    (c) Accumulation of variations over several generations.
    (d) Movement of individuals from one habitat to another.
34. The presence of which of the following types of organs in two animals indicates that they are not derived from a common ancestor?
    (a) Reproductive organs (b) Excretory organs
    (c) Homologous organs (d) Analogous organs
35. One of the following has not been produced from wild cabbage by the process of artificial selection. This one is:
    (a) Kohlrabi (b) Cabbage (c) Kale (d) Spinach
36. Human beings belong to species
    (a) Homo sapiens (b) Homo erects
    (c) Homo habilis (d) None of these
37. Human evolution started in
    (a) Africa (b) America (c) Asia (d) Australia

38. The sex of a child in humans is determined by
    (a) Duration of pregnancy
    (b) Period of time between ovulation and copulation
    (c) Presence of X chromosome in the egg
    (d) Presence of Y chromosome in sperm
39. Vestigial organs are
    (a) Primordial organs
    (b) Primitive organs
    (c) Organs reduced due to disuse
    (d) Organs marked only in embryonic stage
40. The visible characteristic of an organism is known as
    (a) Phenotype (b) Genotype (c) Prototype (d) Stereotype
41. Genetic drift operates only in
    (a) Smaller populations (b) Larger populations
    (c) Mendelian populations (d) Island populations
42. Which human organ is vestigial?
    (a) Nose (b) Ear (c) Wisdom tooth (d) Finger
43. Which of the following are not homologous organs?
    (a) Forelimbs of humans and lizard (b) Forelimbs of frog and lizard
    (c) Forelimbs of human and frog (d) Wings of bat and butterfly
44. The maleness of a child is determined by
    (a) The X chromosome in the zygote
    (b) The Y chromosome in zygote.
    (c) The cytoplasm of germ cell which determines the sex.
    (d) Sex is determined by chance.
45. A zygote which has an X-chromosome inherited from the father will develop into a
    (a) Boy
    (b) Girl
    (c) X-chromosome does not determine the sex of a child
    (d) Either boy or girl
46. Under certain conditions the scientists have obtained cell-like structure but not true organization of a cell. These are referred as
    (a) Eubionts (b) Protists (c) Microbes (d) Coacervates

47. The gases used in the spark discharge apparatus of Miler and Urey were
    (a) NH_3, CO_2
    (b) NH_3, NH_4 and O_2
    (c) NH_3, H_2, CH_4
    (d) NH_3, CO_2, CH_4
48. The first generation of offspring or hybrids produced by a cross between genetically different individuals called parents
    (a) F_1 Generation
    (b) F_2 Generation
    (c) Pure line
    (d) Hybrid
49. The alleles which does not express itself physically and remains unexpressed due to the presence of dominant allele is called
    (a) Recessive Alleles
    (b) Hybrids
    (c) Traits
    (d) Pure line
50. The alleles which express itself in hybrid is called as
    (a) Recessive Alleles
    (b) Dominant Alleles
    (c) Traits
    (d) Pure line

ANSWER KEY

1. (b)	2. (b)	3. (b)	4. (a)	5. (a)	6. (a)	7. (b)	8. (b)
9. (b)	10. (d)	11. (d)	12. (a)	13. (c)	14. (d)	15. (c)	16. (b)
17. (c)	18. (a)	19. (b)	20. (c)	21. (b)	22. (b)	23. (a)	24. (c)
25. (b)	26. (c)	27. (b)	28. (b)	29. (c)	30. (b)	31. (c)	32. (a)
33. (c)	34. (d)	35. (d)	36. (a)	37. (a)	38. (d)	39. (b)	40. (a)
41. (a)	42. (c)	43. (d)	44. (b)	45. (b)	46. (d)	47. (d)	48. (a)
49. (a)	50. (b)						

5

CHAPTER

SOURCES OF ENERGY

Different Forms of Energy, Conventional and Non-conventional Sources of Energy: Fossil Fuels, Solar Energy; Biogas; Wind, Water and Tidal energy; Nuclear Energy, Renewable vs. Non-renewable Sources of Energy.

5.1 ENERGY RESOURCES

Energy is defined by physicists as the capacity to do work. Energy is found on our planet in a variety of forms, some of which are immediately useful to do work, while others require a process of transformation. The sun is the primary energy source in our lives. The sun's rays power the growth of plants, which form our food material, give off oxygen which we breathe in and take up carbon dioxide that we breathe out. Energy from the sun evaporates water from oceans, rivers and lakes, to form clouds that turn into rain. Today's **fossil fuels** were once the forests that grew in prehistoric times due to the energy of the sun. **Chemical energy**, contained in chemical com-pounds is released when they are broken down by animals in the presence of oxygen. In India, manual labour is still extensively used to get work done in agricultural systems, and domestic animals used to pull carts and ploughs. **Electrical energy** produced in several ways, powers trans-port, artificial lighting, agriculture and industry. This comes from hydel power based on the water cycle that is powered by the sun's energy that supports evaporation, or from thermal power stations powered by fossil fuels. **Nuclear energy** is held in the nucleus of an atom and is now harnessed to develop electrical energy. At present almost 2 billion people worldwide have no access to electricity at all. While more people will require electrical energy, those who do have access to it continue to increase their individual requirements. It is estimated that the currently used methods of using renewable energy and non-renewable fossil fuel sources together will be insufficient to meet foreseeable global demands for power generation beyond the next 50 to 100 years.

5.2 GROWING ENERGY NEEDS

Electricity is at present the fastest growing form of end-use energy worldwide. By 2005 the Asia-Pacific region is expected to surpass North America in energy consumption and by 2020 is expected to consume some 40% more energy than North America. For almost 200 years, coal was the primary energy source fuelling the industrial revolution in the 19th century. At the close of the 20th century, oil accounted for 39% of the world's commercial energy consumption, followed by coal (24%) and natural gas (24%), while nuclear (7%) and hydro/renewables (6%) accounted for the rest. Among the commercial energy sources used in India, coal is a predominant source accounting for 55% of energy consumption estimated in 2001, followed by oil (31%), natural gas (8%), hydro (5%) and nuclear (1%). In India, biomass (mainly wood and dung) accounts for almost 40% of primary energy supply. While coal continues to remain the dominant fuel for electricity generation, nuclear power has been increasingly used since the 1970s and 1980s and the use of natural gas has increased rapidly in the 80s and 90s.

India is the world's third largest energy consuming country. Energy use has doubled since 2000, with 80% of demand still being met by coal, oil and solid biomass. On a per capitata basis, India's energy demand was projected to increase by almost 50% between 2019 and 2030, but growth over this period is now closer to 35% in the STEPs (Stated Policies Scenario) and 25% in the delayed recovery scenario. Global energy demand is set to increase by 4.6% in 2021, more than of setting the 4% contractions in 2020 and pushing demand above 2019 levels (IEA Report). Global coal demand in 2021 is set to exceed 2019 levels, rising 4.5% with more than 80% of the growth concentrated in Asia. Global oil demand is forecast to rise 6.2% in 2021, and natural gas demand is expected to grow 3.2% in 2021, while electricity demand is set to register its fastest growth in more than ten years in 2021. The demand would expand by 4.5% in 2021.

India has achieved a historic milestone of 100 Gigawatt's installed renewable energy capacity. With this presently India stands at 4th position in the world in terms of installed renewable energy capacity, 5th in solar and 4th in wind energy.

The sun is the primary energy source and the sun's rays power the growth of plants, which form our food material, give off oxygen which we breathe in and take up CO_2 that we breathe out. Energy is found on our planet in a variety of forms (chemical, electrical and nuclear energy), some of which are immediately useful to do work, while others require a process of transformation. Chemical energy, contained in chemical compounds is released when they are broken down by animals in the presence of oxygen. Electrical energy produced by powers transport, artificial lighting, agriculture and industries. This comes from hydel power or thermal power stations powered by fossil fuels. Nuclear energy is held in the nucleus of an atom and is now harnessed to develop electrical energy.

5.3 CONVENTIONAL ENERGY

Conventional sources of energy are the natural energy resources which are regularly used for many years and are accepted as fuel to produce heat, light, food and electricity. The energy sources include firewood, fossil fuels, cow dung cake etc. Fossil fuel is the greatest conventional source among these resources wherein fossil implies the remains of plants and animals, which got buried under the earth and transformed into rocks over the years. These fossil fuels are coal, oil (petroleum), and natural gas. Conventional sources of energy are generally non-renewable sources of energy as the accumulation or creation of conventional sources of energy takes years, once they are exploited or consumed. As these sources are used on a large scale, the reserves have been depleted, and their alternative is hard to find.

5.3.1 Conventional Energy Sources:

The energy sources which cannot be compensated, once these are used (after their exploitation) are termed as conventional energy sources. Some of the important conventional energy sources are discussed here:

5.3.1.1 Coal

Coal is a major conventional energy sources. It was formed from the remains of the trees and ferns grew in swamps around 500 million years ago. The bacterial and chemical decomposition of such plant debris (which remained buried under water or clay) produced an intermediate product known as peat which is mainly cellulose $(C_6H_{10}O_5)n$. Out of the 6000 billion tons coal stocks under earth crust, 200 tons have been exploited the present society. The coal reserves are found in the states like Jharkhand, Orissa, West Bengal, M.P. and A.P. Some important coal fields are such as Talcher, Raniganj, Jharia, Bokaro, Panch Konkam, Signoulli, Chanda etc. Coal is the world's single largest contributor of greenhouse gases and is one of the most important causes of global warming (Fig. 5.1). Many coal-based power generation plants are not fitted with devices such as electrostatic precipitators to reduce emissions of suspended particulate matter (SPM) which is a major contributor to air pollution. Burning coal also produces oxides of sulphur and nitrogen which, combined with water vapour, lead to acid rain. This kills forest vegetation, and damages architectural heritage sites, pollutes water and affects human health.

Fig. 5.1 Coal

5.3.1.2 Oil:

India's oil reserves which are being used at present lie off the coast of Mumbai and in Assam. Most of our natural gas is linked to oil and, because there is no distribution system, it is just burnt off. This wastes nearly 40% of available gas. The processes of oil and natural gas drilling, processing, transport and utilization have serious environmental consequences, such as leaks in which air and water are polluted and accidental fires that may go on burning for days or weeks before the fire can be controlled. Oil slicks are caused at sea from offshore oil wells, cleaning of oil tankers and due to shipwrecks.

Petroleum is a complex mixture of hydrocarbons, mostly alkanes and cycloalkanes. It occurs below the earth crust entrapped under rocky strata. In its crude form, the viscous black liquid is known as petroleum and a gas in contact with petroleum layer which flows naturally from oil wells is termed as natural gases. The composition of natural gas is a mixture of mainly methane (95.0%), small amounts of ethane, propane and butane (3.6%) and traces of CO_2 (0.48%) and N_2 (1.92%). A liquid mixture of propane and butane can be obtained from natural gas or refinery gases at room temperature under a pressure of 3-5 atmospheres (Fig. 5.2). This is stored and distributed in 40-100 litre capacity steel cylinders. The crude petroleum after being refined and purified, are available as petrol, diesel, kerosene, lubricating oil, plastic etc. for commercial and domestic use. In India, the oil deposits are found at Ganga-Brahmaputra Valley, Bombay high, Plains of Gujarat, Thar Desert of Rajasthan (Fig. 5.3A) and area around Andaman Nicobar islands (Fig. 5.3B) . Running petrol vehicles with unleaded fuel has been achieved by adding catalytic converters on all the new cars, but unleaded fuel contains benzene and butadene which are known to be carcinogenic compounds. Delhi, which used to have serious smog problems due to traffic, has been able to reduce this health hazard by changing a large number of its vehicles to CNG, which contains methane.

Fig. 5.2 Petroleum Crude Oil

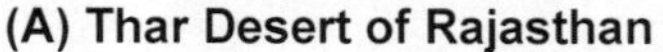

(A) Thar Desert of Rajasthan **(B) Adaman Nicobar Islands**

Fig. 5.3 Oil deposits in India

Conventional sources of energy are the natural energy resources which are regularly used for many years and are accepted as fuel to produce heat, light, food and electricity. The energy sources include firewood, fossil fuels, cow dung cake etc. The average formula of coal is $(C_3/H_4/)n$. Coal is a major conventional energy sources. Coal was formed from the remains of the trees and ferns grew in swamps around 500 million years ago. The coal reserves are found in the various states like Jharkhand, Orissa, West Bengal, M.P. and A.P etc. Some important coal fields are such as Talcher, Raniganj, Jharia, Bokaro, Panch Konkam, Signoulli and Chanda etc.

Coal is the world's single largest contributor of greenhouse gases and is one of the most important causes of global warming. Many coal-based power generation plants are not fitted with devices such as electrostatic precipitators to reduce emissions of suspended particulate matter (SPM) which is a major contributor to air pollution. Burning coal also produces oxides of sulphur and nitrogen which, combined with water vapour, lead to acid rain. This kills forest vegetation, and damages architectural heritage sites, pollutes water and affects human health.

Petroleum is a complex mixture of hydrocarbons (mostly alkanes and cycloalkanes). It occurs below the earth crust entrapped under rocky strata. The composition is a mixture of mainly methane (95.0%), small amounts of ethane, propane and butane (3.6%) and traces of CO_2 (0.48%) and N_2 (1.92%). In India, the oil deposits are found at Ganga-Brahmaputra Valley, Bombay high, Plains of Gujarat, Thar Desert of Rajasthan and area around Andaman Nicobar islands.

5.3.1.3 Fuel Woods:

The rural peoples require fuel wood or fire wood for their day to day cooking which are obtained from natural forests and plantations (Fig. 5.4). Due to rapid deforestation, the availability of fire wood or fuel wood becomes difficult. This problem can be avoided by massive plantation (afforestation) on degraded forest land, culturable waste land, barren land grazing land etc.

Fig. 5.4 Fuel woods

5.3.1.4 Hydropower:

Energy obtainable from water flow or water falling from a higher potential to lower potential, is known is hydropower. It is a conventional and renewable form of energy which can be transmitted to long distance through cables and wires. This uses water flowing down a natural gradient to turn turbines to generate electricity known as hydroelectric power by constructing dams across rivers. In order to produce hydel electricity, high-rise dams are constructed on the river to obstruct the flow of water and thereby collect water in larger reservoirs. The water level rises and in this process the kinetic energy of flowing water gets transformed into potential energy. The water from the high level in the dam is carried through pipes, to the turbine, at the bottom of the dam (Fig. 5.5). Since the water in the reservoir would be refilled each time it rains (hydro power is a renewable source of energy).

Hirakund Dam

Fig. 5.5 Hydro Power Plant

Drawbacks: Although hydroelectric power has led to economic progress around the world, it has created serious ecological problems such as:

- To produce hydroelectric power, large areas of forest and agricultural lands are sub-merged. These lands traditionally provided a livelihood for local tribal people and farmers. Conflicts over land use are inevitable.
- Water is required for many other purposes besides power generation. These include domestic requirements, growing agricultural crops and for industry.
- Silting of the reservoirs especially as a result of deforestation reduces the life of the hydroelectric power installations.
- The use of rivers for navigation and fisheries becomes difficult once the water is dammed for generation of electricity.
- Resettlement of displaced persons is a problem for which there is no ready solution.
- In certain regions large dams can induce seismic activity which will result in earth quakes. There is a great possibility of this occurring around the Tehri dam in the Himalayan foothills. Shri Sunderlal Bahuguna, the initiator of the Chipko Movement has fought against the Tehri Dam for several years.

Narmada Project:

The Narmada Bachao Andolan in India is an example of a movement against large dams. The gigantic Narmada River Project has affected the livelihoods of hundreds of extremely poor forest dwellers (Fig. 5.6). The rich landholders downstream from the Sardar Sarovar dam will derive the maximum economic benefit, whereas the poor tribal people have lost their homes and traditional way of life. The dam will also destroy the livelihood of fishermen at the estuary. The disastrous impact that this project has on the lives of the poor, and the way in which they are being exploited, need to be with large dams causing social problems, there has been a trend to develop small hydroelectric generation units. Multiple small dams have less impact on the environment. China has the largest number of these 60,000, generating 13,250 mw, i.e. 30% of China's electricity. Sweden, the US, Italy and France also have developed small dams for electrical power generation. The development of small hydroelectric power units could become a very important re-source in India, which has steeply falling rivers and the economic capability and technical resources to exploit them. In India, hydroelectric power is generated by a number of multipurpose river valley projects **e.g.** Hydropower Project Hirakud, Bhakra mangal project, Narmada Valley Project, Nagarjun Sagar Project, and Sardar Sarovar Project etc.

Fig. 5.6 Narmada Project

The rural peoples require fuel wood or fire wood for their day to day cooking which are obtained from natural forests and plantations. Due to rapid deforestation, the availability of fire wood or fuel wood becomes difficult. This problem can be avoided by massive plantation (afforestation) on degraded forest land, culturable waste land, barren land, grazing land etc.

Energy obtainable from water flow or water falling from a higher potential to lower potential, is known is hydropower. Hydropower is a conventional and renewable form of energy which can be transmitted to long distance through cables and wires. Hydroelectric power has led to economic progress around the world, and also created serious ecological and human problems. In India, hydroelectric power is generated by a number of multipurpose river valley projects i.e. Hydro Power Project Hirakud, Bhakra mangal Project, Narmada Valley Project, Nagarjun Sagar Project and Sardar Sarovar Project etc.

5.3.1.5 Nuclear Energy:

A small amount of radioactive substance (U^{235}) can produce a lot of energy through the process of nuclear fission. For example, one ton of uranium can provide energy which is much higher than three million tons of coal or 12 million barrels of oil. In order to obtain nuclear energy, nuclear reactors are required. There are around 300 nuclear reactors all over the world. India has only 04 nuclear power reactors). The nuclear energy can be used in production of electrical energy, as a fuel for marine vessel and space crafts and for the generation of heat in chemical processing plants. In India, Uranium deposits are found at different parts of Rajasthan and Singhbum of Jharkhand (Fig. 5.7). Thorium is recovered from monazite sand found in the state of Kerala. Due to the higher energy releasing tendency of these radioactive substances, they can be used in nuclear reactors to release energy crisis. But the radioactive substances are exhaustible and can be used to develop nuclear weapons of mass destruction.

Fig. 5.7 Nuclear Energy Plant

Currently all commercial nuclear reactors are based on nuclear fission. But there is another possibility of nuclear energy generation by a safer process called nuclear fusion. Fusion means joining lighter nuclei to make a heavier nucleus, most commonly hydrogen or hydrogen isotopes to create helium, such as 2H + 2H ® 3He (+ n). The hydrogen bomb is based on thermonuclear fusion reaction. A nuclear bomb based on the fission of uranium or plutonium is placed at the core of the hydrogen bomb. This nuclear bomb is embedded in a substance which contains deuterium and lithium. When the nuclear bomb (based on fission) is detonated, the temperature of this substance is raised to 10^7 K in a few microseconds. The high temperature generates sufficient energy for the light nuclei to fuse and a devastating amount of energy is released.

> A small amount of radioactive substance (U^{235}) can produce a lot of energy through the process of nuclear fission. Fusion means joining lighter nuclei to make a heavier nucleus, most commonly hydrogen or hydrogen isotopes to create helium, such as 2H + 2H ® 3He (+ n). The hydrogen bomb is based on thermonuclear fusion reaction. The nuclear energy can be used in production of electrical energy, as a fuel for marine vessel and space crafts and for the generation of heat in chemical processing plants.

5.4 NON-CONVENTIONAL ENERGY

An alternative to conventional sources of energy is the non-conventional sources of energy that achieved popularity in recent years, after the oil crisis in 1973 and since then they are in use on a large scale. The energy can be obtained from various sources such as the sun, wind, biological wastes, hot springs, tides, etc. to generate heat and power. These are not only renewable sources of energy but are also pollution free. These sources are present in abundance in nature, and they are constantly generated, so it cannot be exhausted easily, and used again and again.

5.4.1 Non-conventional Energy Sources

The conventional energy sources are exhaustible and in some cases, installation of plants to get energy is highly expensive. In order to meet the energy demand of increased population, the scientists developed alternate non-conventional natural resources sources of energy which should be renewable and provide a pollution free environment. Some non-conventional, renewable and inexpensive energy sources are described as:

5.4.1.1 Solar Energy

The sun has been radiating an enormous amount of energy at the present rate for nearly 5 billion years and will continue radiating at that rate for about 5 billion years more. Only a small part of solar energy reaches the outer layer of the earth's atmosphere. Nearly half of it is absorbed while passing through the atmosphere and the rest reaches the earth's surface. The energy from sun collected per unit time by a unit area at the outer surface of earth's atmosphere is 1.40×10^6 ergs/cm^2. It is called solar constant. The entire area of the earth intercepts 1.78×10^{24} ergs of this energy. About 35% of incident solar energy is reflected back into space as visible light. About 12% is absorbed in the earth's atmosphere on the average and is converted into thermal energy. The total energy falling on the ground per second is about 9.44x1023 ergs. Solar energy, a primary energy source, is non-polluting and inexhaustible. There are three methods to harness solar energy such as:

- Converting solar energy directly into electrical energy in solar power stations using photo cells or photovoltaic cells or silicon solar cell.
- Using photosynthetic and biological process for energy trapping. In the process of photosynthesis, green plants absorb solar energy and convert it into chemical energy, stored in the form of carbohydrate.
- Converting solar energy in to thermal energy by suitable devices which may be subsequently converted into mechanical, chemical or electrical energy.

Since solar energy is non-ending and its conversion to some other energy form is nonpolluting, attention should be paid for the maximum utilization of solar energy. Today we have developed several methods of collecting this energy for heating water and generating electricity.

5.4.1.2 Solar Heating

Modern housing that uses air conditioning and/or heating are extremely energy dependant. A passive solar home or building is designed to collect the sun's heat through large, south-facing glass windows. In solar heated buildings, sunspaces are built on the south side of the structure which act as large heat absorbers. The floors of sunspaces are usually made of tiles or bricks that absorb heat throughout the day, and then release heat at night when it's cold. Traditional bungalows had high roofs and ventilators that permitted hot air to rise and leave the room. Cross ventilation where wind can drive the air in and out of a room keeps it cool. Large overhangs over windows prevent the glass from heating the room inside. Double walls are used to prevent heating. Shady trees around the house help reduce temperature.

5.4.1.3 Solar Water Heating (Solar Collector and Storage Tank)

Most solar water-heating systems have two main parts such as **(i)** the solar collector and the **(ii)** storage tank. The solar energy collector heats the water, which then flows to a well-insulated storage tank. Solar water heating can reduce domestic water heating costs by as much as 70% (Fig.5.8). Solar domestic hot water systems are technically mature and available particularly all over the world. Today several million homes and business use solar water heating systems (Fig. 5.9). These systems are providing consumers a cost effective and reliable choice for hot water. These systems are also being used in schools, hospitals, restaurants, agricultural, farms and different processing industries for hot water production.

Fig. 5.8 Solar Collector and Storage Tank

Fig. 5.9 Solar Water Heater

5.4 .1.4 Solar Cookers

A solar cooker is a metal box which is black on the inside to absorb and retain heat. The lid has a reflective surface to reflect the heat from the sun into the box. The box contains black vessels in which the food to be cooked is placed (Fig. 5.10). India has the world's largest solar cooker programme and an estimated 2 lakh families that use solar cookers. Although solar cookers reduce the need for fuel wood and pollution from smoky wood fires, they have not caught on well in rural areas as they are not suitable to traditional cooking practices. However, they have great potential if marketed well.

Fig. 5.10 Solar Cooker

5.4 .1.5 Photovoltaic Energy

The solar technology which has the greatest potential for use through-out the world is that of solar photo voltaic cells which directly produce electricity from sunlight using photovoltaic (PV). PV cells are commonly used today in calculators and watches (Fig. 5.11). They also provide power to satellites, electric lights, and small electrical appliances such as radios and for water pumping, highway lighting, weather stations, and other electrical systems located away from power lines.

Fig. 5.11 Photovoltaic Energy

Non-conventional energy can be obtained from various sources such as the sun, wind, biological wastes, hot springs, tides, etc. to generate heat and power. These are not only renewable sources of energy but are also pollution free. These sources are present in abundance in nature, and they are constantly generated, so it cannot be exhausted easily, and used again and again. The conventional energy sources are exhaustible and in some cases,

Solar energy is non-ending and its conversion to some other energy form is non-polluting, attention should be paid for the maximum utilization of solar energy. Recently, we have developed several methods of collecting this energy for heating water and generating electricity. Solar water heating can reduce domestic water heating costs by as much as 70%. India has the world's largest solar cooker programme and an estimated 2 lakh families that use solar cookers. Photovoltaic cells also provide power to satellites, electric lights, and small electrical appliances such as radios and for water pumping, highway lighting, weather stations, and other electrical systems located away from power lines.

5.4 .1.6 Wind Energy:

Wind is air in motion. The movement of air takes place due to the convection current set out in the atmosphere which is again due to heating of earth's surface by solar radiation, rotation of earth etc. (Fig. 5.12A). The movement of air occurs both horizontally and vertically. The average annual wind density is 3 kW/m^2/day along costal lines of Gujarat, Western Ghat, Central Parts of India, which may show a seasonal variation (i.e., in winter it may go up to 10kW/m^2/day). Over the past two decades, a great deal of technical progress has been made in the design, siting, installation, operation, and maintenance of power-producing wind mills (turbines). Wind energy is an environment-friendly and efficient source of renewable energy. It requires no recurring expenses for the production of electricity, but there are many limitations in harnessing wind energy.

Fig. 5.12 Largest Wind Farm Plant at Kanyakumari

Wind energy farms can be established only at those places where wind blows for the most part of a year. The wind speed should also be higher than 15 km/h to maintain the required speed of the turbine. Furthermore, there should be some back-up facilities (like storage cells) to take care of the energy needs during a period when there is no wind. Establishment of wind energy farms requires large area of land. The farm needs about 2 hectares of land for one 1 MW generator. The initial cost of establishment of the farm is quite high. Moreover, since the tower and blades are exposed to the vagaries of nature like rain, Sun, storm and cyclone, they need a high level of maintenance.

Environmental Impacts of Wind Energy:

Wind power has few environmental impacts, as there are virtually no air or water emissions, or radiation, or solid waste production. The principal problems are bird kills, noise, effect on TV reception, and aesthetic objections to the sheer number of wind turbines that are required to meet electricity needs. Although large areas of land are required for setting up wind farms, the amount used by the turbine bases, the foundations and the access roads is less than 1% of the total area covered by the wind farm. The rest of the area can also be used for agricultural purposes or for grazing. Siting windmills offshore reduces their demand for land and visual impact.

Wind energy is an environment-friendly and efficient source of renewable energy. It requires no recurring expenses for the production of electricity, but there are many limitations in harnessing wind energy. The average annual wind density is

3 kW/m^2/day along coastal lines of Gujarat, Western Ghat, Central Parts of India, which may show a seasonal variation (i.e., in winter it may go up to 10kW/m^2/day). Wind power has few environmental impacts, as there are virtually no air or water emissions, or radiation, or solid waste production. Siting windmills offshore reduces their demand for land and visual impact.

5.4.1.7 Tidal Energy

The energy associated with the tides of the Ocean canbe converted in to electrical energy. France constructed the first tidal power plant in 1966. India could take up Ocean thermal energy conversion (OTEC) and by the process it will be capable of generating 50,000 mW of electricity, to meet the power requirements of remote oceanic islands and coastal towns. In India, Gujarat and Tamilnadu have windmills (Fig. 5.13). The largest wind farm has been set at Kanyakumari which generates 380mW of electricity. Generating technologies for deriving electrical power from the ocean include tidal power, wave power, ocean thermal energy conversion, ocean currents, ocean winds and salinity gradients. Of these, the three best developed technologies are tidal power, wave power and ocean thermal energy conversion.

Fig. 5.13 Tidal Energy

Tidal Power:

Due to the gravitational pull of mainly the moon on the spinning earth, the level of water in the sea rises and falls. If you live near the sea or ever travel to some place near the sea, try and observe how the sea-level changes during the day. This phenomenon is called high and low tides and the difference in sea-levels gives us tidal energy. Tidal energy is harnessed by constructing a dam across a narrow opening to the sea. A turbine fixed at the opening of the dam converts tidal energy to electricity. Tidal Power is tapped by placing a barrage across an estuary and forcing the tidal flow to pass through turbines. In a one-way system the in-coming tide is allowed to fill the basin through a sluice, and the water so collected is used to produce electricity during the low tide. In a two-way system power is generated from both the incoming as well as the outgoing tide.

Tidal power stations bring about major ecological changes in the sensitive ecosystem of coastal regions and can destroy the habitats and nesting places of water birds and interfere with fisheries. A tidal power station at the mouth of a river blocks the flow of polluted water into the sea, thereby creating health and pollution hazards in the estuary. Other drawbacks include offshore energy devices posing navigational hazards. Residual drift current could affect spawning of some fish, whose larvae would be carried away from spawning grounds. They may also affect the migration patterns of surface swimming fish.

Wave Power:

The earth's surface is 70% water. By warming the water, the sun, creates ocean currents and wind that produces waves. It is estimated that the solar energy absorbed by the tropical oceans in a week could equal the entire oil reserves of the world 1 trillion barrels of oil. The energy of waves in the sea that crash on the land of all the continents is estimated at 2 to 3 million megawatts of energy. Similarly, the kinetic energy possessed by huge waves near the seashore can be trapped in a similar manner to generate electricity. The waves are generated by strong winds blowing across the sea. Wave energy would be a viable proposition only where waves are very strong. A wide variety of devices have been developed to trap wave energy for rotation of turbine and production of electricity. Wave power converts the motion of waves into electrical or mechanical energy. For this, an energy extraction device is used to drive turbo-generators. Electricity can be generated at sea and transmitted by cable to land. This energy source has yet to be fully explored. The largest concentration of potential wave energy on earth is located between latitudes 40 to 60 degrees in both the northern and southern hemispheres, where the winds blow most strongly.

The energy associated with the tides of the ocean can be converted in to electrical energy. Generating technologies for deriving electrical power from the ocean include tidal power, wave power, ocean thermal energy conversion, ocean currents, ocean winds and salinity gradients. Of these, the three best developed technologies are tidal power, wave power and ocean thermal energy conversion.

Tidal power is tapped by placing a barrage across an estuary and forcing the tidal flow to pass through turbines. Tidal power stations bring about major ecological changes in the sensitive ecosystem of coastal regions and can destroy the habitats and nesting places of water birds and interfere with fisheries.

The waves are generated by strong winds blowing across the sea. Wave energy would be a viable proposition only where waves are very strong. A wide variety of devices have been developed to trap wave energy for rotation of turbine and production of electricity.

5.4.1.7 Ocean Thermal Energy

Another developing concept harnesses energy due to the differences in temperature between the warm upper layers of the ocean and the cold deep sea water. These plants are known as Ocean Thermal Energy Conversion (OTEC). This is a high tech installation which may prove to be highly valuable in the future (Fig. 5.14). The water at the surface of the sea or ocean is heated by the Sun while the water in deeper sections is relatively cold (Fig. 5.15). This difference in temperature is exploited to obtain energy in ocean-thermal-energy conversion plants. These plants can operate if the temperature difference between the water at the surface and water at depths up to 2 km is 20 K (20°C) or more. The warm surface-water is used to boil a volatile liquid like ammonia. The vapours of the liquid are then used to run the turbine of generator. The cold water from the depth of the ocean is pumped up and condenses vapour again to liquid. The energy potential from the sea (tidal energy, wave energy and ocean thermal energy) is quite large, but efficient commercial exploitation is difficult.

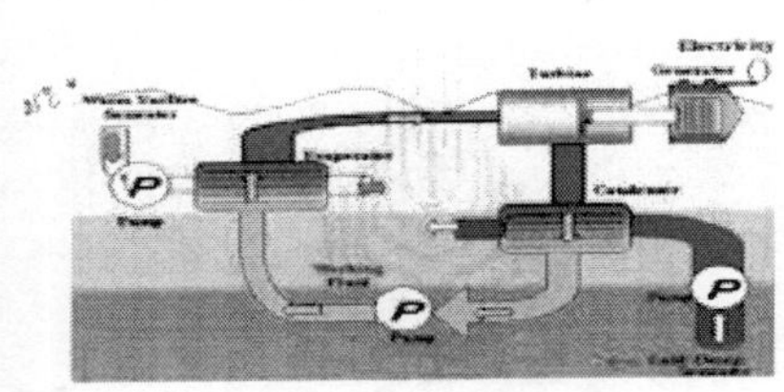

Fig. 5.14 Ocean Thermal Energy

5.4.1.8 Geothermal Energy

The geothermal energy may be defined as the heat energy obtainable from hot rocks present inside the earth crust. At the deeper region of earth crust, the solid rock gets melted in to magma, due to very high temperature. The magma layer is pushed up due to some geological changes and get concentrated below the earth crust. The places of hot magma concentration at fairly less depth are known as hot spots. These hot spots are known as sources of geothermal energy. Now a days, efforts are being made to use this energy for generating power and creating refrigeration etc. There are a quite few number of methods of harnessing geothermal energy. Different sites of geothermal energy generation are Puga (Ladakh) (Fig. 5.15), Tattapani (Suraguja, M.P.) and Cambay Basin (Alkananda Valley, Uttaranchal).

Fig. 5.15 Geothermal Energy Plant at Puga (Ladakh)

Another developing concept harnesses energy due to the differences in temperature between the warm upper layers of the ocean and the cold deep sea water. These plants are known as Ocean Thermal Energy Conversion (OTEC). This is a high tech installation which may prove to be highly valuable in the future.

Geothermal energy may be defined as the heat energy obtainable from hot rocks present inside the earth crust. There are a quite few number of methods of harnessing geothermal energy. Different sites of geothermal energy generation are Puga (Ladakh), Tattapani (Suraguja, M.P.) and Cambay Basin (Alkananda Valley, Uttaranchal).The waves are generated by strong winds blowing across the sea.

Wave energy would be a viable proposition only where waves are very strong. A wide variety of devices have been developed to trap wave energy for rotation of turbine and production of electricity.

5.4 .1.9 Bio-mass based Energy

The organic matters originated from living organisms (plants and animals) like wood, cattle dung, sewage, agricultural wastes etc. are called as biomass. These substances can be burnt to produce heat energy which can be used in the generation of electricity. Thus, the energy produced from the biomass is known as biomass energy (Fig. 5.16). There are three forms of biomass such as:

Fig. 5.16 Bio-mass based Energy

(a) **Biomass in traditional form:** Energy is released by direct burning of biomass (e.g. wood, agricultural residue etc.)

(b) **Biomass in nontraditional form:** The biomass may be converted in to some other form of fuel which can release energy. For example carbohydrate can be converted into methanol or ethanol which may be used as a liquid fuel.

(c) **Biomass for domestic use:** When organic matters like cow dung, agricultural wastes, human excreta etc. subjected to bacterial decomposition in presence of water in absence of air, a mixture of CH_4, CO_2, H_2, H_2S etc. are produced. These gases together are known as biogas. The residue left after the removal of biogas is a good source of manure and biogas is used as a good source of non-polluting fuel.

Organic Municipal Solid Waste includes paper, food wastes, and other organic non-fossil-fuel derived materials such as textiles, natural rubber, and leather that are found in the waste of urban areas (Fig. 5.17). Currently, in the US, approximately 31% of organic waste is recovered from municipal solid waste via recycling and composting programs, 62% is deposited in landfills, and 7% is incinerated. Waste material can be converted into electricity by combustion boilers or steam turbines. Note that like any fuel, biomass creates some pollutants, including carbon dioxide, when burned or converted into energy. In terms of air pollutants, biomass generates less relative to fossil fuels. Biomass is naturally low in sulphur and therefore, when burned, generates low sulphur dioxide emissions. However, if burned in the open air, some biomass feed stocks would emit relatively high levels of nitrous oxides (given the high nitrogen content of plant material), carbon monoxide, and particulates.

Fig. 5.17 Organic Municipal Solid Waste

5.4 .1.10 Biogas:

Biogas is produced from plant material and animal waste, garbage, waste from house-holds and some types of industrial wastes, such as fish processing, dairies, and sewage treatment plants. It is a mixture of gases which includes methane, carbon dioxide, hydrogen sulphide and water vapour. In this mixture, methane burns easily. With a ton of food waste, one can produce 85 Cu. M of biogas (Fig. 5.18). Once used, the residue is used as an agricultural fertilizer. Denmark produces a large quantity of biogas from waste and produces 15,000 megawatts of electricity from 15 farmers' cooperatives. Lon-don has a plant which makes 30 megawatts of electricity a year from 420,000 tons of municipal waste which gives power to 50,000 families. In Germany, 25% of landfills for garbage produce power from biogas. Japan uses 85% of its waste, and France about 50%. Biogas plants have become increasingly popular in India in the rural sector. The biogas plants use cow dung, which is converted into a gas which is used as a fuel. It is also used for running dual fuel engines. The reduction in kitchen smoke by using biogas has reduced lung conditions in thousands of homes. The fibrous waste of the sugar industry is the world's largest potential source of biomass energy. Ethanol produced from sugarcane molasses is a good automobile fuel and is now used in a third of the vehicles in Brazil.

Fig. 5.18 Biogas plant

The organic matters originated from living organisms (plants and animals) like wood, cattle dung, sewage, agricultural wastes etc. are called as biomass. These substances can be burnt to produce heat energy which can be used in the generation of electricity. Thus, the energy produced from the biomass is known as biomass energy. Biomass is a mixture of gases which includes methane, carbon dioxide, hydrogen sulphide and water vapour. Biomass is categorized in to three types such as biomass in (a) traditional form (b) non-traditional form and (c) domestic use.

Biogas is produced from plant material and animal waste, garbage, waste from house-holds and some types of industrial wastes, such as fish processing, dairies, and sewage treatment plants. It is a mixture of gases which includes methane, carbon dioxide, hydrogen sulphide and water vapour. The fibrous waste of the sugar industry is the world's largest potential source of biomass energy.

5.4.1.11 Petro Plants (Petroleum Plants)

There are some species of certain families which accumulate the photosynthetic products (hydrocarbons) of high molecular weight (10,000). They are commonly known as petroplants or petroleum plants. In 1979, Dr. M. Calvin of the University of California reported for the first time the collection and use of photosynthetically produced hydrocarbons from the plants Furthermore, he suggested it as a substitute for conventional petroleum sources. The petroplants have lactiferous canals in their stem and secrete milky latex. The latex can be either continuously tapped like *Hevea* latex and stored or extracted from the biomass by using the organic solvents. The product rich in hydrocrackable hydrocarbon is called as ‹biocrude›. Biocrude yields about 70.6% energy; out of which 22% as kerosene and 44.6% as gasoline. About 400 plant species, belonging to different families are known which grow in different part of the country. It is hoped that petroplants can yield petroleum more than 40-45 barrel/acre. Plants producing rubber and other hydrocarbons are given in Table 5.1 and (Fig. 5.19).

(A) Algae (*Chlorella pyrenoidosa*)

(B) Hevea Rubber (*Hevea brasiliensis*)

(C) Jatropha (*Jatrophacurcas*)

(D) Russian dandelion (*Taraxacum koksaghy*)

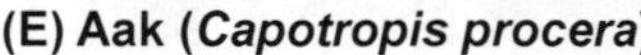

(E) Aak (*Capotropis procera*) **(F) Samprani (*Hardwicika pinnata*)** **(G) Gurjun (*Dipterocarpus turbinates*)**

Fig. 5.19 Petroleum Plants

Table 5.1 Plants Producing Hydrocarbons

Plant Group/Families	Common names	Botanical names
Algae (Chlorophyta)	-	*Botryococcus sp.* *Chlorella pyrenoidosa*
Euphorbiaceae	Hevea rubber	*Hevea brasiliensis*
	Rubber plant	*Euphorbia abyssinica*
	Rubber plant	*E. resinifera, E. lathyris,*
	Sehund	*E. tirucalli*
	Jatropha	*Jatropha curcas*
Compositae	Guayule	*Parthenium argentatum*
	Russian dandelion	*Taraxacum koksaghyz*
Asclepiadaceae	Aak	*Calotropis procera*
Leguminoseae	-	*Copaifera langsdorfii, C. mutijuga*
	Samprani	*Hardwickia pinnata*
Dipterocarpaceae	Gurjun	*Dipterocarpus turbinatus*
Myristicaceae	-	*Dialynthera otoba*
Pittosporaceae	-	*Pittosporum resiniferum*

At present, there is need to encourage cultivation of petroplants, in waste lands. There is need for joint efforts of botanists, phytochemists, engineers, economists, and geneticists to make successful research on these lines.

Petro Plants (Petroleum Plants) belongs to some species of certain families like Euphorbiaceae, Compositae, Asclepiadaceae, Leguminoseae, Dipterocarpaceae, Myristicaceae and Pittosporaceae, which accumulate hydrocarbons having high molecular weight (10,000) i.e. Hevea rubber (*Hevea brasiliensis*), Jatropha (Jatropha curcas), Sehund (*E. tirucalli*), Aak (*Calotropis procera*) and Gurjun (Dipterocarpus turbinatus) etc.

First of all, Dr. M. Calvin (in 1979) reported the collection and use of photosynthetically produced hydrocarbons from the plants. The product rich in hydrocrackable hydrocarbon is called as 'biocrude'. Biocrude yields about 70.6% energy; out of which 22% as kerosene and 44.6% as gasoline. Petroplants have lactiferous canals in their stem and secrete milky latex. It is hoped that petroplants can yield petroleum more than 40-45 barrel/acre. At present, there is need to encourage cultivation of petroplants, in waste lands.

5.4.1.12 Dendrothermal Energy (Energy Plantation)

Due to rapid deforestation and overgrazing, a number of denuded wastelands are formed (Fig. 5.20). On these wastelands, fast growing trees and shrubs may be planted which will provide fuel wood, charcoal, fodder, etc. Through gasification, these plants can produce a lot of energy-

Fig. 5.20 Dendrothermal Energy

5.4 .1.13 Bagasse-based Plants

Bagasse is generated as a waste product in sugar mills. This can be utilized to produce electrical energy (Fig. 5.21). As per available data, the sugar mills in India can generate about 2000 mW surplus electricity during crushing season.

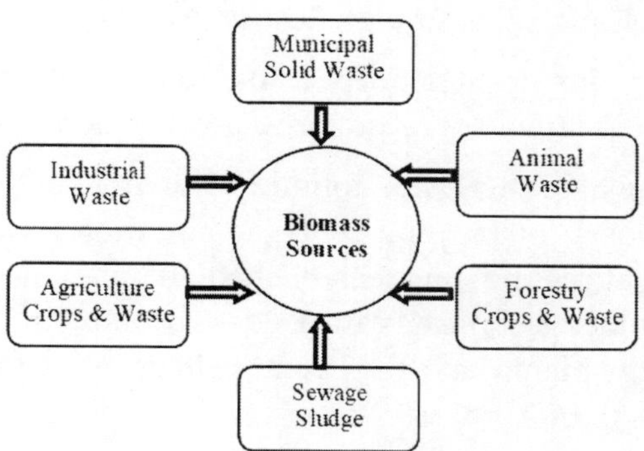

Fig. 5.21 Biomass sources

5.4.1.14 Energy from Urban Waste

Sewage and solid municipal wastes can also generate energy on their suitable treatments.

Conventional sources of energy are those sources which are widely used all around the world since ages. Conventional sources of energy are limited in nature, and their formation takes millions of years, they can be exhausted one day. Equally, non-conventional sources of energy are the sources that are in abundance in the environment and are easily renewable, so they are inexhaustible.

Conventional sources of energy pollute the environment, while non-conventional sources of energy are environment-friendly, so they do not harm the nature.. Conventional sources of energy are costly because they are scarce but their uses are unlimited. In contrast, non-conventional sources of energy are less expensive, because of their enormous presence in nature.

5.5 RENEWABLE RESOURCES

5.5.1 Definition of Renewable Resources:

Renewable resources are the natural assets that can be replenished in the future. The resources can be used over and over again, as they can be revived naturally. Biomass, oxygen, water, sunlight are some of the common examples of renewable resources. These resources are replaced at a rate faster than or equal to the rate of usage, in the sense that the resources regenerate itself by the time it is consumed. This type of resources are enduring in nature, which never gets extinct, i.e. they are believed to have constant supplies over the years and includes solar energy, wind energy, hydroelectricity, etc. However, there are some renewable resources which revive itself at a sustainable rate like wood, oxygen, bio-energy, oils from plants and seeds, etc.

5.5.2 Advantages of Renewable Energy Resources:

- Renewable energies are not burned like fossil fuels, they do not release pollutants into the atmosphere and provide a cleaner, healthier environment.
- Sources of renewable energy are found everywhere in the world and cannot be depleted. Costs for tapping renewable energies are decreasing as the technology advances and, once established, maintenance costs are generally low. Because trained technicians are needed to maintain the equipment, some renewable energy plants have the potential to generate more jobs than highly mechanized fossil fuel plants.
- Most importantly, there are little or no greenhouse gas emissions associated with renewable energies that are contributing to driving up the planet's temperature.

5.5.3 Disadvantages of Renewable Energy Resources

- Renewable energy plants are often quite high and require careful planning and implementation.

- Renewable energies like solar and wind require large tracts of land to produce energy quantities competitive with fossil fuel burning.
- Renewable sources of energy are also affected by weather, reducing their reliability.

5.6 NON-RENEWABLE RESOURCES

5.6.1 Definition of Non-renewable Resources:

Non-renewable resources represent the resources which do not revive itself at a substantial scale, for enduring economic extraction in the specified period. These natural resources are available in finite quantity, which is once used, cannot be replenished. Examples of non-renewable resources are coal, fossil fuel, crude oil, nuclear energy, etc.

5.6.2 Advantages of Non-renewable Energy Resources

- Fossil fuels are the world's traditional energy sources and electrical power plants, vehicles and various industrial plants are built around using them.
- Much non-renewable energy is more reliable than most renewables and is not subject to weather conditions. They provide continuous not intermittent, weather-dependent energy. New technologies, such as carbon, capture and storage (CCS) are emerging that may allow fossil fuel use with less harmful effects to the environment.

5.6.3 Disadvantages of Non-renewable Energy Resources:

- Fossil fuels are in a limited supply and will one day be depleted.
- Processes for extracting and transporting fossil fuels have caused widespread environmental damage from strip mining and accidental oil spills.
- Most importantly, burning fossil fuels releases harmful greenhouse gases into the atmosphere, primarily CO_2.
- Incorporating Carbon Capture and Storage (CCS) technologies into existing fossil fuel plants to prevent CO_2 emissions is extreme costly.
- Nuclear power plants do not release CO_2, but pose other risks such as potential radiation leaks and waste storage problems.

5.7 DIFFERENCES BETWEEN RENEWABLE AND NON-RENEWABLE RESOURCES:

(a) Renewable resources are the resources which can be revived, through the natural process, over time. On the other hand, non-renewable resources are the natural resources which are not going to be replenished, shortly.

(b) Renewable resources do not become extinct easily, and so they are sustainable in nature. Conversely, non-renewable resources deplete over time, i.e. they are exhaustible in nature, which can be ended when they are prone to large scale consumption.

(c) Renewable resources exist in nature in infinite quantity, but non-renewable resources are present in limited quantity.

(d) Renewable resources emit less carbon as compared to non-renewable resources. Hence, they are pollution free and environment-friendly.

(e) As renewable resources are available to us in abundance, they are relatively cheaper than non-renewable resources, which are costlier. However, the maintenance cost of renewable resources is higher than that of non-renewable resources.

(f) The renewable resources regenerate/reproduce itself, faster than it is used up by the living organisms. In contrast, the non-renewable resources either take centuries to regenerate itself, or they get extinct, so the scale at which it is regenerated is lower than that of its consumption.

(g) Rate of renewal for renewable resources is greater than the rate of consumption and vice versa for non-renewable resources

Renewable resources are the resources which can be revived, through the natural process, over time. Biomass, oxygen, water, sunlight are some of the common examples of renewable resources. This type of resources are enduring in nature, which never gets extinct, i.e. they are believed to have constant supplies over the years and includes solar energy, wind energy, hydroelectricity, etc. On the other hand, non-renewable resources are the natural resources which are not going to be replenished, shortly. Examples of non-renewable resources are coal, fossil fuel, crude oil, nuclear energy, etc.

Renewable resources emit less carbon as compared to non-renewable resources. Hence, they are pollution free and environment-friendly. Non-renewable resources represent the resources which do not revive itself at a substantial scale, for enduring economic extraction in the specified period. These natural resources are available in finite quantity, which is once used, cannot be replenished. Rate of renewal for renewable resources is greater than the rate of consumption and vice versa for non-renewable resources.

5.8 DIFFERENCES BETWEEN CONVENTIONAL AND NON-CONVENTIONAL SOURCES OF ENERGY:

Energy implies the capacity or ability to work vigorously. It has a significant role to play in our day to day life, in fact, it is predominantly required in every field, whether it is a household, industry, communication, transport, defence or agriculture. Energy resources are broadly classified as Conventional and Non-conventional sources of energy. Conventional sources of energy are not present in the environment in abundance, however there uses are unlimited. On the contrary, non-conventional energy sources are the sources present in the environment in large quantity but are used for limited purposes only. The differences between conventional and non-conventional sources of energy are given as:

(a) Conventional sources of energy, as the name suggests, are those sources which are widely used all around the world since ages. On the contrary, non-conventional sources of energy are described as the energy sources, whose evolution have been done in the recent past and have gained popularity since then.

(b) As the conventional sources of energy are limited in nature, and their formation takes millions of years, they can be exhausted one day. Conversely, non-conventional sources of energy are the sources that are in abundance in the environment and are easily renewable, so they are inexhaustible.

(c) Conventional sources of energy pollute the environment on a large scale through the smoke and hazardous waste emitted from the power plants. However, the energy produced from running water does not pollute the environment. On the other hand, non-conventional sources of energy are environment-friendly, so they do not harm the nature, by polluting it.

(d) The energy produced from conventional sources are highly used for industrial and commercial purposes. As against, the energy generated out of non-conventional sources are used for domestic purposes.

(e) Conventional sources of energy are costly because they are scarce but their uses are unlimited. In contrast, non-conventional sources of energy are less expensive, because of their enormous presence in nature.

Table 5.2 Difference between Conventional and Non-conventional Sources of Energy

Basis for Comparison	Conventional Sources of Energy	Non-conventional Sources of Energy
Meaning	Conventional sources of energy are the sources that are commonly in use since long time.	Non-conventional sources of energy refer to the sources that are identified few decades ago.
Exhaustible	They can be exhausted due to over consumption	They cannot be exhausted.
Pollution	They pollute environment, on a large scale and adds to global warming	They are environment friendly sources, which does not causes pollution
Use	They are primarily used for industrial and commercial purposes	They are mainly used for domestic purposes.
Expense	Costly	Comparatively less expensive

Table 5.3 A Summarized Table of Energy Sources

Sr No.	Energy Source	Production	Advantages	Disadvantages
1.	Coal fired Electricity	Burning coal heats water that drives turbines to make electricity	Readly available	Non-renewable, Insufficient, High watere use and greenhouse gases
2.	Hydro-electricity	Running water drives turbines that generate electricity	Renewable. Readly available	Set up generates greenhouse gases and damages environment
3.	Natural Gas	Gas piped to consumer and burnt on site	Efficient. Lowe greenhouse gases output than coal fired electricity	Not available everywhere. Non-renewable. Limited applications.
4.	LPG	Gas supplied in cylinders and tanks and burnt on site,	Widely available. Low greenhouse gas emissions	Expensive. Transportation to site uses additional energy.
5.	Solar	Solar energy converted to power using photo tactic cells	Abundant, free renewable No greenhouse gas emissions.	Not yet available in all areas. Expensive to setup
6.	Wind	Wind drives turbines that produce electricity	Free, clean. No greenhouse gas emissions	Expensive to setup. Turbines obtrusive and may endanger birds.
7.	Wave/Tidal	Waves or tides power turbines that produce electricity	Source readly available. Renewable and close to cities	Difficult and expensive to harness wave power efficiency.
8.	Biomass	Plant matter is buried to power electricity generations	Potentially renewable. Recycles agricultural wastes.	Cultivation and burning of fuel on yield low levels of pollutants.
9.	Landfill	Methane from rotting rubbish is burned to power generators	Recycle waste. Helps prevent methane build up in atmosphere	Expensive. Requires large quantities of waste
10.	Geothermal	Heat, steam or hot water from the earth's interior drives generators	Inexpensive once set-up efficient	Limited to areas of geothermal activity or appropriate geology.

REMEMBER SOCKETS

- Energy is defined by physicists as the capacity to do work. Energy is found on our planet in a variety of forms, some of which are immediately useful to do work, while others require a process of transformation.
- The sun is the primary energy source in our lives. The sun's rays power the growth of plants, which form our food material, give off oxygen which we breathe in and take up carbon dioxide that we breathe out.

- Chemical energy, contained in chemical compounds is released when they are broken down by animals in the presence of oxygen.
- Electrical energy produced in several ways, powers transport, artificial lighting, agriculture and industry. This comes from hydel power based on the water cycle that is powered by the sun's energy that supports evaporation, or from thermal power stations powered by fossil fuels.
- Nuclear energy is held in the nucleus of an atom and is now harnessed to develop electrical energy.
- Conventional sources of energy are the natural energy resources which are regularly used for many years and are accepted as fuel to produce heat, light, food and electricity. The energy sources include firewood, fossil fuels, cow dung cake etc.
- Coal is a major conventional energy sources. It was formed from the remains of the trees and ferns grew in swamps around 500 million years ago. The coal reserves are found in the states like Jharkhand, Orissa, West Bengal, M.P. and A.P. Some important coal fields are such as Talcher, Raniganj, Jharia, Bokaro, Panch Konkam, Signoulli, Chanda etc.
- Coal is the world's single largest contributor of greenhouse gases and is one of the most important causes of global warming.
- Petroleum is a complex mixture of hydrocarbons, mostly alkanes and cycloalkanes. It occurs below the earth crust entrapped under rocky strata.
- The composition of natural gas is a mixture of mainly methane (95.0%), small amounts of ethane, propane and butane (3.6%) and traces of CO_2 (0.48%) and N_2 (1.92%).
- In India, the oil deposits are found at Ganga-Brahmaputra Valley, Bombay high, Plains of Gujarat, Thar Desert of Rajasthan and area around Andaman Nicobar islands.
- Energy obtainable from water flow or water falling from a higher potential to lower potential, is known is hydropower.
- Hydropower is a conventional and renewable form of energy which can be transmitted to long distance through cables and wires. Hydroelectric power has led to economic progress around the world, and also created serious ecological and human problems.
- The Narmada Bachao Andolan in India is an example of a movement against large dams. The gigantic Narmada River project has affected the livelihoods of hundreds of extremely poor forest dwellers.
- In India, hydroelectric power is generated by a number of multipurpose river valley projects e.g. Hydropower project Hirakud, Bhakra Nangal project, Narmada valley project, Nagarjun Sagar project, Sardar Sarovar project etc.

- A small amount of radioactive substance (U235) can produce a lot of energy through the process of nuclear fission. For example, one ton of uranium can provide energy which is much higher than three million tons of coal or 12 million barrels of oil.
- Fusion means joining lighter nuclei to make a heavier nucleus, most commonly hydrogen or hydrogen isotopes to create helium, such as 2H + 2H ® 3He (+ n). The hydrogen bomb is based on thermonuclear fusion reaction.
- India has the world's largest solar cooker programme and an estimated 2 lakh families that use solar cookers.
- Photovoltaic cells are commonly used today in calculators and watches. They also provide power to satellites, electric lights, and small electrical appliances such as radios and for water pumping, highway lighting, weather stations, and other electrical systems located away from power lines.
- Wind energy is an environment-friendly and efficient source of renewable energy. It requires no recurring expenses for the production of electricity, but there are many limitations in harnessing wind energy.
- The geothermal energy may be defined as the heat energy obtainable from hot rocks present inside the earth crust. Different sites of geothermal energy generation are Puga (Ladakh), Tattapani (Suraguja, M.P.) and Cambay Basin (Alkananda Valley, Uttaranchal).
- Petro Plants (Petroleum Plants) belongs to some species of certain families which accumulate hydrocarbons having high molecular weight (10,000) i.e. Hevea rubber (Hevea brasiliensis), Jatropha (Jatropha curcas), Aak (Calotropis procera) and Gurjun (Dipterocarpus turbinatus) etc.
- First of all, Dr. M. Calvin of the University of California (in 1979) reported the collection and use of photosynthetically produced hydrocarbons from the plants.
- Petroplants have lactiferous canals in their stem and secrete milky latex.
- Biomass is a mixture of gases which includes methane, carbon dioxide, hydrogen sulphide and water vapour.
- Biomass is categorized in to three types such as biomass in (a) traditional form (b) non-traditional form and (c) domestic use.
- Renewable resources are the resources which can be revived, through the natural process, over time. On the other hand, non-renewable resources are the natural resources which are not going to be replenished, shortly.
- Renewable resources emit less carbon as compared to non-renewable resources. Hence, they are pollution free and environment-friendly.
- Conventional sources of energy are those sources which are widely used all around the world since ages. On the conflicting, non-conventional sources

of energy are described as the energy sources, whose evolution have been done in the recent past and have gained popularity since then.

- Conventional sources of energy are limited in nature, and their formation takes millions of years, they can be exhausted one day. Conversely, non-conventional sources of energy are the sources that are in abundance in the environment and are easily renewable, so they are inexhaustible.
- Conventional sources of energy pollute the environment. On the other hand, non-conventional sources of energy are environment-friendly, so they do not harm the nature, by polluting it.

VERY SHORT ANSWER TYPE QUESTIONS:

1. What are resources?
2. What are natural resources?
3. Definition of renewable resources.
4. Definition of non-renewable resources.
5. Name any one renewable and non-renewable natural resource.
6. What type of fuel combustion is needed to increase the efficiency of engines and to reduce air pollution?
7. Name any two products formed when fossil fuels are burnt.
8. Define the term chemical energy.
9. Define the term electrical energy.
10. Define the term nuclear energy.
11. What is conventional energy?
12. What is non- conventional energy?
13. Give one example of conventional energy sources.
14. Give one example of non-conventional energy sources.
15. Define the term natural gas.
16. Full form of CNG.
17. What is hydropower?
18. What is nuclear power?
19. Give the use of uranium.
20. What is nuclear fusion?
21. What is photovoltaic energy?
22. Give the role of wing energy.
23. Full form of OTEC.
24. Give two sites of geothermal energy generation.

25. What is biomass energy?
26. What is biogas?
27. Give the Composition of biogas.
28. What is petroleum plants?
29. Who discovered the petro plants?
30. Give two examples of petroleum plants.
31. Give the significance of Jatropha.
32. What is dendrothermal energy?

SHORT ANSWER TYPE QUESTIONS

1. How does formation of coal and petroleum occur?
2. Name the four elements that constitute fossil fuels.
3. Give the differences between conventional and non- conventional energy.
4. What are fossil fuels? How are they formed?
5. List any four methods of conserving fossil fuels.
6. Write a short note on any one (a) growing of energy needs (b) Coal (c) Petroleum (d) Fossil fuels (e) Narmada Project (f) Solar Energy (g) Solar Cooker (h) Solar Collector (i) Tidal Power (j) Jatropha
7. Mention the drawback of hydropower plant.
8. Define the term photovoltaic energy and give uses of this energy.
9. Write on environmental impacts of wind energy.
10. Write on Ocean Thermal Energy.
11. Write on different forms of biomass.
12. Write on petroleum plants with uses.
13. Give the advantages of renewable or non-renewable energy resources.
14. Give the disadvantages of renewable or non-renewable energy resources.
15. Differences between renewable and non-renewable resources.
16. Differences between conventional and non-conventional sources of energy.

LONG ANSWER TYPE QUESTIONS

1. What are natural resources? State two factors that work against an equitable distribution of these resources.
2. How can we as individuals help in reducing the use of fossil fuels? Mention two ways.
3. How has the use of fossil fuels led to the deterioration of environment? State any two reasons.

4. Define the term hydropower energy and explain with the example of Narmada Project
5. What is solar energy? Explain any two uses of solar energy in daily life
6. Explain the conventional or non-conventional energy sources.
7. Why should use coal and petroleum. State the reasons.
8. List four ways in which management of fossil fuels differ from natural resources of energy.
9. How is the increase in demand of fossil fuels affecting the environment adversely? Suggest two ways to minimize the problem.
10. The construction of large dams lead to (i) Social and (ii) Environmental problems. List two problems in each case.
11. Under what conditions is carbon monoxide released when fossil fuels are burnt?
12. List two gases released on burning carbon. With their effects.
13. How are fossil fuels formed in nature? What are the products formed when combustion takes place in insufficient air?
14. Write the harmful effects of using plastic bags on the environment. Suggest alternatives to plastic bags.
15. Why is sustainable management of natural resources necessary?

MULTIPLE CHOICE QUESTIONS:

1. Coal and Petroleum are
   (a) Used in biogas production
   (b) Fossil Fuel
   (c) Fuel Wood
   (d) Inorganic Fuel resources
2. Which of the following is not a natural resource
   (a) Natural gases
   (b) Water
   (c) Electricity
   (d) Soil
3. Example of Petro Plants
   (a) Hevea rubber (*Hevea brasiliensis*)
   (b) Jatropha (*Jatropha curcas*)
   (c) Aak (*Calotropis*)
   (d) All of these
4. First of all, who reported the collection and use of photosynthetically produced hydrocarbons from the plants
   (a) M. Calvin
   (b) A. Chakraborty
   (c) A.K. Banerjeee
   (d) Amrita Devi

5. Which have lactiferous canals in their stem and secrete milky latex.
   (a) Petroleum Plants (b) Algae
   (c) Fungi (d) Protozoas
6. Biomass is a mixture of
   (a) Methane and CO_2 (b) CO_2 and H_2S
   (c) Metane and H_2S (d) Methane, CO_2 and H_2S
7. Which is the following combinations has no fossil duel
   (a) Wind, Ocean, Water (b) Wood, Sun, Petroleum
   (c) Kerosene, Tide, Wind (d) Wood, Wind, Sun
8. Which type of resources emit less carbon
   (a) Renewable (b) Non-renewable
   (c) Both (d) None of these
9. Which type of energy sources pollute the environment
   (a) Conventional (b) Non-conventional
   (c) Both a and b (d) None of these
10. Which type of energy sources are environment friendly
   (a) Conventional (b) Non-conventional
   (c) Both a and b (d) None of these

ANSWER KEY

1. (b) 2. (c) 3. (d) 4. (a) 5. (a) 6. (d) 7. (a) 8. (a)
9. (a) 10. (b)

❖❖❖

6

CHAPTER

OUR ENVIRONMENT

Eco-system, Environmental Problems, Ozone Depletion, Waste Production and their Solutions, Biodegradable and Non-biodegradable Substances

The environment of any organism means the physical and biological conditions in which it lives. The word environment was derived from the word "environ" meaning "to encircle" or "surround". The environment includes the physical surroundings like water bodies, air (or atmosphere), soil (land) and all the organisms such as plants, animals, human beings and micro-organisms (Algae, Fungi, Bacteria and Viruses etc.). All these elements of the environment are dependent on each another. Therefore, all the constituents of environment interact with one another and maintain a balance in the environment. Human beings are the only organisms who change/disturb the natural environment to fulfill their demands/needs such as food, clothing, housing, transport and industry etc. In fact, the uninhibited actions of human beings are damaging the balanced and healthy environment more and more. A change in any of these conditions can affect the organism. To understand how, we need to look at the different ways in which an organism interacts with other organism and with its surroundings.

6.1 ECOSYSTEM

The various communities of living organisms (plants and animals) interact among themselves as well as with their physical environment (like soil, air and water). The living organisms interact with one another through their food chains in which one organism consumes another organism. The living organisms like plants interact with soil to get essential nutrients like N (nitrogen), P (phosphorus), etc. which are to get carbon dioxide (CO_2) and also with water bodies, for carrying out the process of photosynthesis. Thus, the various communities of living organisms (called biotic communities) like plants and animals alongwith soil, air and water of that region form a self-sustaining or functional unit of the living world. This 'functional unit' or 'system' made up of living and non-living components which are capable of independent existence is called an ecosystem. The ecosystem includes all the communities of an area (all the plants and animals of an area) functioning with their non-living environment like soil, air and water. **The term "ecosystem" was first used in 1935 in a publication by British ecologist Arthur Tansley (Fig 6.1).**

Fig 6.1 British Ecologist-Arthur Tansley

(15 August 1871 to 25th November 1955)

6.1.1 Definition of Ecosystem

An ecosystem is a self-contained unit of living things such as plants, animals and decomposers, and their non-living environment (soil, air and water). An ecosystem needs only the input of sunlight energy for its functioning. Thus an ecosystem is a functional unit consisting of all the living beings of an area and the non-living components of their environment, interacting to form a stable ecosystem. Or

An ecosystem may be defined as a structural and functional unit of the biosphere comprising living organism and their non-living environment which interact by means of food chains and biogeochemical cycles resulting in energy-flow, biotic diversity and material cycling to form a stable and self-supporting.

6.1.2 Types of Ecosystem

In the biosphere, ecosystems may be classified on the basis of their nature, duration and size.

6.1.2.1 On the Basis of Nature

Ecosystems are classified into two types such as:

(i) **Natural Ecosystems:** These ecosystems operate in the nature by themselves without any human interference. Common examples of natural ecosystems are such as pond ecosystem, Lake Ecosystem, desert ecosystem, grassland ecosystem, forest ecosystem and oceanic ecosystem etc.

(ii) **Artificial Ecosystems:** These are maintained by man and hence are also termed man-engineered or man-made ecosystems. In these ecosystems, man preserves or disrupts the natural balance by the addition of energy and planned manipulations as crop lands, orchards, gardens and aquarium etc.

6.1.2.2 On the Basis of Duration

Ecosystems are classified into:

(i) **Temporary Ecosystems**: These are short-lived ecosystems which may be natural or man-made as rainfed pond and laboratory culture of protozoans.

(ii) **Permanent Ecosystems:** These are self-supporting natural ecosystems that maintain themselves for relatively long duration e.g., lake, forest, desert etc.

6.1.2.3 On the Basis of Size

Ecosystems are classified into:

(i) **Small Ecosystems:** Small-sized ecosystems are also termed micro ecosystems, e.g., a flowerpot, a site under a stone etc.

(ii) **Large Ecosystems:** Very large-sized ecosystems are also termed macro ecosystems, e.g an ocean, a forest and a desert etc.

The term ecosystem was first used in 1935 by British Ecologist Sir Arthur Tansley. An ecosystem may be defined as a structural and functional unit of the biosphere comprising living organism and their non-living environment which interact by means of food chains and biogeochemical cycles resulting in energy-flow, biotic diversity and material cycling to form a stable, self-supporting. Ecosystems may be classified on the basis of their nature, duration and size. On the basis of nature, ecosystems are classified into two types such as (a) Natural ecosystems (Pond and Lake Ecosystem etc.) and (b) Artificial ecosystems (Croplands and Aquarium etc.).

6.1.3 Components of an Ecosystem:

All the ecosystems are made up of two main components such as (a) Abiotic and (ii) Biotic components. Abiotic components means non-living components and biotic components mean living components. Thus, we can now say that an ecosystem consists of non-living environment and the living biological community.

6.1.3.1. Abiotic Components (Non-living Components)

Abiotic components or non-living components of an ecosystem include the non-living physic-chemical factors of the environment. These components not only affect the distribution and structure of organisms but also their behavior and interrelationships. Abiotic factors include inorganic substance, organic compounds and climate factors.

(i) **Inorganic Substances:** Inorganic substances, e.g., carbon, nitrogen, oxygen, calcium, phosphorus etc. and their compounds (water, carbon dioxide, etc.) constitute the main abiotic components. These occur either in the form of compounds dissolved in water, in the soil or in free state in the air.

(ii) **Organic Compounds:** These include carbohydrates, proteins, lipids, nucleic acids etc. These are present in living organisms and dead remains of them. The dead organic matter is broken down by the action of decomposers (e.g., bacteria and fungi) into inorganic substances for their recycling.

(iii) **Climatic Factors:** These include light, temperature, humidity, wind, rainfall, water and edaphic factors (e.g., soil and substrate, topography, minerals, pH) etc.

6.1.3.2 Biotic Components (Living Components)

The biotic or living component of an ecosystem is a community of organisms (like plants and animals), which is made up of many different inter-dependent populations. The biotic community (or living community) of an ecosystem includes three types of organisms such as:

- **Producer Organisms (or Autotrophs):** They synthesize (make) their own food. They are also called autotrophs (In Greek, autos=self, trophe=nutrition). All the green plants and certain blue green algae (BGA) are act as producers.
- **Consumer Organisms (or Heterotrophs):** These are dependent on other organisms for food. They are also called as heterotrophs (In Greek hetero = other). All the animals which eat plants or other animals are consumers.
- **Decomposer Organisms (or Saprotrophs):** They consume the dead remains of other organisms or feed on dead plants and animals called decomposers. They are also known as Saprophytes on saprotrophs (In Greek, sapros = rotten). Certain bacteria, fungi and worms are decomposers.

6.1.4 Functions of an Ecosystem

The ecosystem has non-living components like soil, water and air which contain inorganic nutrient elements, and the living components called producers, consumers and decomposer organisms. All these components make the ecosystem function in this manner as from the nutrient pool of the earth (soil, water and air), carbon dioxide and water are absorbed by the producer organisms (green plants). With the help of sunlight energy, the producer organisms convert these inorganic substances into organic compounds (like carbohydrates) which act as a food. Thus, producers trap the solar energy and then provide the basic food or energy for all other live forms in the ecosystem. The consumers (animals) derive their energy needs, directly or indirectly, from producers (plants). When the producers (plants) and consumers (animals) die, then the decomposer organisms act on their dead bodies to return the various elements back to the nutrient pool (soil, water and air). Thus, an ecosystem involves input of energy and matter which are exchanged between living and non-living components in a cyclic process.

6.1.5 Producers, Consumers and Decomposers

According to the manner in which they obtain their food from the environment, all the organisms can be divided into three groups such as:

6.1.5.1 Producers

Those organisms which produce food are called producers. Producers are the organisms which can prepare their own food from simple inorganic substances like carbon dioxide and water by using sunlight energy in the presence of chlorophyll (Fig 6.2). The green plants synthesize their food during photosynthesis by taking raw materials from the earth and energy from the sun. The green plants produce carbohydrates by photosynthesis and also synthesize proteins and fats. Thus, the green plants are called producers in the living world. Producers are the autotrophic organisms (self-feeder organisms) in the ecosystem upon which other organisms depend for food. Thus, producers (like green plants) are autotrophs.

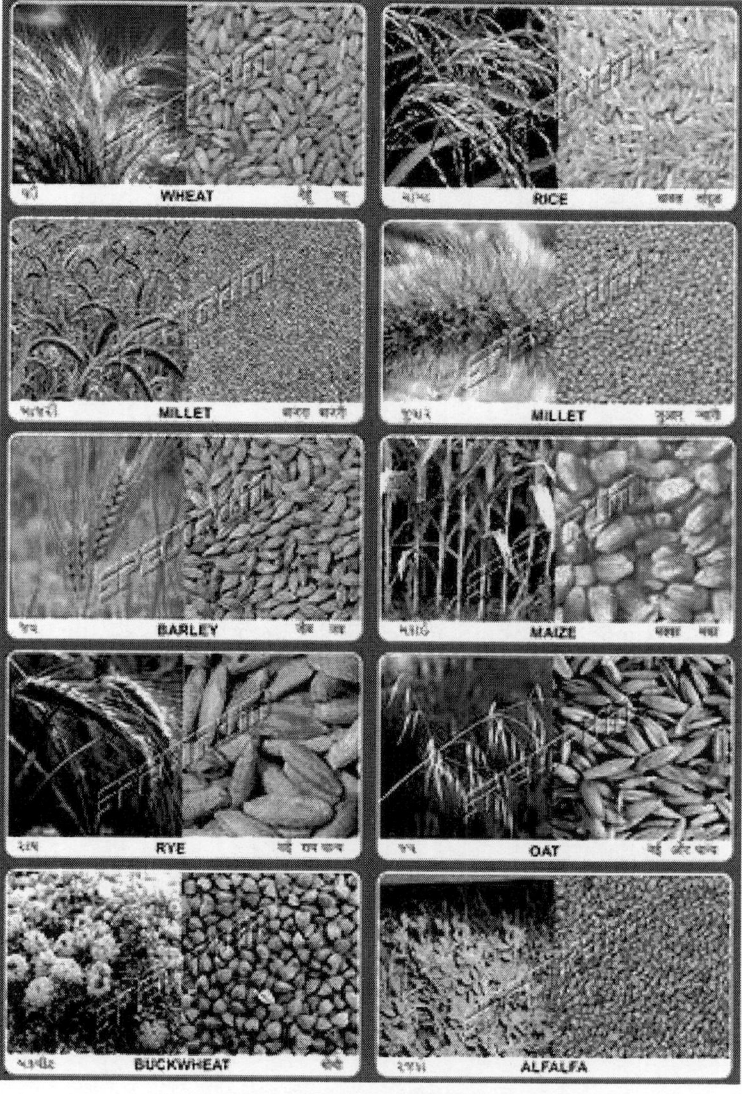

Fig. 6.2 Diagram shows the producers

6.1.5.2 Consumers

The consumers depend on producers for food, directly or indirectly. The consumers get their food by eating other organisms or their products. In most simple words, consumers are the organisms that eat other organisms. All the animals are consumers. Even the microscopic animal lives of the water called protozoa are consumer organisms. The examples of common consumer organisms are man, goat, deer, fish, lion, cow and buffalo, etc. The cow and buffalo eat green grass and other green fodder because green grass and other green plants are producers of food. The bio-mass of grass and plants supplies food and energy to these animals like cow and buffalo. It should be noted that the consumer organisms like animals cannot prepare food from simple inorganic substances through photosynthesis. The consumers need ready-made food for their survival which they get from producers (green plants), either directly or indirectly. If an animal eats grass or other green plants or their products itself we say that it gets the food from producers directly. For example, a goat gets the food from producers directly when it eats grass. On the other hand, if an animal eats the meat of another animal (which eats grass), then we say that it gets the food from producer indirectly. For example, a lion gets food by eating goat which in turn eats grass. So, in this case the lion gets its food indirectly from producer grass (through the goat). Consumer organisms are also called heterotrophs. Consumers can be further divided into three categories such as (i) Herbivores (ii) Carnivores and (iii) Omnivores.

(i) Herbivores or Herbivorous animals:

Some animals eat only plants or plant products. Those animals which eat only plants are called herbivores. The herbivores may eat grasses, leaves, grains, fruits or the bark of trees. Some of the examples of herbivores are as Cow, Buffalo, Goat, Sheep, Horse, Deer, Camel, Ass, Ox, Elephant, Monkey, Squirrel, Rabbit and Hippopotamus (Fig 6.3). The animals which get their food by eating the producers (plants) directly are called Primary Consumers. Since herbivores obtain their food directly from plants (or producers), therefore, herbivores (like cattle, deer, goat, etc.) are primary consumers.

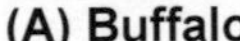

(A) Buffalo

(B) Goat

(C) Sheep

(D) Deer (E) Cow

Fig. 6.3 Herbivorous Animals

(ii) Carnivores or Carnivorous animals:

Some animals eat only other animals. They do not eat plant food at all. Those animals which eat only other animals as food are called carnivores. The carnivores eat the meat (or flesh) of other animals. So, we can also say that those animals which eat only the meat (or flesh) of other animals are called carnivores. Some of the examples of the carnivores are as Lion, Tiger, Frog, Vulture, Kingfisher, Lizard, Wolf, Snake and Hawk (Fig 6.4). Lion is called a carnivore because it eats only the meat (or flesh) of other animals like deer, rabbit and goat, etc. Carnivores are also known as carnivorous animals.

(A) Tiger (B) Lion

Fig. 6.4 Carnivorous Animals

The carnivores are further categorized in to two types such as (i) small carnivores and (ii) large carnivores. **The small carnivores** which feed on herbivores (Primary Consumers) are called Secondary consumers i.e. a bird, frog, lizard and fox, etc., are secondary consumers. **The large carnivores** (or top carnivores) which feed upon the small carnivores (secondary consumers) are called tertiary consumers i.e. lion, tiger and birds of prey (such as hawk)

(iii) Omnivores or Omnivorous Animals:

Some animals eat both, plants as well as other animals. Those animals which eat both, plants and animals are called omnivores. In other words, the omnivores eat plant food as well as the meat (or flesh) of other animals. Some of the examples of omnivores are man (human beings), Dog, Bear, Crow, Sparrow, Mynah and Ant. Man is called an omnivore (Fig 6.5) because he eats both, plant food (such as grains, pulses, vegetables and fruits) as well as meat of animals (such as goat, chicken and fish). Omnivores are also called omnivorous animals or Tertiary Consumers.

Fig. 6.5 Omnivorous Animal

We will now describe another type of producers and consumers which are extremely small. These are called planktons. Planktons are very minute or microscopic organisms freely floating on the surface of water in a pond, lake, river or ocean. Planktons are of two types: such as (i) Phytoplanktons and (ii) Zooplanktons. **The microscopic aquatic plants freely floating on the surface of water are called phytoplanktons** (Fig. 6.6A). The free-floating algae are an example of phytoplankton. Phytoplanktons are capable of producing food by the process of photosynthesis. **The microscopic aquatic animals freely floating on water are called zooplanktons.** The freely-floating protozoa are an example of zooplankton (Fig. 6.6B). A very, very small fish is also a zooplankton. Planktons float near the surface of water and provide food for many fish and other aquatic animals.

(a) Phytoplanktons

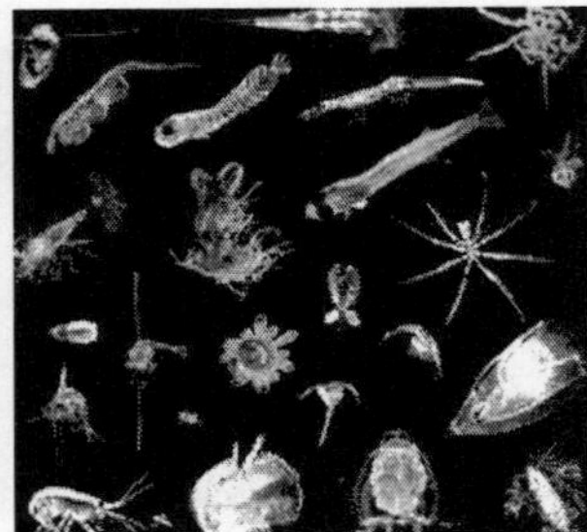

(b) Zooplanktons

Fig. 6.6 Phyto and Zooplanktons (Microscopic Aquatic Plants and Animals)

6.1.5.3 Decomposers:

The micro-organisms like some bacteria and fungi which break down the complex organic compounds present in dead organisms like dead plants and animals and their products like faeces, urine, etc., into simpler substances are called decomposers. The bacteria which act as decomposers are called *putrefying bacteria*. The bacteria and fungi act as decomposers by the secretions of their body surfaces which decompose the organic matter present in dead plants and animals into simpler substances and liberate ammonia, carbon dioxide etc. They absorb some of these simpler substances for their own maintenance and release the remaining into the soil, water and air to be used by the producers again in this way, decomposers help in the recycling of materials in ecosystem. The decomposers are also known as micro-consumers or saprotrophs (Fig 6.7).

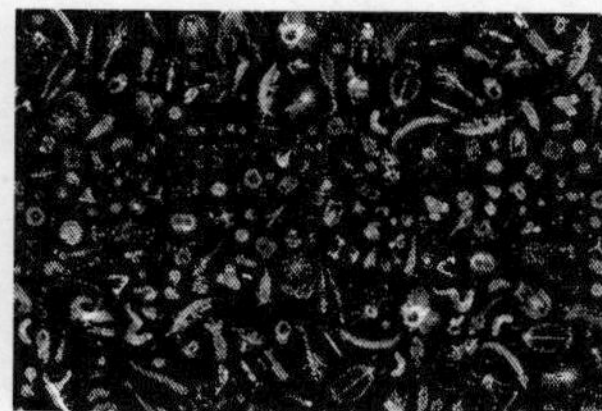

Fig. 6.7 Decomposers

Importance of Decomposers:

The decomposers act as cleansing agents of environment due to decomposing the dead bodies of plants and animals. The decomposers also help in putting back the various elements of which the dead plants and animals are made, back into the soil, air and water for re-use by the producers like crop-plants. This maintains the fertility of soil and the soil would continue to support crops again and again. For example, the decomposers like putrefying bacteria and fungi decompose the dead plants and animal bodies into ammonia (and other simpler substances). This ammonia is converted into nitrates by the nitrifying bacteria present in soil. These nitrates act as fertilizer in the soil and are again absorbed by the plants for their growth. Thus, it is only due to the presence of decomposers that the various nutrient elements which were initially taken by plants from the soil, air and water are returned to after the death of plants and animals. If, however, there were no decomposers, then the dead bodies of plants and animals would keep lying as such and the elements of which plant and animal bodies are made, would never be returned to their original pools like soil, air and water. In that case, the cyclic process of life and death would be disrupted. This is because in the absence of decomposers the soil, air and water would not be replenished by elements from the bodies of dead organisms. All the nutrients present in soil, air and water would soon be exhausted and evolution of life would come to an end. Thus, the decomposer organism's help in recycling the materials in the ecosystem so that the process of life may go on and on like an unending chain.

6.18 EXAMPLES OF ECOSYSTEM:

6.1.6.1 Garden Ecosystem:

It is an artificial, terrestrial ecosystem. It comprises of following components (Fig 6.8A) such as:

Abiotic components:

They include soil, its chemical (pH, nutrient concentration etc.) and physical (soil texture, soil water, aeration, porosity etc.) characters, sunlight, air etc.)

Biotic components*:*

(i) **Producers:** Grasses, flower bearing plants (e.g., rose, jasmine etc.) and trees are producers.

(ii) **Consumers:** Animals such as frogs, insects and birds are consumers. All three types of consumers i.e., primary (e.g., insects), secondary (e.g., frogs) and tertiary (e.g., kites) are present in a garden ecosystem. All these living organisms interact with each other and their growth, reproduction and other activities are influenced by the abiotic components of ecosystem.

(iii) **Decomposers:** Decomposers are microorganisms present in the soil. These cannot be seen with naked eye. These provide raw materials back to the environment, by decomposition of dead, decaying matter, for their reuse by the producers. They maintain fertility of the garden soil.

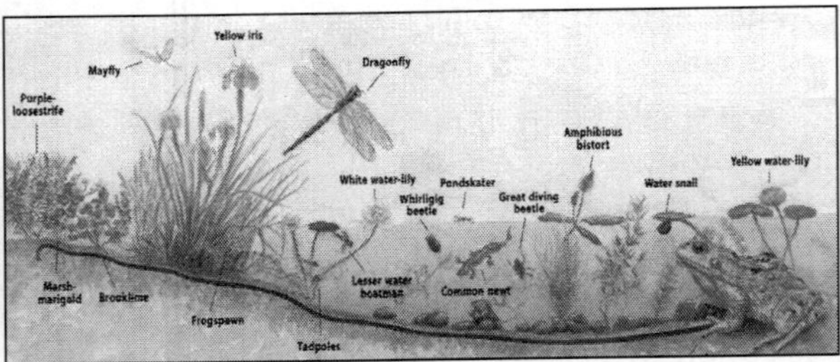

(a) Garden Ecosystem **(b) Pond Ecosystem**

Fig. 6.8 Example of Ecosystems

6.1.6.2 Pond Ecosystem

A pond is a classic example of an aquatic ecosystem (Fig 6.8B). The pond ecosystem comprises of the following abiotic and biotic components such as:

(i) **Abiotic Components:** These include water and the physical and chemical environment of water. Physical environment comprises of temperature, light intensity, pH, of water etc., while the chemical environment constitutes the basic elements, minerals, dissolved gases, organic and inorganic compounds.

(ii) **Biotic Components:** Following are the biotic components of pond ecosystem.

(iii) **Producers (Autotrophs):** These include rooted and floating plants as well as phytoplanktons.

(iv) **Consumer:** These include the heterotrophs mainly animals that feed upon other organisms. The primary consumers, (C_1) are minute floating fauna (zooplanktons), water fleas and larvae of insects. Consumers of the second order (C_2) are crustaceans, rotifers, small fish and frogs. The tertiary consumers (C_3) are large fish and water birds.

(v) Decomposers: These include aquatic bacteria and fungi. Although they are distributed throughout the pond, they are abundantly present in the mud at the bottom of the pond. Producers (plants) receive solar-energy to form organic materials through photosynthesis.

Producers are eaten by primary consumers (insect larva, rotifers, molluscs) which in turn are eaten by secondary consumers (insects, fish, frogs). The secondary consumers become the prey of tertiary consumers (large fish, snakes, water birds). At some stage or the other, producers and different grades of consumers die. Their dead bodies are acted upon by decomposers who are responsible for breaking the complex organic substances into simple substances to be utilized again by producers. In this way, pond is an ideal example of an ecosystem.

All the ecosystems are made up of two main components such as (a) Abiotic (non-living components) and (ii) Biotic (living components) components. Abiotic factors include inorganic substance, organic compounds and climate factors. The biotic community (or living community) includes three types of organisms such as producer, consumer and decomposer.

Consumers can be further divided into three categories such as (i) Herbivores (ii) Carnivores and (iii) Omnivores. Animals eat only plants or plant products are called herbivores. Animals which eat only other animals as food are called carnivores, and animals which eat both, plants and animals are called as omnivores.

Planktons are of two types such as (i) Phytoplanktons and (ii) Zooplanktons. The microscopic aquatic plants freely floating on the surface of water are called Phytoplanktons (free-floating algae), whereas freely floating aquatic animals on water are called zooplanktons (freely-floating protozoa).

6.2 FOOD CHAINS

Food acts as a fuel to provide energy to do work. It is true for all living organisms. Anything which we eat to live is called food. The food (or energy) can be transferred from one organism to the other through food chains. In any given ecosystem, all living organisms are linked in a systematic chain with respect to their mode of manufacturing food/feeding habits. The interactions among various components of the environment involve flow of energy from one component of the ecosystem to other. The sequence of living organisms in a community, in which one organism consumes another organism to transfer food energy, is called a food chain. In simple words, a list of organisms (living beings) showing "who eats whom" is called a food chain (Fig 6.9). For example, in a grassland or forest, there is a lot of grass (which are green plants). This grass is eaten up by animals like cows and deer, when in turn are then eaten up by a lion. This simple food chain operating in a grassland or forest can be represented as:

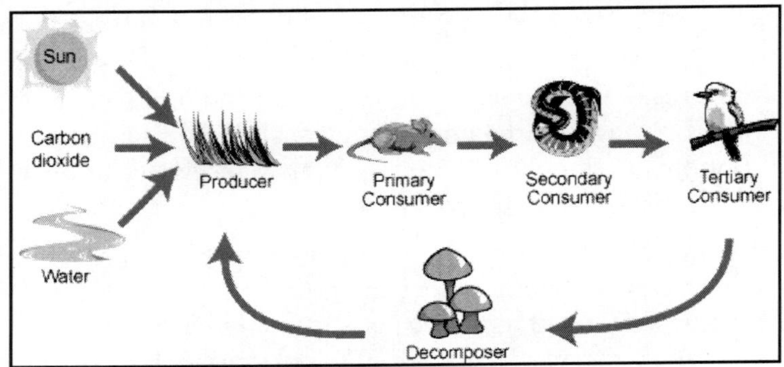

Fig. 6.10 Digram show Food Chains

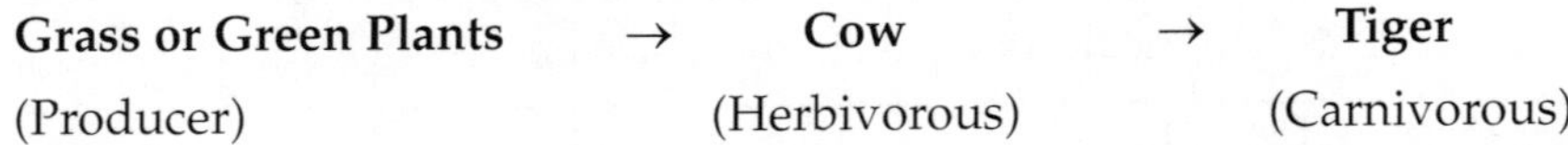

Grass or Green Plants → **Cow** → **Tiger**

(Producer) (Herbivorous) (Carnivorous)

Thus, food chain is a sequence of organisms in a biotic community through which food passes with members of a step becoming food of the next step of the sequence. Each step in food chain which is characterized by a particular method of obtaining food is called trophic level. A food chain usually possesses 3-4 trophic levels in a terrestrial ecosystem and 4-5 trophic levels in aquatic ecosystem.

1. **Producers:** They constitute the base or beginning of a food chain. Producers trap solar energy, manufacture organic food from inorganic raw materials and release oxygen. They constitute the first trophic level (T_1) of a food chain.
2. **Herbivores or First Order Consumers (Primary Consumers):** They are animals which feed on plants or plant products, e.g., Grasshopper, Rabbit, Deer, Elephant, herbivores constitute second trophic level (T_2).
3. **First Order (Primary) Carnivores or Second Consumers (Secondary Consumers):** They are animals which prey upon herbivores, e.g., Frog, Wild Cat and Fox. These animals form the third trophic level (T_3).

Table 6.1 Differences between First and Third Trophic Level

Sr. No.	Particulars	First Trophic Level	Third Trophic Level
1.	Status	The organisms belonging to first trophic level are producers.	They are first order carnivores.
2.	Nutrition	The members of first trophic level prepare their own food from inorganic raw materials in the process of photosynthesis.	The members of third trophic level prey upon second trophic level organisms or herbivores.
3.	Energy	Energy is obtained from solar radiations. The same is converted into chemical energy. Example. Plants, Phytoplankton.	They energy is obtained from herbivore biomass. It is chemical energy. Examples. Fox, Wild Cat, Frog.

4. **Second Order (Secondary) Carnivores or Third Order Consumers (Tertiary Consumers):** They are larger carnivores which prey upon first order carnivores or secondary consumers. The second order carnivores are third order consumers which form fourth trophic level (T_4), e.g., Hawk, Kingfisher.

The last orders of consumers or carnivores which are not further preyed upon the other animals are called top carnivores. They may belong to T_3, T_4 or T_5. The last trophic level (generally T_5 or T_6) is always that of decomposers.

Table 6.2 Differences between Second Trophic Level and Fourth Trophic Level

Sr. No.	Particulars	Second Trophic Level	Fourth Trophic Level
1.	Status	The members of second trophic level are herbivores.	The members of fourth trophic level are second order carnivores.
2.	Nutrition	They feed on plants.	The feed on primary carnivores.
3.	Energy	The number is large. Examples. Grasshopper, Deer, Rabbit	The number is small. Examples. Hawk, Peacock.

(a) The plants are producers (or autotrophs) and constitute the first trophic level. They fix up the sun's energy and make it available for consumers (or heterotrophs).

(b) Herbivores (which feed upon plants) constitute the second trophic level.

(c) Carnivores (that feed upon herbivores) constitute the third trophic level.

(d) Large carnivores or Top carnivores (which feed upon small carnivores), constitute the fourth trophic level.

6.2.1. On the Basis of Habitat Types of Food Chains

Depending upon the habitat, food chains are of two types, terrestrial (on land) and aquatic (in water). A food chain has 2-6 trophic levels or steps. Every sustainable food chain is producer based. Each higher trophic level exerts a biocontrol over the lower trophic level. Biomagnifications of non-degradable chemicals occurs with the rise of trophic levels. A two-step food chain occurs in those forests where elephants are found. It ends at the level of herbivore. A three step food chain occurs in Gir forest where lions are found. Four steps of food chain are more common.

Terrestrial Food Chains:

1. Two Step Food Chain: Vegetation → Elephant
   (Producer) (Herbivore)
2. Three Step Food Chain: Grass → Deer → Lion
   (Producer) (Herbivore) (Carnivore)
3. Four Step Food Chain: → Grass → Insect → Insectivorous Bird → Hawk
   (Producer) (Herbivore) (Carnivore I) (Carnivore II)

4. Five Step Food Chain:

   Grass (Producer) → Insect Herbivore → Frog Carnivore (I) → Snake Carnivore (II) → Peacock Carnivore (III)

6.2.3 Types of Food Chains

In nature, basically two types of food chains are recognized such as (i) Grazing food chain and (ii) Detritus food chain.

1. **Grazing Food Chain:** This type of food chain starts from the living green plants goes to grazing herbivores, and on to carnivores (Fig 6.10A). Ecosystems with such type of food chain are directly dependent on an influx of solar radiation. Thus, this type of food chain depends on autotrophic energy capture and the movement of this captured energy to herbivores. Most of the ecosystems in nature follow this type of food chain. **The Phytoplanktons → Zooplanktons → Fish Sequence or the Grasses → Rabbit → Fox sequences are the examples of the grazing food chain.**

2. **Detritus Food Chain:** This type of food chain goes from dead organic matter into micro-organisms and then to organisms feeding on detritus (detrivores) and their predators. Thus, such ecosystems are less dependent on direct solar energy (Fig 6.10B). These depend chiefly on the influx of organic matter produced in another system. For example, such type of food chain operates in the decomposing accumulated litter in a temperate forest.

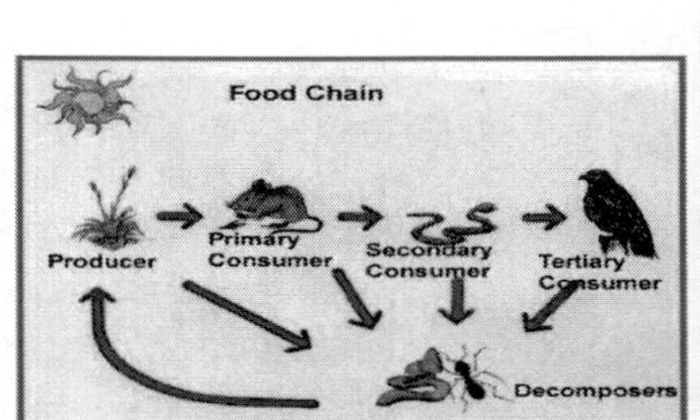

(A) Grazing Food Chain (B) Detritus Food Chain

Fig. 6.10 Types of Food Chains

6.2.4 Length of Food Chains

In ecosystems different food chains may have two, three, four or maximum five trophic levels. Accordingly, a food chain may end at the **(i)** Herbivore (Primary consumer level) **(ii)** Primary carnivore (Secondary consumer level), **(iii)** Secondary carnivore (Tertiary consumer level) or **(iv)** Tertiary carnivore (Quaternary consumer level). First trophic level in all of these food chains will be the producers.

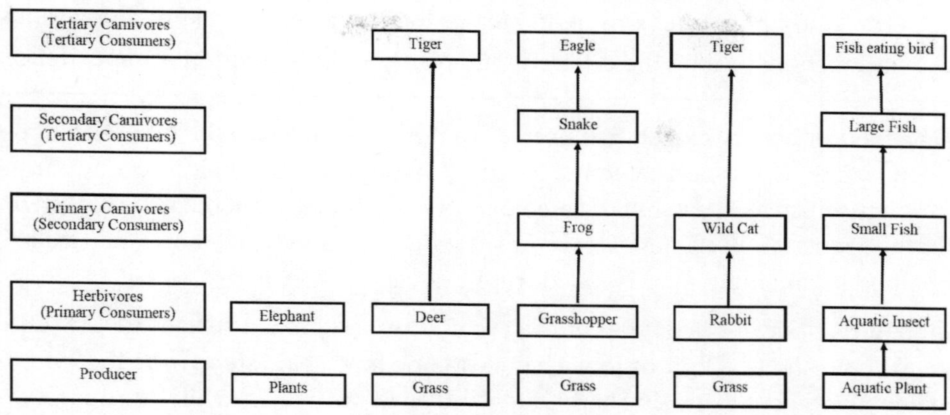

Fig. 6.11 Food Chain may consist of various levels

6.2.5 Characteristics of Food Chain

(a) A food chain involves a nutritive interaction between the living organisms (biotic components) of an ecosystem. In a food chain, there occurs repeated eating, i.e., each group eats the other group and subsequently is eaten by some other group of organisms.

(b) A food chain is always straight and proceeds in a progressive straight line.

(c) In a food chain, there is unidirectional flow of energy from sun to producers and subsequently to series of different types of consumers.

(d) Usually, there are 3 or 4 trophic levels in the food chain. In few chains, there may be a maximum of 5 trophic levels.

(e) Some organisms are omnivores. These occupy different trophic positions in different food chains.

(f) Generally 80-90% of energy is lost at each transfer, as heat in accordance with second law of thermodynamics.

6.2.6 Significance of Food Chain

- They also help us to appreciate the energy flow mechanism and from one species to another at various biotic levels matter circulation in ecosystem and understand the movement of toxic substances in the ecosystem.
- The study of food chain helps us to understand the problems of bio-magnifications.
- Food chain helps in studying feeding relationships between organisms.
- A food chain represents a single directional or unidirectional transfer of energy. It shows the transfer of energy takes place from grass to cow and then to lion. It cannot take place in the reverse direction from lion to cow to grass.

The study of food chains in an area or habitat helps us in knowing various interactions among the different organisms and also their interdependence.

The food or energy can be transferred from one organism to the other through food chains. Thus, food chain is a sequence of organisms in a biotic community through which food passes with members of a step becoming food of the next step of the sequence. Each step in food chain which is characterized by a particular method of obtaining food is called trophic level.

In nature, basically two types of food chains are recognized such as (i) Grazing food chain and (ii) Detritus food chain. Grazing food chain starts from the living green plants goes to grazing herbivores, and on to carnivores, while detritus food chain goes from dead organic matter into micro-organisms and then to organisms feeding on detritus (detrivores) and their predators.

6.3 FOOD WEB

A large number of food chains exist in a community of living organisms in an ecosystem such as grassland, forest, pond or crop-field. Many of these food chains are inter-connected by species (organisms) which occur in more than one food chain. A food chain is a linear arrangement of animals. It does not give a complete picture of the feeding relationship among the different organisms of an ecosystem because many of them eat more than one kind of food. The inter-connected food chains functioning in an ecosystem which establish a network of relationships between various species, is called a food web (Fig. 6.12). In simple words, the network of a large number of food chains or interconnected food chains existing in an ecosystem is called a food web. The food web has many intercrosses and linkages among the various species (producers and consumers) present in it. It means that the various food chains in and ecosystem do not operate in isolation (or alone).

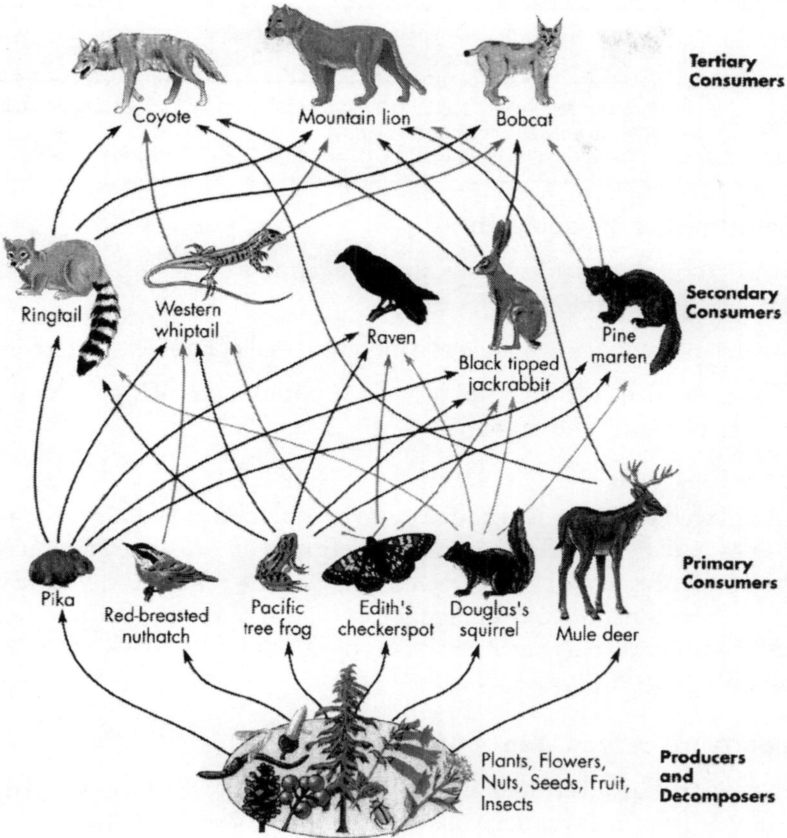

Fig. 6.12 Food Web (A food web consists of many interconnect food chains)

In this food web, we can see a network of numerous pathways along which the food (or energy) flows within grassland community. This food web starts from the plants which is a producer and ends in top carnivore hawk (baaz).

Table 6.3 Examples of the Food Chain

Food Chain	Particulars
1st Food chain	Plants are eaten by rabbit and then rabbit is eaten by hawk Plants → Rabbit → Hawk
2nd Food chain	Plants are eaten by mice (or rats) and the mice are eaten by hawks Plants → Mice → Hawk
3rd Food chain	Plants are eaten by mice; mice are eaten by snakes and then snakes are consumed by hawks Plants → Mice → Snake → Hawk
4th Food chain	Plants are eaten by seed-eating birds and the seed-eating birds are consumed by hawks Plants → Seed-eating Bird → Hawk

5th Food chain	Plants are eaten up by grasshopper and the grasshopper is consumed by hawks Plants → Grasshopper → Hawk
6th Food chain	Plants are eaten by grasshopper, grasshopper is eaten by frog, frog is eaten by snake and then snake is consumed by hawk Plants → Grasshopper → Frog → Snake → Hawk

6.3.1 Characteristics of Food Web

(a) Unlike food chains, food webs are never straight. Instead, each food web is formed by interlinking of food chains.

(b) Greater alternatives available in a food web make the ecosystem more stable.

(c) Food webs also help in checking the overpopulations of highly reproductive species of plants and animals.

(d) Food webs also help in ecosystem development.

(e) A food web provides alternative pathways of food availability. For example, if a particular species of producer is destroyed by a disease in the ecosystem, the herbivores of that area can feed on other species of producers. Similarly, secondary consumers (e.g., predatory birds) may feed on rats or mice in the event of decrease in population of rabbits in that area on which they commonly feed.

6.3.2 Importance of Food Web

1. **Starvation:** Food web does not allow any population to starve when members of lower trophic level decrease in number.
2. **Checking Overpopulation:** Food web does not allow a species to overgrow as increased availability will increase the chances of higher number of its predations.
3. **Endangered Population:** It allows endangered population to grow in size.
4. **Stability:** Food webs provide stability to ecosystems.

The network of a large number of food chains or interconnected food chains existing in an ecosystem is called a food web. The food web has many intercrosses and linkages among the various species (producers and consumers) present in it. It means that the various food chains in and ecosystem do not operate in isolation (or alone).

Table 6.4 Differences between Autotrophs and Heterotrophs

Sr. No.	Autotrophs	Heterotrophs
1.	They are producer organisms.	They are consumer organisms.
2.	They prepare their organic nutrients themselves	They do not prepare their organic nutrients themselves.

3.	They get only inorganic materials from outside, using which they synthesize organic materials inside their body.	They get both organic and inorganic materials from outside.
4.	These obtain energy from sunlight or inorganic chemical reactions.	These obtain energy from organic nutrients.
5.	These usually add O_2 to the environment.	These add CO_2 to the environment.
6.	These include plants and blue green algae.	These include animals, many protists, bacteria and fungi.
7.	These possess photosynthetic pigments (e.g., chlorophyll).	Photosynthetic pigments are lacking in them.
8.	They constitute the first trophic level.	They belong to second and higher trophic levels.

Table 6.5 Differences between Herbivores and Carnivores

Sr. No.		Herbivores	Carnivores
1.	Food	They are animals which feed on plants.	Carnivores are animals which feed on other animals.
2.	Canines	The animals do not have well developed canines.	Canines are well developed to tear flesh of the prey.
3.	Cellulose	Herbivores harbor microorganisms which degrade cellulose for them.	Carnivores do not degrade cellulose.
4.	Ecological Status	Herbivores are primary consumers or second trophic level organisms. Examples. Deer, Goat, Cow, Mouse.	Carnivores are higher level consumers. They belong to third or still higher trophic level. Examples. Frog. Fox, Snake.

Table 6.6 Differences between Carnivores and Omnivores

Sr. No.		Carnivores	Omnivores
1.	Food	It feeds on other animals.	It eats both plant and animal food.
2.	Teeth	Canine teeth are well developed. Incisors are less developed.	Both canine and incisor teeth are well developed.
3.	Claws	Claws are well developed.	Claws are less developed.
4.	Trophic Level	It lies at a fixed trophic level.	It operates at two or more trophic levels.
5.	Population	Population of a carnivore depends upon population of its prey. Examples. Peacock and Lion.	Population of an omnivore is less dependent upon population of its preys. Examples. Cockroach, Humans and Domestic Dog.

Table 6.7 Differences between Producers and Consumers

Sr. No.		Producers	Consumers
1.	Manufacture of Food	Producers manufacture their own food through the process of photosynthesis.	They cannot manufacture their own food.

2.	Chlorophyll.	They possess chlorophyll.	Photosynthesis is absent.
3.	Energy	Producers trap solar energy and change the same into chemical energy.	They cannot trap solar energy. Chemical energy is obtained from outside.
4.	Nutrients	They obtain inorganic nutrients from outside.	They obtain organic nutrients from outside.
5.	Trophic Level	Trophic Level. They constitute the first trophic level.	Consumers belong to second and higher trophic levels.
6.	Oxygen	Producers release oxygen during manufacture of food.	They consume oxygen in their respiration.

Table 6.8 Differences between Producers and Decomposers

Sr. No.		Producers	Decomposers
1.	Organisms	They are green plants.	They are microorganisms.
2.	Nature	Producers are autotrophs.	Decomposers are heterotrophs.
3.	Inorganic Nutrients	They require inorganic nutrients.	They liberate inorganic nutrients.
4.	Function	Producers supply food to all other organisms.	They liberate raw materials and create space for newer generations.
5.	Trophic Level	They constitute the first trophic level.	They constitute the last trophic level.

Table 6.9 Differences between Food Chain and Food Web

Sr. No.		Food Chain	Food Web
1.	Units	It consists of a single unit of food relations.	It is a complex formed by several units of food relations.
2.	Sequence	It is a straight sequence of organisms.	It is a network of many linkages among the connected food chains.
3.	Populations	It has a maximum of 4-6 populations of different species.	A food web consists of numerous populations of different species.
4.	Food	Only one type of organism is used as food by a particular type of organism.	An organism can use two to several types of organisms as food.
5.	Disturbance	A disturbance in food chain is difficult to overcome.	A disturbance is overcome after some time.
6.	Endangered Species	It does not help in restoring population of endangered species. Rather it may decline further.	A food web helps in increasing population of endangered species.

6.4 ECOLOGICAL PYRAMID:

The term ecological pyramid was discovered by Elton, 1927 (Fig. 6.13). An ecological pyramid (also trophic pyramid, eltonian pyramid, energy pyramid, or sometimes food pyramid) is a graphical representation designed to show the biomass or bio-productivity at each trophic level in a given ecosystem. Biomass is the amount of living or organic matter present in an organism. Biomass pyramids show how much biomass is present in the organisms at each trophic level, while productivity pyramids show the production or turnover in biomass. Ecological pyramids begin with producers on the bottom (such as plants) and proceed through the various trophic levels (such as herbivores that eat plants, then carnivores that eat herbivores, then carnivores that eat those carnivores, and so on). The highest level is the top of the chain. An ecological pyramid of biomass shows the relationship between biomass and trophic level by quantifying the biomass present at each trophic level of an ecological community at a particular time. Thus, it is a graphical representation of biomass (total amount of living or organic matter in an ecosystem) present in unit area in different tropic levels. Typical units are grams per meter, or calories per meter.

Fig. 6.13 Charles S. Elton (Eminent Ecologist)

6.4.1 Kinds of Pyramids

There are 3 kinds of ecological pyramids such:

(a) Pyramid of Energy

(b) Pyramid of Numbers

(c) Pyramid of Biomass

6.4.1.1 Pyramid of Energy:

The pyramid of energy describes the overall nature of the ecosystem. During the flow of energy from one organism to other, there is considerable loss of energy in the form of heat. More amount of energy available in the primary producers like autotrophs, and the least energy in the tertiary consumers (Fig 6.14).Thus, shorter food chain has more amount of energy available even at the highest trophic level.

- Energy pyramid shows the flow of energy at different trophic levels.

- Energy pyramid always upright and vertical.
- Energy pyramid depicts the energy is minimum as the highest trophic level and is maximum at the lowest trophic level.
- At each trophic level, there is successive loss of energy in the form of heat and respiration, etc.
- To compare the functional roles of the trophic levels in an ecosystem, an energy pyramid is most suitable.

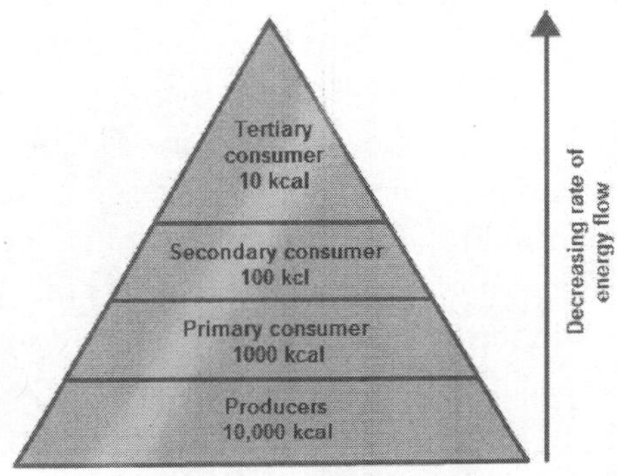

Fig. 6.14 Pyramid of Energy

6.4.1.2 Pyramid of Numbers

The pyramid of numbers depicts the relationship in terms of the number of producers, herbivores and the carnivores at their successive trophic levels. There is a decrease in the number of individuals from the lower to the higher trophic levels. The number pyramid varies from ecosystem to ecosystem. Depending upon the size, the pyramid of numbers may not always be upright, and may even be completely inverted. It is very difficult to count all the organisms, in a pyramid of numbers and so the pyramid of number does not completely define the trophic structure for an ecosystem. There are three of pyramid of numbers (Fig1.5A,B,C)

(a) Upright pyramid of number

(b) Partly upright pyramid of number and

(c) Inverted pyramid of number

Upright Pyramid of Number:

This type of pyramid number is found in the aquatic and grassland ecosystem, in these ecosystems there are numerous small autotrophs which support lesser herbivores which in turn support smaller number of carnivores and hence this pyramid is upright.

- In this pyramid, the number of individuals is decreased from lower level to higher trophic level.
- This type of pyramid can be seen in **grassland ecosystem** and **pond ecosystem**.
- The grasses occupy the lowest trophic level (base) because of their abundance.
- The next higher trophic level is primary consumer – herbivore (example – grasshopper).
- The individual number of grasshopper is less than that of grass. The next energy level is primary carnivore e.g. rat.
- The number of rats are less than grasshopper, because, they feed on grasshopper. The next higher trophic level is secondary carnivore e.g. snakes. They feed on rats.
- The next higher trophic level is the top carnivore e.g. Hawk.
- With each higher trophic level, the number of individual decreases.

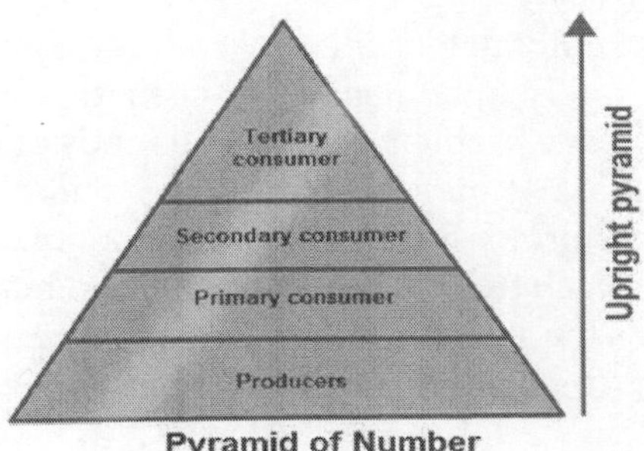

Fig. 6.15A Upright Pyramid of Numbers (Pond Ecosystem)

Inverted Pyramid of Number

This type of ecological pyramid is seen in parasitic food chain where one primary producer supports numerous parasites which support more hyperparasites. In this pyramid, the number of individuals is increased from lower level to higher trophic level. e.g. Tree ecosystem.

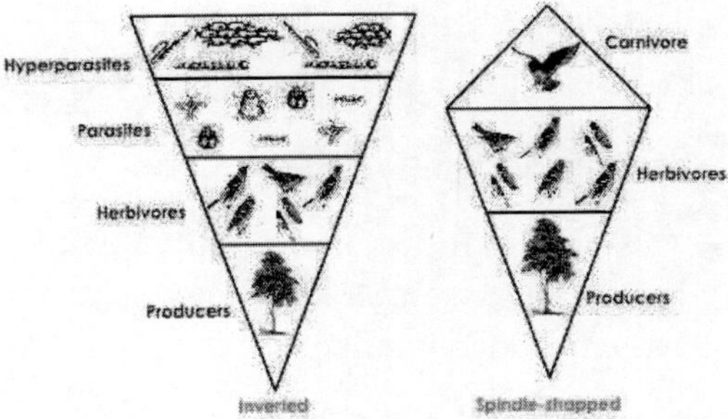

Fig.6.15B Inverted Pyramid of Numbers (Tree Ecosystem)

Partially Upright pyramid of Number

It is seen in the forest ecosystem where the number of producers are lesser in number and support a greater number of herbivores and which in turn support a fewer number of carnivores. This type of pyramid also called as spindle-opal pyramid

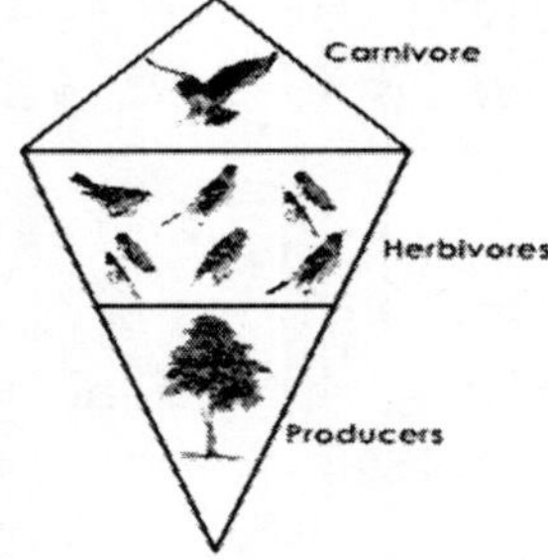

Fig 6.15C Partially Upright Pyramid of Number (Forecast Ecosystem)

6.4.1.3 Pyramid of Biomass:

Pyramid of biomass is usually determined by collecting all organisms occupying each trophic level separately and measuring their dry weight (Fig.1.16). This overcomes the size difference problem because all kinds of organisms at a trophic level are weighed. Biomass is measured in g/m^2. The biomass of a species is expressed in terms of fresh or dry weight. Measurement of biomass in terms of dry weight is more accurate. Each trophic level has a certain mass of living material at a particular time called as the **standing crop**. The standing crop is measured as the mass of living organisms (biomass) or the number in a unit area.

Upright Pyramid of Biomass:

- For most ecosystems on land, the pyramid of biomass has a large base of primary producers with a smaller trophic level perched on top.
- The biomass of producers (autotrophs) is at the maximum. The biomass of next trophic level i.e. primary consumers is less than the producers. The biomass of next higher trophic level i.e. secondary consumers is less than the primary consumers. The top, high trophic level has very less amount of biomass.

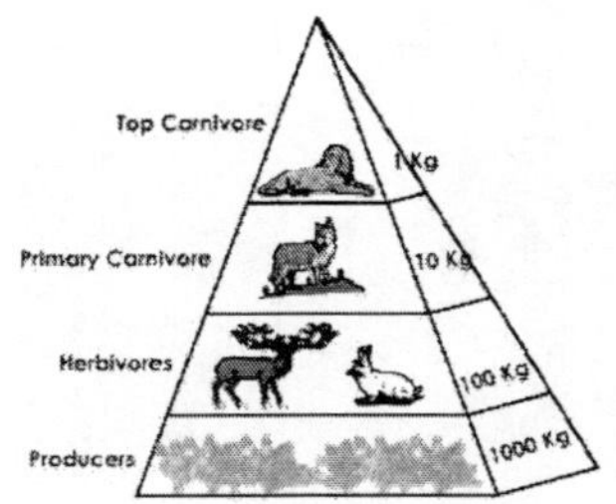

(A) Upright pyramid of biomass (Forest Ecosystem)

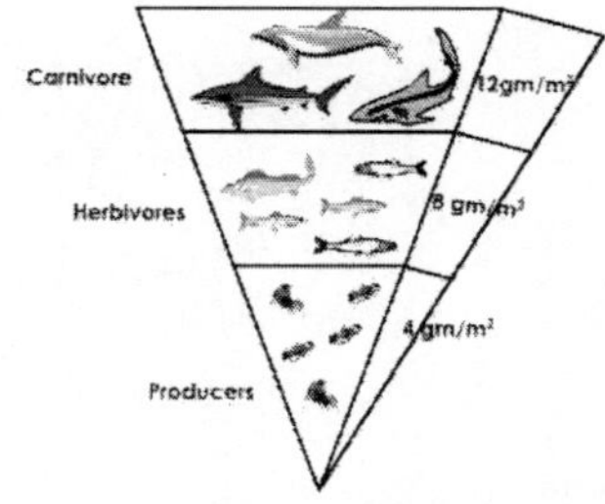

(B) Inverted pyramid of biomass (Pond Ecosystem)

6.16 Pyramid of Biomass

Pyramid of Biomass – Inverted

- In contrast, in many **aquatic ecosystems**, the pyramid of biomass may assume an inverted form. [Pyramid of numbers for aquatic ecosystem is upright]
- This is because the producers are tiny phytoplankton that grow and reproduce rapidly.
- Here, the pyramid of biomass has a small base, with the consumer biomass at any instant actually exceeding the producer biomass and the pyramid assumes inverted shape.

6.4.5 Limitations of Ecological Pyramids:

- It does not take into account the same species belonging to two or more trophic levels.
- It assumes a simple food chain, something that almost never exists in nature; it does not accommodate a food web.
- Moreover, saprophytes (plant, fungus, or microorganism that lives on decaying matter) are not given any place in ecological pyramids even though they play a vital role in the ecosystem.

6.5 FLOW OF MATERIALS IN ECOSYSTEM IS CYCLIC BUT FLOW OF ENERGY IS UNIDIRECTIONAL:

The materials like water, carbon (as carbon dioxide) and nitrogen (as minerals) are taken up by the up by the plants from soil, air and water bodies, etc, and made into food. This food is then passed on to the animals like herbivores and carnivores in a food chain. After the death and decay of plants and animals, the materials like water, carbon and nitrogen present in their bodies are returned to soil, air and water, from where they were taken originally. These materials can then be reused for the growth of new plants. In this way, the same materials are used again and again, the materials are not lost from the environment. So, the flow of materials like water, carbon and nitrogen, etc., in the ecosystem is said to be cyclic. This is not so in the case of energy. The flow of energy in the ecosystem is unidirectional (or-directional). The energy enters the plants (from the sun) through photosynthesis during the making of food. This energy is then passed on from one organism to another in a food chain. Energy given out by the organisms as heat is lost to the environment, it does not return to be used by the plants again (Fig. 6.17). This makes the flow of energy in ecosystem 'unidirectional'. Thus, the flow of energy in the ecosystem is said to be unidirectional because the energy lost as heat form the living organisms of a food chain cannot be reused by plants in photosynthesis.

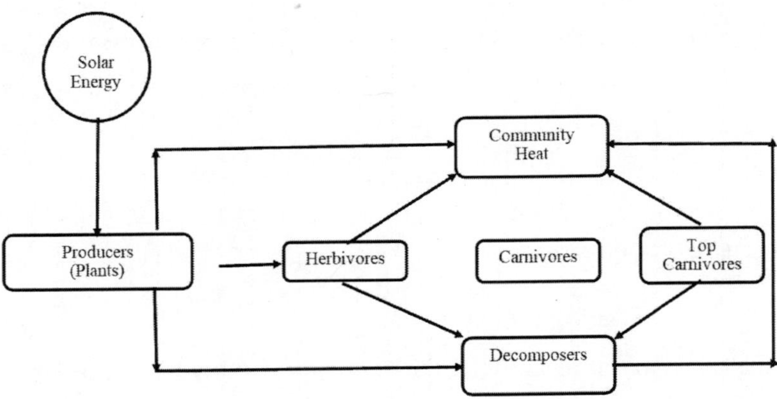

Fig. 6.17 Energy transfer in a food chain

6.6 TEN PER CENT (10%) LAW

The 10 percent law which was given by R. Lindeman in the year of 1942 (Fig 6.18) is one of the most useful generalizations about the magnitude of loss of energy in food chains. It is also termed as Second Law of Thermodynamics or Law of Entropy. During the transfer of energy through successive trophic levels in an ecosystem, there is a loss of energy all along the path. No transfer of energy is 100%. The study of transfer of energy in different food chains in a large number of ecosystems has revealed a uniform pattern of transfer of energy, which is given by 10 percent law. According to this law, only 10 percent of the energy entering a particular trophic level of organisms is available for transfer to the next higher trophic level. All the energy transfers in food chain follow the 10% law which in simple terms means that the energy available at each successive trophic level is 10 percent of the previous level. Thus, there is a progressive decline (gradual reduction) in the amount of energy available as we go from producer level to the higher trophic levels of organisms. Let us take one example to understand the 10 percent law more clearly.

Fig. 6.18 R. Lindeman, discovered 10% Law of Energy

Presume 1000 joules of light energy emitted by the sun falls on the plants (called producers). We know that the plants convert only 1% of the light energy falling on them into chemical energy of food. So, the energy which will be available in plant matter as food will be only 1% of 1000 Joules (J), which comes to 10 Joules. The remaining 1000 – 10 = 990 Joules of light energy of solar energy which is not utilized by the plants is reflected back into the environment (Fig. 6.19). Please note that the 10 percent law will not apply at this stage. It will apply only in the transfer of energy in the food chain.

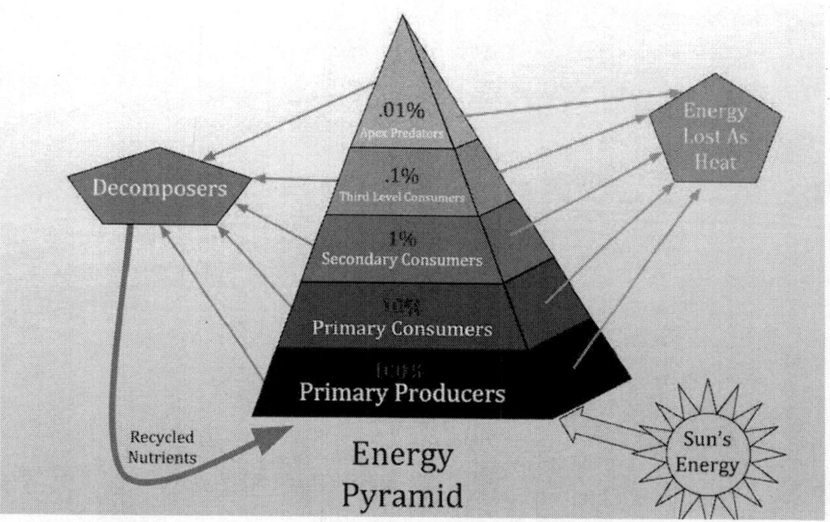

Fig. 6.19 Diagram shows transfer of energy (as per 10% law of Energy)

We will now apply the 10 per cent law to the food chain: Plants → Herbivores Carnivores (represented by plants, deer and lion). The plants or first trophic level has 10 Joules of energy in it. Now, according to ten per cent law, only 10% of 10 Joules of energy (which is 1 joule) will be available for transfer at the next trophic level, so that the herbivore (deer) will have only 1 joule of energy stored as food at the second trophic level. Applying the 10% law again we find that 10% of the remaining 1 Joule (which is 0.1 Joule) will be transferred to third trophic level of carnivore (lion). So, the energy available in the lion as food will be only 0.1 Joule.

The 10 percent law, given by R. Lindeman (in 1942) is one of the most useful generalizations about the magnitude of loss of energy in food chains. It is also termed as second law of thermodynamics or law of entropy. During the transfer of energy through successive trophic levels in an ecosystem, there is a loss of energy all along the path. No transfer of energy is 100%. The study of transfer of energy in different food chains in a large number of ecosystems has revealed a uniform pattern of transfer of energy, which is given by 10 percent law.

Interdependence with Ecosystem:

The biotic components of any ecosystem are dependent on one another for their requirements of food and nutrition. The producers prepare food from inorganic components of an ecosystem. Consumers depend directly or indirectly on producers for their food and energy. When producers and consumers die, the decomposers feed on them and decompose their bodies releasing the various elements back into the nutrient pool (Soil, Water and Air). Thus, cycling of nutrients within the ecosystem is maintained which is vital for the life on earth to continue. A constant supply of nutrients like carbon, hydrogen, oxygen, sulphur and phosphorus is essential for living organisms to make complex organic compounds like proteins, fats and carbohydrates. Thus, decomposers are a vital link between living components of the environment.

6.7 HUMAN'S ACTIVITIES ON ENVIRONMENT

Human beings are an important part of the environment. Our activities change the environment around us. And the changes in environment then affect us. Here, we will now discuss two environmental problems caused by our activities such as (i) Ozone Layer and their depletion (ii) Management of Solid Waste

6.7.1 Ozone Layer and their Depletion :

The ozone layer forms a thick layer in stratosphere, encircling the earth that has large amount of ozone in it. The ozone layer protects life on earth from strong ultraviolet (UV) radiation that comes from the sun. The ozone layer was discovered in 1913 by the French physicists Charles Fabry and Henri Buisson.

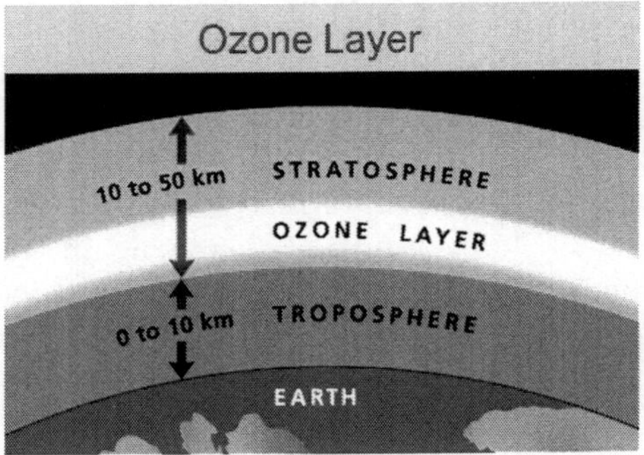

Fig. 6.20 Ozone Layers

The ozone layer has the capability to absorb almost 97-99% of the harmful ultraviolet radiations that sun emit and which can produce long term devastating effects on humans beings as well as plants and animals. Ozone layer depletion leads to decrease in ozone in the stratosphere (The stratosphere is the second major layer of Earth's atmosphere, just above the troposphere, and below the mesosphere. About 20% of the atmosphere's mass is contained in the stratosphere) and increase in ozone present in the lower atmosphere. Presence of ozone in the lower atmosphere is considered as a pollutant and a greenhouse gas. Ozone in the lower atmosphere contributes to global warming and climate change. The depletion of ozone layer has trickle down effects in the form of global warming, which in turn leads to melting of polar ice, which will lead to rising sea levels and climatic changes around the world. September 16 was designated by the United Nations General Assembly as the International Day for the Preservation of the Ozone Layer, and celebrates at World Ozone Day.

6.7.1.1 Depletion of Ozone Layer :

Ozone layer depletion is one of the most serious problems faced by our planet earth (Fig 6.22). It is also one of the prime reasons which are leading to global warming. Ozone is a colorless gas which is found in the stratosphere of our upper atmosphere. The layer of ozone gas is what which protects us from the harmful ultraviolet radiations of the sun. The ozone layer absorbs these harmful radiations and thus prevents these rays from entering the earth's atmosphere. Ultraviolet radiations are high energy electromagnetic waves emitted by the sun which if enters the earth's atmosphere can lead to various environmental issues including global warming, and also a number of health related issues for all living organisms. Thanks to the ozone layer which protects us from these harmful rays.

Oxygen is O_2 Oxygen molecule is made up of 2 atoms of oxygen combined together. Oxygen gas is essential for life because it is needed in respiration. Ozone is O_3. Ozone molecule is made up of 3 atoms of oxygen combined together. Ozone is also a gas but it is poisonous in nature (if inhaled). Ozone is formed high up in the atmosphere by the action of ultraviolet (UV) radiation on oxygen gas.

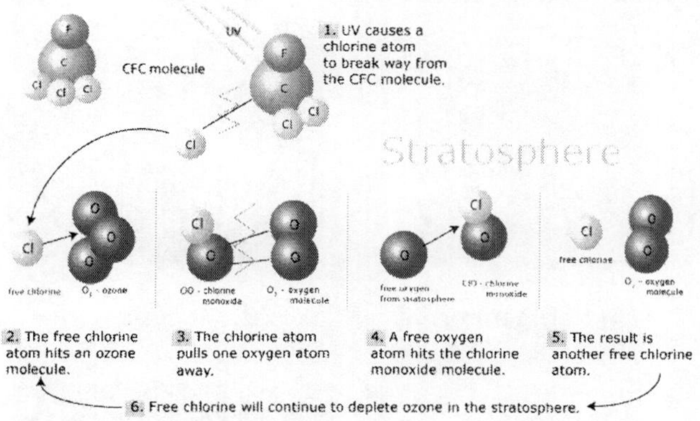

Fig. 6.21 Depletion of Ozone layer

The high energy ultraviolet radiation (UV radiation) coming from the sun splits oxygen gas into free oxygen atoms.

$$\underset{\text{Oxygen molecule}}{O_2} \xrightarrow[\text{(From sun)}]{\text{UV radiations}} \underset{\text{Oxygen atom}}{O} + \underset{\text{Oxygen atom}}{O}$$

The free oxygen atoms thus produced are very reactive. One oxygen atom reacts with an oxygen molecule to form an ozone molecule:

$$\underset{\text{Oxygen molecule}}{O_2} + \underset{\text{Oxygen atom}}{O} \rightarrow \underset{\text{Ozone molecule}}{O_3}$$

At a height between 15km and 60km, there is a layer of ozone gas (O_3) in the upper atmosphere. This ozone layer is very important for the existence of life on earth because it absorbs most of the harmful ultraviolet radiations coming from the sun and prevents them from reaching the earth. The ultraviolet radiations have extremely harmful effects on human beings, other animals as well as plants. For example, ultraviolet rays can cause skin cancer. They also damage the eyes by causing an eye disease called cataract. Ultraviolet rays immune system by lowering the body's resistance to diseases.

Thus, it is the ozone layer in the upper atmosphere which protects us from these diseases by absorbing ultraviolet rays coming from the sun.

It has now been found that the amount of ozone is getting depleted (or reduced) due to which the ozone layer in the upper atmosphere is becoming thinner and thinner day by day. The depletion of ozone layer is due to the use of chemicals called chlorofluorocarbons.

Chlorofluorocarbon (CFC) is the chemicals which are widely used in refrigeration (refrigerators and air conditioners) as a coolant; in fire extinguishers and in aerosol sprayers (Fig 6.22). Chlorofluorocarbons released into the air react with ozone gas present in the ozone layer and destroy it gradually. Due to this, the ozone layer in the upper atmosphere has become thinner, allowing more ultraviolet rays to pass through it to the earth. Thus, due to the depletion of ozone layer caused by Chlorofluorocarbons, more ultraviolet radiations reach the earth.

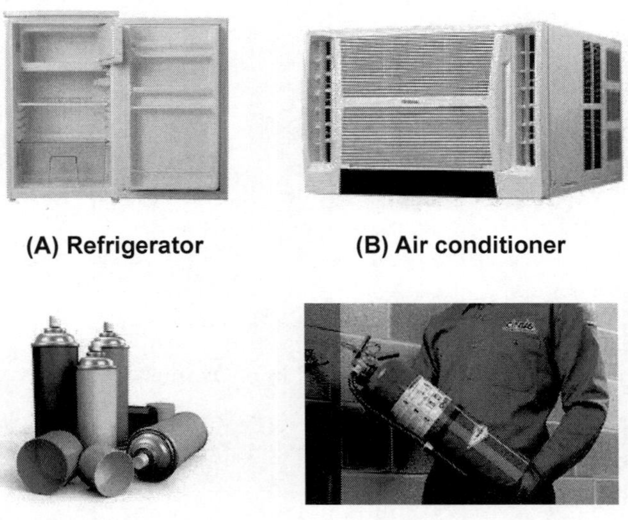

(A) Refrigerator **(B) Air conditioner**

(C) Aerosol cans **(D) Free extinguishers**

Fig. 6.22 Chlorofluorocarbons producing Machines

It the ozone layer in the atmosphere disappears completely, then all the extremely harmful ultraviolet radiations coming from the sun would reach the earth. These ultraviolet radiations would cause skin cancer and other ailments in men and animals, and also damage the plants. All the life on the earth would be destroyed gradually. In 1987, in an attempt to protect ozone layer, the United Nations Environment Programme (UNEP) forged an agreement among its member countries to freeze CFC production at 1986 levels.

6.7.1.2 Ozone depleting substances:

Ozone depleting substances or ODS are those substance which react with ozone present in the stratosphere and destroy the same. The main ozone depleting substances are chlorofluorocarbons (CFCs), halons, nitrous oxide, methane, carbon tetrachloride and chlorine. Chlorofluorocarbons are highly stable, odourless, synthetic gaseous substances used as aerosol propellants, coolants, refrigerants, blowing agents, etc. Halons are bromine containing gases that have been in use in fire extinguishers. Some of the ODS are released in stratosphere by jets and rockets. Others pass gradually from earth to its stratosphere. Ozone depleting substances react with ozone and convert the same into oxygen. Chlorofluorocarbons split up in the presence of ultraviolet radiations to release atomic chlorine. The letter reacts with ozone to produce oxygen. Chlorine monoxide is formed. The same reacts with ozone again to form oxygen and release chlorine (Molina and Rowland), 1974). Chlorine repeats the process. It is estimated that a chlorine atom formed in the stratosphere can persist in stratosphere for 100 years and destroy 1,00,000 ozone molecules.

UV-radiations

$$CCI_2F_2 \rightarrow CCIF_2 + CI$$

$$2CI + O_3 \rightarrow CI_2O + O_2$$

$$2CI_2O \rightarrow 4CI + O_2$$

6.7.1.3 International Efforts to Check Ozone Depletion:

(i) **Montreal Protocol:** In 1987, UNEP or United Nations Environment Programme succeeded in forging an agreement between industrialized nations to limit production of chlorofluorocarbons to half the level of 1986.

(ii) **Helsinki Declaration:** In 1989 most nations pledged to phase out CFCs and halons by 2000. Production of the two ODS has been stopped since then. CFCs have been replaced with hydro fluorocarbon (HFC) and hydro chlorofluorocarbon (HCIFC). While the chlorine content is low in hydro chlorofluorocarbon (HCIFC), it is absent in hydro chlorofluorocarbon (HFC). Because of these efforts, ozone hole over Antarctica has come down from 28.3 million sq. km in 2000 to 27.5 million sq. km in 2006, 25.0 million sq. km in 2011, and 21.0 million sq. km in 2013.

6.7.1.4 Major Impacts of Depletion of Ozone Layer :

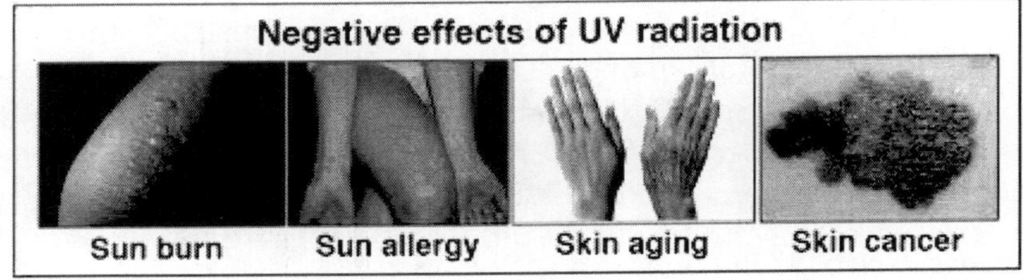

(A) Impact on skin

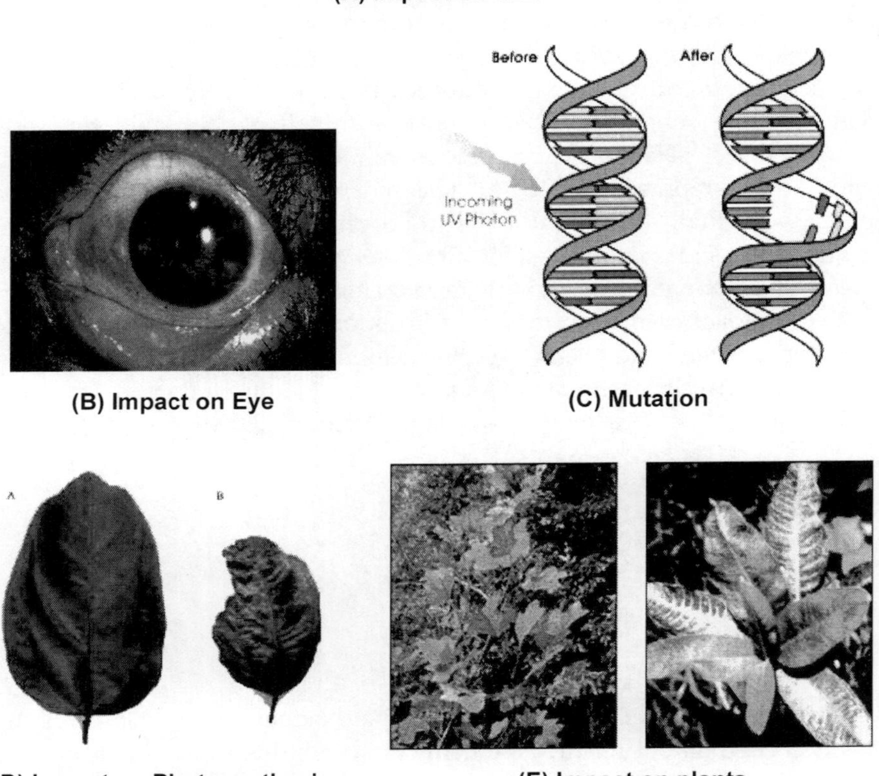

(B) Impact on Eye

(C) Mutation

(D) Impact on Photosynthesis

(E) Impact on plants

Fig. 6.23 Some examples of adverse effect of depletion of ozone layer

(a) **Skin Cancer:** exposure to UV rays from sun can lead to increased risk for developing of several types of skin cancers. Malignant melanoma, basal and squamous cell carcinoma are the most common cancers caused by exposure to UV rays.

(b) **Eye Damage:** UV rays are harmful for our eyes too. Direct exposure to UV rays can lead to Cataract problems, and also Photokeratitis or snow blindness. There will be dimming of eye sight, photo burning and higher incidence of cataracts. A large number of land animals will become blind.

(c) **Immune System Damage:** our immune system is also highly vulnerable to UV rays. Increased exposure to UV rays can lead to weakening of the response of immune system and even impairment of the immune system in extreme cases.

(d) **Aging of Skin:** exposure to UV rays can lead to acceleration of the aging process of your skin. This will result in you looking older than what you actually are. It can also lead to photo allergy that result in outbreak of rashes in fair skinned people

(e) **Death of Young Ones:** (In embryonic state, the young ones are highly susceptible to UV radiations and get killed.

(f) **Mutations:** More mutations will occur. Most of them would be harmful.

(g) **Photosynthesis**: 10-25% decline in photosynthesis would occur.

(h) Global Warming: Reduced photosynthesis will increase CO_2 concentration causing global warming.

(i) **Effect on other Life Forms:** It adversely affects the different species of amphibians and is one of the prime reasons for the declining numbers of the amphibian species. It affects them in every stage of their life cycle; from hampering the growth and development in the larvae stage, deformities and decreases immunities in some species and to even retinal damage and blindness in some species.

(j) **Effect on Marine Ecosystem:** It adversely affects the planktons which plays a vital role in the food chain and oceanic carbon cycle. Affecting phytoplankton will in turn affect the whole ocean ecosystem.

(k) **Effect on the Plants:** UV radiations can alter the time of flowering in some plant species. It can also directly affect the plant growth by altering the physiological and developmental processes of the plants.

In humans, exposure to UV rays can also lead to difficulty in breathing, chest pain, and throat irritation and can even lead to hampering of lung function.

6.7.1.5 Ultra Violet Rays :

Although the concentration of the ozone in the ozone layer is very small, it is vitally important to life because it absorbs biologically harmful ultraviolet (UV) radiation coming from the sun. Extremely short or vacuum UV (10–100nm) is screened out by nitrogen. UV radiation capable of penetrating nitrogen is divided into three categories, based on its wavelength; these are referred to as UV-A (400–315nm), UV-B (315–280nm), and UV-C (280–100 nm). UV-C, which is very harmful to all living things, is entirely screened out by a combination of dioxygen (< 200nm) and ozone (> about 200nm) by around 35kms altitude. UV-B radiation can be harmful to the skin and is the main cause of sunburn; excessive exposure can also cause cataracts, immune system suppression, and genetic damage, resulting in problems such as skin cancer UV light transmission through eye.

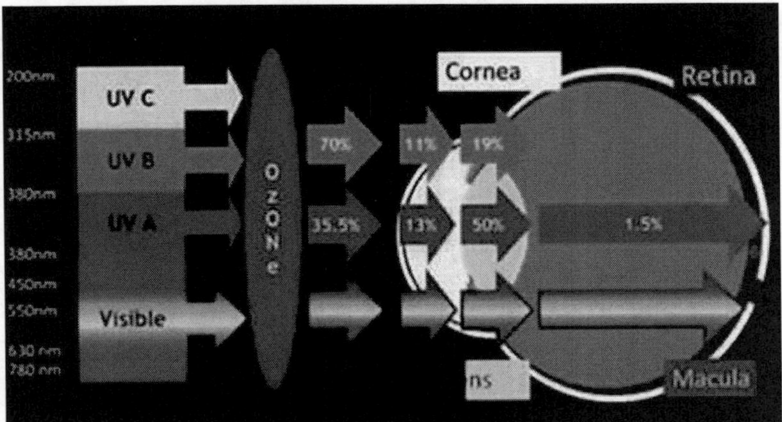

Fig. 6.24 UV Light transmission through eye

The ozone layer (which absorbs from about 200 to 310nm with a maximal absorption at about 250nm is very effective at screening out UV-B; for radiation with a wavelength of 290nm, the intensity at the top of the atmosphere is 350 million times stronger than at the Earth's surface. Nevertheless, some UV-B, particularly at its longest wavelengths, reaches the surface, and is important for the skin's production of vitamin D.

Ozone is transparent to most UV-A, so most of this longer-wavelength UV radiation reaches the surface, and it constitutes most of the UV reaching the Earth. This type of UV radiation is significantly less harmful to DNA, although it may still potentially cause physical damage, premature aging of the skin, indirect genetic damage, and skin cancer.

The ozone layer forms a thick layer in stratosphere, encircling the earth that has large amount of ozone in it. The ozone layer protects life on earth from strong ultraviolet (UV) radiation that comes from the sun. The ozone layer was discovered in 1913 by French physicists Charles Fabry and Henri Buisson. The ozone layer has the capability to absorb almost 97-99% of the harmful UV radiations. Ozone depleting substances (ODS) are those substances which react with ozone present in the stratosphere and destroy the same. The main ODS are chlorofluorocarbons (CFCs), halons, nitrous oxide, methane, carbon tetrachloride and chlorine.

Ozone is transparent to most UV-A, so most of this longer-wavelength UV radiation reaches the surface, and it constitutes most of the UV reaching the Earth. Thinning of ozone layer allows more UV radiations falling on the earth. These cause majorly such as (a) skin cancer, (b) increased chances of cataract disease in eyes, (c) damage to immune system in human beings and animals (d) increased embryonic mortality, (e) decreased photosynthesis and (f) contribute towards global warming. September 16 was designated by the United Nations General Assembly as the International Day for the Preservation of the Ozone Layer, and celebrates at World Ozone Day.

6.7.2 Bioaccumulation Vs. Biomagnification:

Food chains are important interrelations in the ecosystems. It starts with a producer (Plants), which can produce food for themselves. The second level is occupied with herbivores and the upper levels are occupied with carnivores. The food produced at the bottom level is passed to the upper levels. Therefore, whatever the substances in the lower tropic levels can be passed onto the upper levels along with the nutrients. Bioaccumulation and the biomagnification of substances are considered to be important if those substances are harmful to organisms. If the substances are not harmful, their accumulation will not do any problem to them. With the industrial development and high population density, a lot of pollutants are released into the environment. Some of them are not easily degradable, and they are toxic. When they enter into the food chains and circulate inside organisms, it is a huge problem. So it is important to have an idea about the concepts of bioaccumulation and biomagnification as well as about the substances which are likely to be bio accumulative.

6.7.2.1 Bioaccumulation:

Bioaccumulation is the accumulation of substances heavy metals, pesticides, or organic chemicals in living organisms. These substances can enter into the systems via water or food. Bioaccumulation is resulted via food chains. The lower tropic levels of the food chain accumulate less concentration of substances than the higher tropic levels. Usually, body has mechanisms to remove all the unwanted and toxic products away from the body. Bioaccumulation results when the accumulating rate is much higher than the removal rate. So, if the life time of the substance is higher, the impact of it also gets higher. Usually, kidneys are responsible for removing the majority of the unwanted substances from the body. Blood carries them into the kidneys and then through filtration and selective reabsorption urine is produced. In order to remove toxics with urine, they should be water soluble.

Bioaccumulative substances are normally fat soluble and cannot be broken down into smaller molecules. Therefore, they tend to remain in the body. Toxic substances are lipophilic or fat-loving, the reason why these substances are deposited and concentrated in the fat tissues of the organisms. The affected organism has a higher concentration of the substance than the concentration in the organism's surrounding environment. The toxic substances are very slowly metabolized or excreted so if the organism keeps on consuming prey or food contaminated with toxic substances, the concentration of the substance will further increase in its body, hence, bioaccumulation results. When a certain threshold level is reached, measured in parts per million (ppm), symptoms due to the type of toxin are demonstrated.

Examples of toxic substances that bioaccumulate are lead, mercury, dichloro diethyl trichloro ethane (DDT) etc. Lead exposure can be serious for young children because their growing bodies absorb lead more easily than adults. They are more susceptible to its harmful effects. Even low level of lead exposure can harm the intellectual development, behaviour, growth, and hearing of infants. High concentrations of lead in pregnant mothers can cause miscarriages and stillbirths. Mercury, as mentioned above, can cause paralysis. DDT at concentration above 236 mg/kg of body weight can cause death among humans. Concentration of 6-10 mg/kg leads to such symptoms as headache, nausea, vomiting, confusion, and tremors.

Bioaccumulation is the accumulation of substances heavy metals, pesticides, or organic chemicals in living organisms. These substances can enter into the systems via water or food. Bioaccumulation is resulted via food chains. Bioaccumulative substances are normally fat soluble and cannot be broken down into smaller molecules. Therefore, they tend to remain in the body. Examples of toxic substances that bioaccumulate are lead, mercury, dichloro diethyl trichloro ethane (DDT) etc. Lead exposure can be serious for young children because their growing bodies absorb lead more easily than adults. Mercury can cause paralysis. DDT at concentration above 236 mg/kg of body weight can cause death among humans.

6.7.2.2 Biomagnification:

Biomagnification is also called Bioamplification. It is simply the increase in concentration of a substance in a food chain, not an organism. Persistent organic pollutants (POPs) are compound that biomagnify. POPs are chemical substances that persist in the environment. These substances bioaccumulate through the food web and attitude risk not only to humans but also other living organisms because of their adverse effects. Thus, the pollutants must be long-lived in order to cause biomagnification. Also, it should be mobile, so that it easily enters into biological systems in means of food or water. If it is not mobile, it may stay inside one organism and will not be passed into the next tropic level. If they are soluble in fat, they tend to remain in bodies of organisms for a longer period

These pollutants consist of pesticides (such as DDT), industrial chemicals (such as polychlorinated biphenyls, PCBs) and unintentional by-products of industrial processes (such as dioxins and furans) (Fig. 6.25). In essence, biomagnification is similar to bioaccumulation but is descriptive of higher level biological processes, not individual. Further, in order for biomagnification to occur, the pollutant must be biologically active. For example, DDT is a chlorinated hydrocarbon which can be bioaccumulated. It is toxic for insects and has a half-life of 15 years. Heavy metals like mercury, lead, cadmium, zinc are also toxic and can be biomagnified.

Fig. 6.25 Pollution by Farmers

Biomagnification is also known as Bioamplification. It is simply the increase in concentration of a substance in a food chain, not an organism. Persistent organic pollutants (POPs) are compound that biomagnify. POPs are chemical substances that persist in the environment. These substances bioaccumulate through the food web and attitude risk not only to humans but also other living organisms because of their adverse effects. For example, DDT is a chlorinated hydrocarbon which can be bioaccumulated. It is toxic for insects and has a half-life of 15 years. Heavy metals like mercury, lead, cadmium, zinc are also toxic and can be biomagnified.

6.7.2.3 Bioaccumulation Vs Biomagnification (Fig. 6.26, 6.27):

Bioaccumulation is increasing the concentration of a substance in one organism whereas biomagnification is increasing the level as you go up in a food chain.

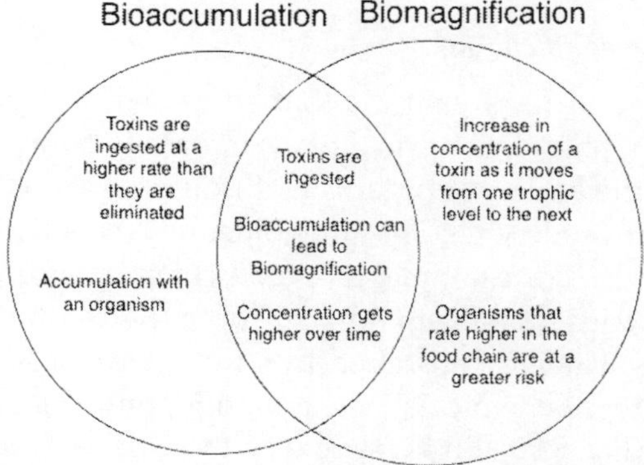

Fig. 6.26 Diagram shows the relationship/differences between Bioaccumulation and Biomagnification

Bioaccumulation occurs within a tropic level and biomagnification occurs between tropic levels.

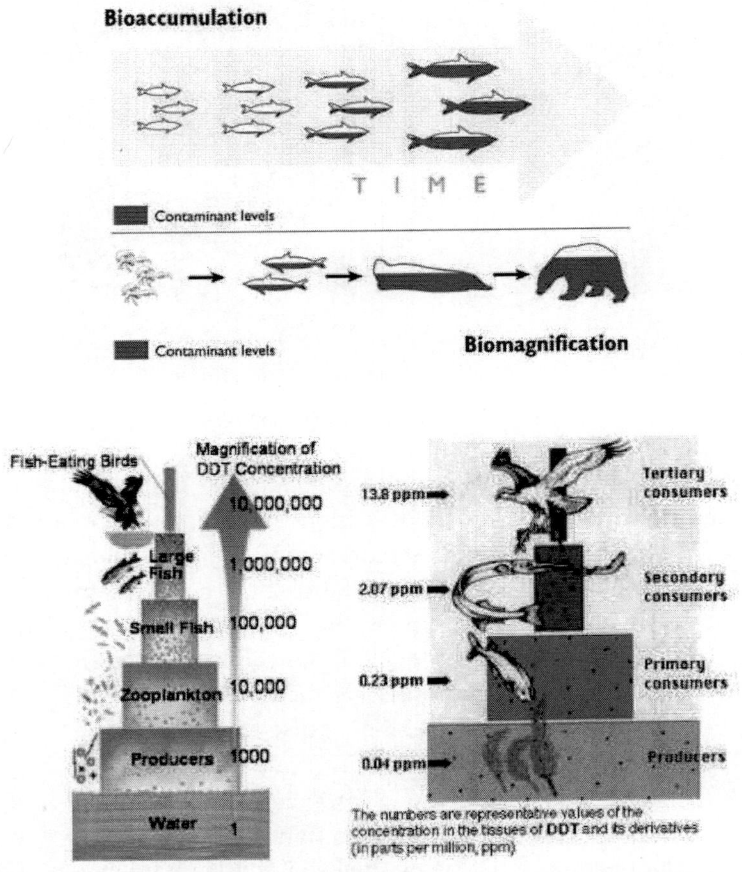

Fig. 6.27 Diagram shows the Bioaccumulation and Biomagnification

6.7.2.4 Example of Biomagnification:

An interesting feature of food chain is that some harmful non-biodegradable chemicals (pesticides, e.g., DDT and heavy metals such as mercury, arsenic, cadmium etc.) enter the bodies of organisms through the food chains and go on concentrating at each trophic level. This phenomenon is called biomagnification or biological magnification. For example, in a food chain operating in a pond, river or lake, the water contains a small amount 0.02 ppm (parts per million) of harmful chemicals. When this water is consumed by phytoplanktons and zooplanktons, the concentration of these chemicals increases to 5 ppm. Fishes feeding on these accumulate 240 ppm (Fig. 6.28). Birds feeding on these fishes were found to contain 1600 ppm of these chemicals. Thus, there is an increase in the concentration of the chemicals at each trophic level.

As human beings usually occupy the highest trophic level in food chain, they get maximum quantity of harmful chemicals.

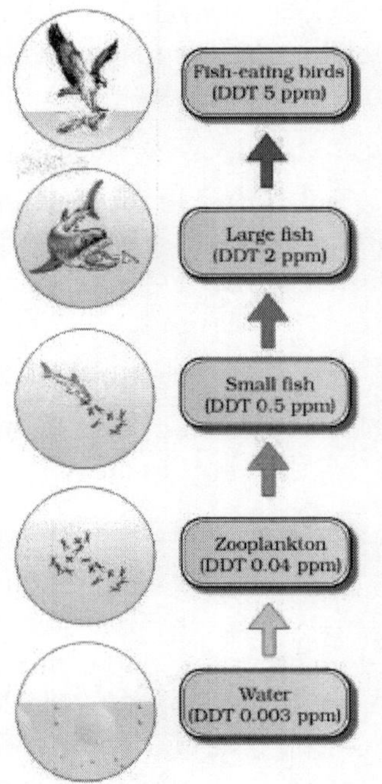

Fig. 6.28 Biomagnification of DDT in an aquatic food chain

6.7.3 Solid Waste Management :

In our daily life, we use a lot of materials and throw away. The useless left over or discarded materials are termed as wastes (Fig 6.29). The waste materials can be gaseous (e.g., automobile exhausts, smoke from chimneys of industries and house), liquid (e.g., effluents from industries, sewage water) or solid (e.g., food waste, cow dung and human excreta, trash and rubbish, farm waste, industrial and chemical wastes). Solid wastes are the main sources of soil pollution.

Fig. 6.29 Agricultural Waste Materials

6.11.3.1 Solid Wastes:

Solid wastes generally comes from households, vegetable and fruit markets, cattle sheds, industries, agricultural fields, and many other places. These can be categorized into following categories.

(i) **Food waste:** It includes waste of vegetable and fruit markets and kitchen (e.g., rotten/left over vegetables and fruits and their peels), waste of slaughter houses and food canning industries.

(ii) Cow dung, human excreta and farm waste.

(iii) **Trash and rubbish :** IT includes dirt, ash, sand and bricks; polythene bags and plastic waste; broken/useless boards card boards, waste paper, waste rubber, and leather articles; waste/broken glass articles; worn-out clothes and metal articles etc.

(iv) **Solid waste** from chemical and other industries.

(v) **Hospital waste:** It includes used cotton, bandages, used/broken syringes and needles, used plastic or glass bottles etc.

6.7.3.2 Disposal of Domestic Wastes or Household Garbage:

The household wastes are called garbage. Every household produces a lot of garbage or wastes daily (Fig. 6.30). This garbage includes left-over food, fruit and vegetable peels, fallen leaves of potted plants, waste paper, unwanted plastic objects (such as plastic bottles, polythene bags, toys, etc.), glass articles (like glass bottles, broken window panes, etc.), metal objects (like aluminum foils, rusted iron grills, etc.), old wooden objects, rags (old, torn clothes), discarded shoes, and sewage. Some of the garbage (or waste) is biodegradable whereas a major part of it is non-biodegradable. If the household garbage or waste is not disposed of properly, it can pollute the environment like soil, water and air. 'Disposal of waste' means 'to get rid of waste'. The disposal of waste should be done in a scientific way. There are different methods of waste disposal. The method to be used depends on the nature of the waste.

Fig. 6.30 Domestic wastage

6.7.3.3 Methods for Waste Disposal:

Some of the important methods of waste disposal are given here in brief such as **(i)** Preparation of Compost **(ii)** Recycling **(iii)** Incineration **(iv)** Landfill, and **(v)** Sewage treatment

Preparation of Compost:

Biodegradable domestic wastes such as left-over food, fruit and vegetable peels, and leaves of potted plants, etc., can be converted into compost by burying in a pit dug into ground, and used as manure. According to this, organic matter of solid wastes is digested an aerobically or aerobically by microbial action and converted into humus and stable mineral compounds. Although aerobic composting is more attractive, it has a drawback that most of the composting plants have unacceptable odour. However, it is of great use, it its products are used as manure in soil (Fig 6.31). The crop yield is improved and there is less need of fertilizers and pesticides. Anaerobic composting is more advantageous due to (i) No need of aeration and (ii) Produces biogas (50-70% methane and 30-40% CO_2) which can be used for heating or generation of electric power.

Fig. 6.31 Preparation of compost

Recycling:

The solid wastes like paper, plastics and metals, etc., are recycled (Fig 6.32). For example, waste paper is sent to paper mills where it is reprocessed to form new paper once again. The broken plastic articles like plastic bags, buckets, bowls, cups, plates, etc., are sent to plastic processing factories where they are melted and remolded to make new articles. Similarly, waste metal articles are sent to metal industries where they are melted and recycled as solid metal for various purposes.

Fig. 6.32 Recycling of the solid wastes

Incineration:

Incineration means 'reducing to ashes'. The burning of a substance at high temperature (of more than 1000°C) to form ash is called incineration. It involves the aerobic burning of the combustible constituents of solid wastes like garbage, rubbish and dead animals in the properly-constructed hearth of furnaces at high temperature (>670°C). This is also the ideal method for medical waste management as it eliminates the infectious organisms. Incineration is used to destroy household waste, chemical waste and biological waste (like that from hospitals). Incineration reduces the volume of the waste. This is because when the large volume of waste material is burned, then only a small amount of ash is left behind which can be disposed of by landfill (Fig. 6.33). Incineration is carried out in and incinerator (which is a kind of furnace). The waste disposal on large scale by using incinerators is done by the Municipality of a City. The solid waste is burned at a high temperature in the incinerator. All the organic matter present in waste is removed as carbon dioxide and water vapour. The ash left behind is removed from time to time.

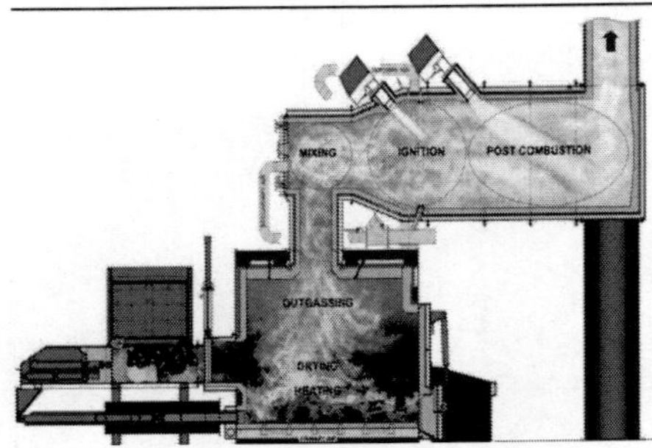

Fig. 6.33 Incineration process

Landfill:

The disposal of wastes by putting it in low-lying areas of ground and covering it with earth is called landfill. A big landfill site can be used to dispose of waste materials (or garbage) for a considerable time. In this, solid wastes are dumped into low lying areas. The refuge is dumped in layers of about 15 meter and each layer is covered by soil of about 20cm thickness (Fig. 6.34). Such landfills can be used for developing parks or other recreational site. So dumping is simple and economical method to manage the urban solid wastes and reclaim the low-lying areas for better use.

Fig. 6.34 Landfilling

Pyrolysis:

It involves anaerobic destructive distillation of the combustible constituents of the solid wastes at high temperature (650° to 1000°C), so as to recover the chemical constituents and chemical energy of organic wastes (Fig. 6.35).

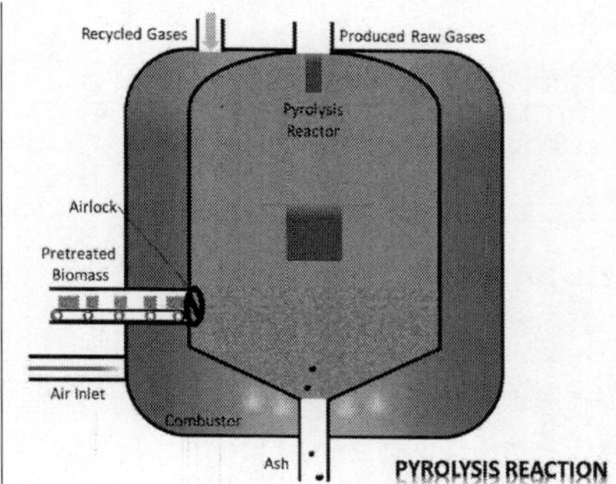

Fig. 6.35 Pyrolysis Reaction

Sewage Treatment:

The dirty drain water containing urine and faeces which is carried from our homes by the underground pipes (called sewers) is called sewage. If untreated sewage is dumped into a river, it can pollute the river waste. Thus, sewage is disposed of by treating it at the sewage treatment plant (or sewage works). The treatment of sewage produces clean water which is discharged into the river (Fig. 6.36). The organic matter present in sewage is 'digested' in the digesters of sewage treatment plant to produce 'sewage gas' (which is a kind of biogas), and 'manure'.

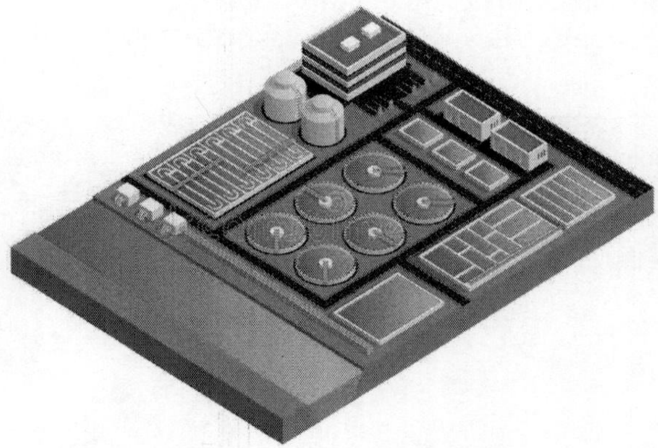

Fig. 6.36 Waste water purification plant

Table 6.11 Differences between Incineration and Pyrolysis

Sr. No.	Characters	Incineration	Pyrolysis
1	Definition	Aerobic burning of combustible components of solid wastes at high temperatures.	Anaerobic burning of combustible components of solid wastes at high temperature.
2	Final prod-ucts	Ashes and clinkers.	Chemical energy and chemical constituents.

Example to illustrate how the use of biodegradable and non-biodegradable materials can make a difference to our environment:

There was a time when tea in trains was served in plastic glasses which had to be returned to the vendor. This was, however, not a hygienic thing to do. Later on disposable plastic cups were introduced (which are used 'once' and then 'thrown away'). Though it was hygienic to use a disposable plastic cup for drinking tea but the disposal of millions of plastic cups on daily basis posed a big problem. Sometime back 'kulhads' (disposable cups made of clay) were introduced for serving tea in trains. It was, however, soon realized that the use of a lot of clay for making millions of kulhads daily led to the loss of fertile top soil from fields (which was used for making kulhads). So, the practice of using 'kulhads' has also been discontinued. These days, disposable paper cups are being used.

6.7.3.4 Paper Vs. Plastic Cups :

- Paper cups are biodegradable. So, even if paper cups are thrown away after use, they will decompose (break down) automatically by the action of micro-organisms in due course of time. On the other hand, plastic cups are non-biodegradable. They will remain as such and pollute the environment (Fig 6.37).

Fig. 6.37 Paper and Plastic Products

- Paper cups can even be disposed of by burning without causing much air pollution. On the other hand, burning of plastic cups produces toxic gases (poisonous gases) which cause too much air pollution.

When accumulated, these heaps of solid wastes make the surroundings dirty and pollute the soil. When the human population was low and technology was in its infancy, the solid wastes, generated due to human activities were easily degraded by decomposers present in nature and these did not create any significant harmful effect on the environment. In the recent time, however, human population has increased tremendously and the technology has become greatly advanced. These two factors have contributed significantly in the deterioration of our environment. Advancement in the technology has resulted in improvements in our life styles and also has changed our attitudes. As a result, we have started using more and more of disposable articles thereby generating huge amounts of waste material.

6.7.3.5 Few examples are given here:

- **Use of disposable plastic/paper cups in trains:** Millions of disposable cups are used and thrown away on daily basis into the environment. Initially, the use of disposable plastic cups in trains was hailed as a step forward for reasons of hygiene, as none at that time, had thought about the impact of this step on the environment. Now, disposable biodegradable paper cups are being used in trains in place of disposable non-biodegradable plastic cups.
- **Use of disposable materials in marriage and other parties:** This is another example where we use and throw away huge quantity of disposable materials such as plastic/paper cups, plates, spoons etc. (biodegradable and non-biodegradable) in the environment without realizing their impact on the environment.
- Some time back, kulhads (disposable cups made of clay) were suggested as an alternative of disposable plastic cups for use in trains. However, a little discouraged this suggestion as making these kulhads on a large scale would have resulted in the loss of the fertile top soil.

6.7.4 Environmental Pollution :

Environmental pollution is defined as an undesirable change in the physical, chemical or biological characteristics of air, water and soil which cause harm to living organisms (Fig 6.41). The environmental pollution is caused due to addition of a number of wastes.

Fig. 6.38 Pollution at India Gate New Delhi

Table 6.12 Example of Polluting Wastes

S.N	Sources of wastes	Types of wastes
1.	Industrial wastes	• CO, CO_2, SO_2, H_2S and hydrocarbons (air released from automobiles also). • Hydrochloric acid, chlorine, nitrogen oxides, etc. • Heavy metals like mercury, lead, cadmium, nickel, etc. • Many organic compounds like phenol, naphtha, aromatic compounds, etc.
2.	Domestic wastes	• Sewage of sanitary facilities of residential, commercial, institutional and other public places. • Sewage with many types of pathogenic organisms like bacteria, encysted protozoans, viruses, eggs of helminthes, etc.
3.	Agricultural wastes	• CO_2 methane (from paddy fields), organopesticides and chlorinated hydrocarbons, etc. • Traces of fertilizers with nitrates, phosphates and sulphates of potassium.
4.	Commercial wastes	• Building wastes, printing press wastes, biomedical wastes of medical institutions and hospitals, etc.

6.7.5 Biodegradable and Non-biodegradable Wastes :

All the waste materials produced by the various activities of man and animals are poisonous to some extent. It is mainly divided into two main groups such as:

Biodegradable Materials

Non-biodegradable Materials

Fig. 6.39 Biodegradable and Non-biodegradable materials

1. Biodegradable Wastes, and
2. Non-biodegradable Wastes

6.7.5.1 Biodegradable wastes:

Those waste materials which can be broken down to non-poisonous substances in nature in due course of time by the action of micro-organisms like certain bacteria, are known as biodegradable wastes. **Biodegradable wastes** usually don't pollute the environment. After sometimes, the biodegradable waste decays (decomposes) naturally and becomes harmless. Biodegradable wastes pollute the environment only when their amount is large which cannot be degraded (or decomposed) into harmless substances in nature at the right time. These wastes can be degraded by biological or microbial actions. These include domestic sewage, livestock wastes, municipal wastes etc. These cause pollution only when they are added in larger amounts than their rate of degradation.

6.7.5.2 Non-biodegradable wastes:

The waste materials which cannot be broken down into harmless (non-poisonous) substances in nature are called non-biodegradable wastes. These include glass and plastic wastes, aluminium cans, radioactive substances, synthetic polymers, heavy metals, pesticides (DDT, BHC, etc.), polyethylene bag, etc.). Metal articles like Aluminum cans, Iron nails, Silver foil and radioactive wastes. All these non-biodegradable wastes cannot be made cannot be made less toxic (less poisonous) easily and hence, they are major pollutants of the environment. These wastes are not acted upon by the microbes or acted very slowly so that they cannot be degraded. These wastes undergo biological magnification in the food chain. The non-biodegradable wastes cannot be decomposed by micro-organisms like bacteria. DDT is a non-biodegradable waste so it can be passed along the food chain from crops to man or other animals and birds and harm them. For this reason, DDT has been banned from use in most countries. Non-biodegradable wastes are the major pollutants of the environment. As the discarded plastic articles, glass articles and metal objects are the non-biodegradable waste materials which cause a lot of pollution in our surroundings. We will now describe a simple experiment to find out whether a given material is biodegradable or non-biodegradable.

Problem: To find out the Biodegradable and Non-biodegradable materials:

Solution:

Procedure: Collect waste material from your homes. This could include all the waste generated during a day, like kitchen waste (spoilt food, vegetable peels, used tea leaves, milk packets and empty cartons), waste paper, empty medicine bottles/ strips/bubble packs, old and torn clothes and broken footwear.

Bury this material in a pit in the school garden or if there is no space available, you can collect the material in an old bucket/flower pot and cover with at least 15 cm of soil. Keep this material moist and observe at 15 day intervals.

Observation:

(i) Polythene milk packets, plastic medicine bottles are the materials which remain unchanged over long periods of time. These materials are known as non-biodegradable materials.

(ii) Vegetable peels, used tea leaves, spoilt food are the materials that change very fast. While old and torn clothes, broken footwear, waste paper, empty cartons are the materials which change their form and structure over time. All these materials are known as biodegradable materials.

Conclusion:

We will find that the piece of paper and the piece of cotton cloth have been partially eaten up (or decomposed) but the plastic bag has remained unaffected, it has not been eaten up (or decomposed). This means that paper and cotton cloth have been decomposed by the micro-organisms present in the soil. So, paper and cotton cloth are biodegradable. On the other hand, the plastic bag has not been decomposed by the micro-organisms present in the soil, therefore, plastic is non-biodegradable. So, the decomposer organisms are not able to decompose plastic into simpler harmless substances.

We will now explain why some materials are biodegradable whereas others are non-biodegradable. The micro-organisms like bacteria and other decomposer organisms (called saprophytes) present in our environment are 'specific' in their action. They break down the natural materials or products made from natural materials (say, paper) but do not break down man-made materials such as plastics. So, it is due to the property of decomposer organisms of being specific in their action that some waste materials are biodegradable whereas others are non-biodegradable. We should use the shopping bags (or carry bags) made of paper, cotton cloth or jute because these are biodegradable materials. On the other hand, plastic bags (or polythene bags) should be avoided because plastic is a non-biodegradable material.

6.7.5.3 Biodegradable Plastics:

- Our whole world seems to be wrapped in plastic. Almost every product we buy, most of the food we eat, and many of the liquids we drink, come encased in plastic. The environmental impact of persistent plastic wastes is evoking more global concern as alternating disposal methods are limited. Incineration (burning) may generate toxic air pollution and satisfactory landfill sites are limited.
- The continuously growing concern has stimulated research interest in biodegradable polymers as alternative to conventional non-biodegradable polymers of petrochemical products such as polyethylene and polystyrene etc.
- The plastics that will decompose in natural environment by microorganisms are known as biodegradable plastics.

- Biodegradable plastics are made from sustainable materials such as corn, hemp and chicken feathers etc. Most of the biodegradable plastics are made from corn. It has been manufactured since 1989. The basic ingredient of corn based plastics is resin polyactide or PLA extracted from corn. Corn based plastics are biodegradable however they can only biodegrade quickly in controlled composting facilities. Composting plants allow compost to reach 140°F for ten days to incite the biodegradation process.
- The conventional plastic made from petroleum products never biodegrade.

Solid wastes generally comes from households, vegetable and fruit markets, cattle sheds, industries, agricultural fields, and many other places. These wastes can be categorized into following categories such as (a) food waste (b) cow dung, (c) human excreta (d) farm waste (e) trash and rubbish, and (f) Hospital waste etc.. Some of the important methods of waste disposal are used such as (i) Preparation of Compost (ii) Recycling (iii) Incineration (iv) Landfill, and (v) Sewage treatment.

All the waste materials produced by the various activities of man and animals are poisonous to some extent. It is mainly divided into two main groups such as

Biodegradable wastes, and Non-biodegradable wastes. Those waste materials which can be broken down to non-poisonous substances in nature in due course of time by the action of micro-organisms like certain bacteria, are known as biodegradable wastes. The waste materials which cannot be broken down into harmless (non-poisonous) substances in nature are called non-biodegradable wastes.

Table 6.13 Differences between Biodegradable and Non-biodegradable Wastes

Sr. No.	Characteristics	Biodegradable Wastes	Non-biodegradable Wastes
1.	Origin	They are biological in origin.	They are commonly man-made.
2.	Degradability	The wastes are degraded by microorganisms	They are not degraded by micro-organisms.
3.	Accumulation	They do not accumulate in nature	The pile up and accumulate in nature.
4.	Biomagnification	They biodegradable wastes do not show biomagnification	The soluble non-degradable wastes enter food chains and undergo biomagnification.
5.	Resources	The wastes can be converted into resource. Examples. Garbage, livestock wastes, sewage	Some wastes can be recycled. Examples. Plastic, polythene, glass, nickel, cadmium, several pesticides.

REMEMBER SOCKETS

- The word environment was derived from the word "environ" meaning "to encircle" or surround. The environment includes the physical surroundings like water bodies, air, soil and all the organisms such as plants, animals, human beings and micro-organisms
- Human beings are the only organisms who change/disturb the natural environment to fulfill their demands/needs such as food, clothing, housing, transport and industry etc.
- The term ecosystem was first used in 1935 by British ecologist Arthur Tansley.
- An ecosystem may be defined as a structural and functional unit of the biosphere comprising living organism and their non-living environment which interact by means of food chains and biogeochemical cycles resulting in energy-flow, biotic diversity and material cycling to form a stable, self-supporting.
- Ecosystems may be classified on the basis of their nature, duration and size.
- On the basis of nature, ecosystems are classified into two types such as (a) Natural ecosystems (Pond and Lake Ecosystem etc.) and (b) Artificial ecosystems (Croplands and Aquarium etc.).
- On the basis of duration, ecosystems are classified into two types such as (a) Temporary ecosystems (Rain-fed pond) and (b) Permanent ecosystems (Forest, desert etc.)
- On the basis of size, ecosystems are classified into two types such as (i) Small (Flowerpot etc.) and (ii) Large ecosystems (Ocean etc.)
- All the ecosystems are made up of two main components such as (a) Abiotic and (ii) Biotic components. Abiotic components mean non-living components and biotic components mean living components.
- The biotic community of an ecosystem includes three types of organisms such as Producer (Autotrophs), Consumer (Heterotrophs) and Decomposer (Saprotrophs).
- Consumers can be further divided into three categories such as (a) Herbivores (b) Carnivores, and (c) Omnivores.
- The microscopic aquatic plants freely floating on the surface of water are called Phytoplanktons (free-floating algae), whereas freely floating aquatic animals on water are called zooplanktons (freely-floating protozoa).
- The decomposers act as cleansing agents of environment due to decomposing the dead bodies of plants and animals.
- In as ecosystem, all living organisms are linked in a systematic chain with respect in their mode of manufacturing food/feeding habits. Such a chain is called food chain.

- The interaction among various components of the environment involves flow of energy from one component of the ecosystem to the other.
- In ecosystem, food chains may have two, three, four or maximum five trophic levels.
- Trophic levels are the distinct sequential steps in the food chain where transfer of energy occurs.
- Quantum of available energy in a food chain successively gets decreased at each trophic level as a result of waste of energy as heat. It restricts the size of the food chain in an ecosystem to usually 3 or 4 steps and rarely maximum of 5 steps.
- In nature, basically two types of food chains are recognized such as (a) Grazing and (b) Detritus food chain.
- Grazing food chain starts from the living green plants goes to grazing herbivores, and on to carnivores, while detritus food chain goes from dead organic matter into micro-organisms and then to organisms feeding on detritus (detrivores) and their predators.
- A food chain is always straight and proceeds in a progressive straight line and involves a nutritive interaction between the living organisms.
- Each food web is a network of food chains which become interconnected at various trophic levels so as to form a number of feeding connections amongst different organisms of a biotic community.
- Food webs provide alternate pathways of food availability.
- Food webs also help in checking the overpopulations of highly reproductive species of plants and animals, and also help in ecosystem development.
- Unlike food chains, food webs are never straight. Instead, each food web is formed by interlinking of food chains.
- On an average, 10% of the amount of organic matter that is present at each step reaches the next trophic level.
- The term ecological pyramid was discovered by Elton, 1927.
- An ecological pyramid (also trophic pyramid, eltonian pyramid, energy pyramid, or sometimes food pyramid) is a graphical representation designed to show the biomass or bio-productivity at each trophic level in a given ecosystem.
- Ecological pyramids are divided in three categories such as (a) pyramid of energy, (b) pyramid of numbers and (c) pyramid of biomass.
- Energy pyramid shows the flow of energy at different trophic levels and always upright and vertical.
- Pyramid of numbers depicts the relationship in terms of the number of producers, herbivores and the carnivores at their successive trophic levels.

There is a decrease in the number of individuals from the lower to the higher trophic levels.

- Pyramid of numbers are further divided in three categories such as such as Upright pyramid, partly upright pyramid and Inverted pyramid.
- Pyramid of biomass is more fundamental, they represent the quantitative relationships of the standing crops. They show a gradual decrease in the biomass from the producers to the higher trophic levels.
- Pyramid of biomass are further divided in two categories such upright and inverted pyramid of biomass
- Flow of materials in ecosystem is cyclic, but flow of energy is unidirectional:
- The 10% law which was given by R. Lindeman (in 1942) and also termed as second law of thermodynamics (Law of Entropy).
- According to 10% law, only 10 percent of the energy entering a particular trophic level of organisms is available for transfer to the next higher trophic level.
- The materials that we generate in daily life and throw them away are termed solid wastes. These generally come from residences, cattle sheds, industries, agricultural field and many other places.
- Recently, increase in human population and advancement in technology have contributed significantly in the deterioration of our environment.
- Solid wastes can be categorized into two types such as (a) biodegradable wastes and (b) non-biodegradable wastes. Biodegradable wastes can easily be degraded by the action of microorganisms (bacteria, fungi of decay) into simpler, harmless substances in due course of time. The latter reach the reservoir pool and are again available to autotrophs for photosynthesis.
- Non-biodegradable wastes cannot be degraded by natural means into simpler, harmless substance. These, therefore, simply persist in the environment and affect human beings and other animals.
- Solid wastes are the main source of soil pollution. These are accumulated as heaps which make the surroundings dirty. Decomposition of waste produces foul smell and wind spreads it in surrounding areas making life more miserable. Flies breed at these places and spread diseases. Solid wastes may block drains which then become breeding places of mosquitoes. The latter are carriers of diseases like malaria and dengue.
- Chemicals and fertilizers reduce the fertility of soil. Certain non-biodegradable pesticides (DDT) enter the food chains, biomagnify and affect various biotic components including man.
- Waste disposal is mainly done by (a) landfills, (b) recycling of wastes, (c) by forming compost, (d) incineration and (e) through biogas plants generating biogas and manure.

- Global environment comprises atmosphere, hydrosphere and lithosphere.
- The atmosphere is formed of air, covering the earth from all sides. Four layers, indentified in the atmosphere, are troposphere, stratosphere, mesosphere and thermosphere.
- Ozone layer was discovered in 1913 by the French Physicists Charles Fabry and Henri Buisson.
- Ozone layer has the capability to absorb almost 97-99% of the harmful UV radiations.
- Ozone layer depletion leads to decrease in ozone in the stratosphere
- Ozone blanket comprises high concentration of ozone some 18-50 km above in the stratosphere. Ozone is a form of oxygen having 3 oxygen atoms (O_3).
- In stratosphere, ozone is being photo dissociated and generated by the absorption of short-wavelength ultraviolet (UV) radiations. By absorbing harmful UV radiations, it acts as a shield to protect all types of life on earth.
- UV radiation is divided into three categories, based on its wavelength such as UV-A (400–315nm), UV-B (315–280nm), and UV-C (280–100 nm).
- Ozone is being depleted by air pollutants such as chlorofluorocarbons (CFCs), methane (CH_4) and oxides of nitrogen (NOx).
- Chlorofluorocarbons (CFCs) are mainly responsible for ozone depletion.
- CFCs are harmful, synthetic chemicals widely used in refrigerators and air conditioners (as coolants) in fire extinguishers, in aerosol sprayers (as propellants).
- September 16 was designated by the United Nations General Assembly as the International Day for the Preservation of the Ozone Layer, and celebrates at World Ozone Day.
- Thinning of ozone layer allows more UV radiations falling on the earth. These cause majorly such as (a) skin cancer, (b) increased chances of cataract disease in eyes, (c) damage to immune system in human beings and animals (d) increased embryonic mortality, (e) decreased photosynthesis and (f) contribute towards global warming.
- Bioaccumulation is the accumulation of substances heavy metals, pesticides, or organic chemicals in living organisms.
- Biomagnification (Bioamplification) is simply the increase in concentration of a substance in a food chain, not an organism.
- Bioaccumulation is increasing the concentration of a substance in one organism whereas biomagnification go up in a food chain.
- Bioaccumulation occurs within a tropic level and biomagnification occurs between tropic levels.

VERY SHORT ANSWER TYPE QUESTIONS:

1. Define the term environment.
2. Define the term ecosystem.
3. Give the types of ecosystem on the basis of nature.
4. Give the types of ecosystem on the basis of nature.
5. Give the difference between organic and inorganic compounds.
6. What percentage of the solar energy is trapped and utilized by the plants?
7. What percentage of energy available at the producer level is transferred at successive trophic levels in a food chain?
8. Give the name of functional unit of the environment comprising of the living and non-living components .
9. Give the name of two natural ecosystems and two artificial ecosystems.
10. Which one of the following is not a terrestrial ecosystem?

    Desert, Aquarium, Grassland,Forest,
11. Why plants are called producers?
12. What term is being used to those organisms which break down the complex organic compounds present in dead animals and plants?
13. Define the term planktons?
14. Define the term zooplanktons.
15. Define the term herbivores.
16. Define the term carnivores.
17. Define the term omnivores.
18. Give the difference between herbivores and carnivores.
19. Define the term food chain.
20. Define the term food web.
21. Which category of organisms forms the starting point of a food chain?
22. Give the name two predators of snakes in a food web operating in a forest ecosystem.
23. Which of the following belong to the same trophic level?

    Spider, Plants, Goat;Rate, Hawk;
24. Which of the following belong to the same trophic level?

    Lizard, Tree, Snake, Frog, Grass;
25. Write the term an aquatic food chain.
26. Name the organisms belonging to the second and the fourth trophic levels in the food chain comprising the following:

    Snakes, Insects, Hawk, Frogs, Plants

27. What are the various steps of food chain called?
28. Construct a food chain comprising the following:

    Plants, Snakes; Hawk, Rats
29. Arrange the following in a food chain:

    Fish, Algae, Small animals, Big Fish
30. Which organisms belong to third and fourth trophic levels in the food chain comprising the following?

    Plants, Rats, Hawk, Snakes,
31. Which one term in the following includes the others?

    Flora, Environment, Sunlight, Air, Soil, Fauna, Water
32. Who discovered the pyramid?
33. Give the types of Pyramid.
34. Which type of pyramid is always upright?
35. Who discovered the term ten percent law?
36. If 5 Joules of energy is available at producer level (plants), then how much energy will be transferred to the lion in the following food chain?

    Plants $\rightarrow$ Goat $\rightarrow$ Lion
37. State whether the following statement is true or false:
38. Only 10% of the light energy given by the sun is available for transfer at each higher trophic level in a food chain.
39. What are wastes?
40. Give any two examples of biodegradable wastes.
41. Which biodegradable waste takes comparatively more time to degrade?
42. Which method of waste disposal is used to remove bulk of waste?
43. What is ozone hole?
44. What do you mean by non-biodegradable wastes?
45. Name any one disease in man which occurs due to germs transferred by flies.
46. Give the name of at least one source which generated solid wastes.
47. Name one ozone depleting substance which is used as coolant in refrigeration and air conditioners.
48. Who first discovered ozone hole over Antarctica and when?
49. Give at least one harmful effect of ozone depletion on man.
50. Which of the following types of wastes are used in biogas plants to generate biogas, biodegradable or non-biodegradable?
51. Name the common wastes used to prepare compost.

52. Expand the abbreviation 'ODS'. Give one example of it.
53. Which protocol was signed in 1987 between nations to limit the production of ozone depleting substances?
54. Name any one chemical which is non-biodegradable and gets biologically magnified.
55. What do you mean by biological magnification?
56. Name any two biodegradable pollutants.
57. List any two non-biodegradable pollutants.
58. List four common waste disposal methods.
59. What is incineration?
60. Name the three major biotic components of an ecosystem.
61. Name the decomposers?
62. Which are more numerous, secondary consumers or tertiary consumers?
63. What is the ultimate source of energy for organisms?
64. Name the radiations that are absorbed by ozone blanket.
65. Name two basic components of an ecosystem.
66. Give reason why ozone layer in the stratosphere is considered useful?
67. Why plastic bags and containers should not be disposed of by burning? Give reason.
68. Give example of autotrophs in a garden ecosystem.
69. Give examples of heterotrophs in an ecosystem.
70. Which of the following are essential components?
71. Expand UNEP.
72. Expand the term ODS.
73. What are biological components of environment?
74. Which two of the following are biodegradable? Tomato leaves, Aluminum wire, synthetic fibre, wood.
75. Which of the following are not biodegradable? Wool, glass, silver foil, leather.
76. What is it necessary to conserve our environment?
77. Which one of the following gases is the major constituent of biogas? Carbon monoxide, Hydrogen, Methane, Carbon dioxide.
78. Why are green plants called producers?
79. Expand the abbreviation IPCC.
80. Where does all the energy in living organisms originate from?

81. Give the full form of DDT.
82. Who discovered the term ozone layer?
83. Which radiations are absorbed by ozone layer?
84. Give the name the group of chemical compounds which damages the ozone layer.
85. Give the full form of UV and CFC.
86. Define the term bioaccumulation.
87. Define the term biomagnification.
88. Give the example of toxic substances.
89. Give the name the process in which a harmful chemical enters the food chain and gets concentrated at each trophic level.
90. In a food chain consisting of grass, frog, bird and insects, where will the concentration of the harmful chemicals be maximum?
91. If a harmful chemical enters a food chain comprising cat, mice and plants, which one of these organisms is likely to have the maximum concentration of the harmful chemical in its body?
92. Define the term incineration.
93. Define the term landfill.
94. What do you understand by pollution?
95. Give the name of two waste materials which can be recycled.
96. Give the name the process by which the volume of solid wastes can be reduced.
97. Give the example of biodegradable and non-biodegradable plastics.
98. Write one word answer for the following statements/definitions:
    - Burning of substances at high temperature (over 1000oC) and converting them into ash.
    - Decline in the thickness of ozone layer over a restricted area.
    - Each level of food chain where transfer of energy takes place.
    - Organisms which depend on the producers either directly or indirectly for food.
    - The physical and biological world where we live in.
    - The physical factors like temperature, rainfall, wind and soil of an ecosystem.
    - The substances which can easily be degraded by biological means into simpler harmless substance.
    - The substances which react with the ozone layer in the stratosphere and destroy it.

- The synthetic chemicals widely used in refrigerators and air conditioners as coolants.
- The useless left over or discarded materials.

SHORT ANSWER TYPE QUESTIONS

1. Define the term environment and ecosystem.
2. Define the terms producer and consumer with examples.
3. Give the differences among autotrophs, heterotrophs and saprotrophs.
4. Define the term decomposers with example.
5. Define the importance of decomposers in the ecosystem?
6. Define the terms as primary consumer, secondary consumer and a tertiary consumer with example of each.
7. Give an example of a four step food chain operating in grassland.
8. Give an example of the secondary consumer in this food chain.
9. Define the term trophic level.
10. Draw the food chain with four trophic levels.
11. Give the difference between the food habits of organisms belonging to the first and the third trophic levels?
12. What is meant by non-biodegradable waste materials? Give two examples of non-biodegradable wastes.
13. Give the differences between biodegradable and non-biodegradable waste with example?
14. Which of the following are biodegradable and which non-biodegradable?
15. Wood, Grass, Cake, Vegetable peels, Cotton Cloth, Wheat, DDT, Glass bottle, Ball point pen refill, Jut bag, Hay, Polythene bag, Paper,
16. What is a food chain? Give one example of a simple food chain.
17. Give the differences between first and third trophic levels
18. Give the differences between second trophic level and fourth trophic level.
19. Write down in brief on types of food chains.
20. Give the characteristics of food chain.
21. Write down on significance of food chain.
22. Give the characteristics of food web.
23. Write down on importance of food web.
24. Give the differences between autotrophs and heterotrophs.
25. Write down on differences between food chain and food web.

26. Write down on pyramid of numbers with examples.
27. Write down on limitations of ecological pyramids.
28. Explain why, some materials are biodegradable but some are non-biodegradable.
29. Describe an activity to show that while paper is biodegradable but plastic (say, polythene) is non-biodegradable.
30. Draw one (a) long food chain, and (b) short food chain, ending with man.
31. Write a common food chain of a pond ecosystem having four links.
32. State one advantage of using jute bags over plastic bags for shopping.
33. Why we do not clean ponds or lakes but an aquarium needs to be cleaned periodically?
34. What will be the consequence of the absence of decomposers in the ecosystem?
35. Give two differences between food chain and food.
36. List various modes of waste disposal. Explain any four of them.
37. Why is improper disposal of waste a curse to environment?
38. Write the common food chain of a pond ecosystem.
39. What are the advantages of cloth bags over plastic bags during shopping?
40. Why are crop fields known as artificial ecosystems?
41. What is incineration? Which types of wastes are generally disposed of by this process?
42. What is meant by recycling of wastes? Explain briefly.
43. How does ozone layer act as protective shield?
44. Give two harmful effects of non-biodegradable wastes.
45. What are CFCs? Where are these widely used?
46. What is montreal protocol? Why was it signed?
47. What is ten percent law? Explain with an example.
48. Write the full form of CFC. Give its one harmful effect.
49. Explain how, harmful ultraviolet radiations of sunlight are prevented from reaching the earth's surface.
50. What are the causes of depletion of ozone layer? Which diseases are likely to be caused if the ozone layer will become thinner?
51. Explain how harmful chemicals enter our bodies.
52. Write down on accumulation of harmful Chemicals in Food chains.
53. Give harmful effects of UV radiations.
54. Give example of food web in a terrestrial ecosystem.

55. Write various characteristics of a food chain.
56. Briefly write various characteristics of a food web.
57. What are trophic levels?
58. Name three important ozone depleting substances.
59. Explain 10 per cent law.
60. What is ecosystem? Explain different components of ecosystem.
61. How does energy flow in an ecosystem?
62. Give any two causes of global warming.
63. Give an example of food chain occurring in grassland.
64. Name two artificial ecosystems.
65. Which type of waste is disposed by composting?
66. Who formulated 10% law?
67. Give the differences among UV-A, UV-B and UV-C rays.
68. Give the differences between Incineration and Pyrolysis.
69. Give the differences between biodegradable and non-biodegradable wastes with examples.
70. What is meant by biological magnification? With the help of a food chain, explain how biological magnification of harmful chemicals can occur.
71. What is meant by bio concentration of pesticides? Which common pesticide has accumulated in human body in considerable amounts?
72. What is garbage? What does garbage consist of?
73. Name the various modes of wastes disposal.
74. How can the wastes such as paper, plastic and metal objects be disposed of?
75. Give a method for the disposal of household wastes such as left-over food, fruit and vegetable peels, and leaves of potted plants.
76. What is meant by incineration? For what purpose is it used?
77. How are most of the solid wastes in urban areas disposed of?
78. State two advantages of using disposable paper cups over disposable plastic cups/
79. What is sewage? How is sewage disposed of?
80. Write the harmful effects of ozone depletion.
81. What would happen if the ozone layer in the atmosphere completely disappears?
82. Write one or two words for each of the following statements/definitions:
    (a) The physical factors like temperature, rainfall, light, soil, are and water of an ecosystem.

(b) The physical and biological world where we live in

(c) Self-contained unit of living things and their non-living environment needing only sunlight for its functioning.

(d) Each level of food chain where transfer of energy takes place.

(e) Organisms which depend on the producers for food either directly or indirectly.

83. Write a short notes on any one (i) Garden ecosystem (ii) Pond ecosystem (iii) Food chain (iv) Ozone layer and their depletion (v) Ozone depleting substances (vi) Helsinki declaration (vii) Aging of skin (viii) Bioaccumulation Vs Biomagnification (ix) Sewage treatment (x) Paper Vs. Plastic cups

LONG ANSWER TYPE QUESTIONS

1. Define the term ecosystem and explain its components.
2. Explain the types of consumers with their role.
3. Describe the garden ecosystem with diagram.
4. What is food chain? Explain the types with example.
5. Describe the grazing food chain with labeled diagram.
6. Describe the detritus food chain with labeled diagram.
7. What is food web Give their characteristic and importance also.
8. Define the term pyramid and brief the types of pyramid.
9. Explain the pyramid of numbers with example and suitable diagram.
10. Explain the pyramid of biomass with example and suitable diagram.
11. Explain why pyramid of energy is always straight.
12. Give the effect of human's activities on ecosystem.
13. Write down on transfer of energy in food chains with diagram.
14. Why flow of materials in ecosystem is cyclic but flow of energy is unidirectional explain it.
15. Explain the role of decomposers in the environment.
16. Indicate the flow of energy in an ecosystem. Why is it unidirectional? Justify.
17. What are decomposers? What will be the consequence of their absence in an ecosystem?
18. Suggest suitable mechanism(s) for waste management in industries.
19. Explain some harmful effects of agricultural practices on the environment.
20. What is ozone hole? Where it was first noticed? Name the equipment and also the unit of measurement of quantity of ozone present at a particular region of the atmosphere.

21. Why proposal of use of kulhads (disposable cups made of clay) in trains was set aside?
22. What kind of disposable cups being used on large scale in trains now a day and why?
23. Explain the role of decomposers in biotic environment.
24. Why natural ecosystem is more stable than man-made artificial ecosystem?
25. What is ozone blanket? How it is being depleted?
26. What do you mean by waste disposal? Explain how is bulk of biodegradable waste removed?
27. Why bacteria and fungi are called decomposers? List any two advantages of decomposers to the environment.
28. Explain the phenomenon of 'biological magnification'. How does it affect organisms belonging to different trophic levels particularly the tertiary consumers?
29. Explain the ten percent (10%) Law with suitable example.
30. Explain on depletion of ozone layer with diagram.
31. Write down on ozone depleting substances and their impact.
32. Define the term bioaccumulation and differentiate from biomagnification with example.
33. Explain any two methods of disposal of domestic wastes or household garbage.
34. Write down on Incineration and land filling with suitable diagram:
35. Define the term environmental pollution and give the types of wastages with examples.
36. Differentiate biodegradable wastes from non-biodegradable wastes with example.

MULTIPLE CHOICE QUESTIONS:

1. The environment of any organism means

   (a) Physical (b) Biological (c) Both a and b (d) None of these

2. World Environment Day is celebrated on every year on

   (a) June 5 (b) June 4 (c) July 5 (d) July 4

3. The term "Ecosystem" was coined by

   (a) Arthur Tansley (b) R. Lindeman

   (c) Watson and Crick (d) Krebs

4. An ecosystem may be defined as a unit of the biosphere comprising living organism and their non-living environment

(a) Structural (b) Functional (c) Both a and b (d) Physical

5. Consumers can be divided into how many categories

(a) 2 (b) 3 (c) 4 (d) 1

6. The microscopic aquatic plants freely floating on the surface of water are called as

(a) Algae (b) Fungi

(c) Nematodes (d) Phytoplankton.

7. The microscopic aquatic animals freely floating on water are called as

(a) Zooplanktons (b) Fungi

(c) Nematodes (d) Phytoplankton

8. An ecosystem includes:

(a) Non-living objects

(b) All living organisms

(c) All living organisms and input of sun's energy

(d) Both living organisms and non-living objects

9. Environment includes

(a) Plants, animals, microbes (b) Light, temperature, rainfall

(c) Land, air, water (d) All of these

10. Abiotic components of environment include

(a) Plants (b) Animals

(c) Microorganisms (d) Water

11. The primary consumers are

(a) Producers (b) Herbivores (c) Carnivores (d) Omnivores

12. The driving force of any ecosystem is

(a) Carbohydrates (b) Solar energy

(c) Biomass (d) ATP

13. Among these, which is a linear arrangement of organisms?

(a) Food chain (b) Food web (c) Trophic level (d) Community

14. The biotic component of an ecosystem consists of

(a) Plants and animals

(b) Algae and fungi

(c) Producers, consumers and decomposers

(d) Air, water and soil

15. Which of the following should not be added in a composting pit to prepare compost?
    (a) Fruits and vegetables peels (b) Wheat plants
    (c) Polythene bags (d) Earthworms
16. Organisms which synthesis carbohydrates from inorganic compounds by using radiant energy are called as
    (a) Producers (b) Herbivores (c) Carnivores (d) Decomposers
17. One of the following is a micro-consumer. This one is:
    (a) Mice (b) Ant (c) Mosquito (d) Fungi
18. Which of the following bacteria act as decomposers in an ecosystem?
    (a) Lactobacillus (b) Rhizobium
    (c) Cyanobacteria (d) Putrefying
19. What will happen if deer is missing in the food chain given below?
    Grass → Deer → Tiger
    (a) The population of grass decreases
    (b) The population of tigers increases
    (c) The population of tigers decreases and the population of grass increases.
    (d) Tigers will start eating grass
20. Organisms of a higher trophic level which feed on several types of organisms belonging to a number of lower trophic level constitute the:
    (a) Food Chain (b) Ecological pyramid
    (c) Food Web (d) None of these
21. Which of the following statement in incorrect?
    (a) All green plants and blue green algae are producers.
    (b) Producers prepare their own food from inorganic compounds.
    (c) Plants convert solar energy into chemical energy.
    (d) Green plants get their food from readymade organic compounds.
22. Which of the following group of organisms are not constituents of a food chain?
    (i) Grass, lion, rabbit, wolf
    (ii) wolf, grass, snake, tiger
    (iii) Frog, snake, eagle, grass, grasshopper
    (iv) plankton, man, fish, grasshopper
    (a) (i) and (iii) (b) (iii) and (iv) (c) (ii) and (iii) (d) (i) and (iv)

23. One of the following helps in the recycling of materials in an ecosystem. This one is:

    (a) Omnivores (b) Autotrophs (c) Carnivores (d) Saprotrophs

25. In nature, which type/s of food chains is/are occurs

    (a) Grazing food chain (b) Detritus food chain

    (c) Both a and b (d) None of these

26. In a food chain, the third trophic level is always occupied by:

    (a) Producers (b) Herbivores (c) Carnivores (d) Decomposers

27. In an ecosystem:

    (a) The flow of materials is unidirectional

    (b) The flow of energy is unidirectional

    (c) The flow of energy is cyclic

    (d) The flow of materials is cyclic

28. A food chain comprises of cat, seed-eating bird, plants, and dog. The organism which will have the maximum concentration of harmful pesticides coming through the food chain is most likely to be:

    (a) Plants (b) Cat

    (c) Dog (d) Seed-eating bird

29. An aquatic food chain comprises of the organisms like tadpoles, weeds, fish and water beetles. The organisms which gets the minimum energy through this food chain is:

    (a) Weeds (b) Water beetles (c) Fish (d) Tadpoles

30. Most of the water surface of a lake is covered with algae, and part of the food chain which also includes small fish, bird, larvae and big fish. In this food chain, which will obtain the maximum energy?

    (a) Larvae (b) Bird (c) Big fish (d) Small fish

31. If a grasshopper is eaten by a frog, then the energy transfer will be from:

    (a) Producer to primary consumer

    (b) Producer to decomposer

    (c) Primary consumer to secondary consumer

    (d) Secondary consumer to tertiary consumer

32. One of the following is not a biodegradable material. This one is:

    (a) Animal bones (b) Cotton

    (c) Wood (d) Aluminum foil

33. Which of the following is not a non-biodegradable material?
   (a) Plastic school bag (b) Jute carry bag
   (c) Polyester clothes (d) Cotton
34. The use of one of the following will pollute the environment. This one is :
   (a) Cotton cloth carry bags (b) Jute carry bags
   (c) Paper carry bags (d) Nylon cloth carry bag
35. Sahara Desert was formed over a period of time due to one of the following uncontrolled activities of man:
   (a) Excessive killing of large carnivores
   (b) Excessive killing of large herbivores
   (c) Excessive use of poisonous chemicals called herbicides
   (d) Excessive cutting down of forest plants and trees
36. Which of the following constitutes a food chain?
   (a) Grass, Fish and Goat (b) Grass, Goad and Human
   (c) Grass, Wheat and Mango (d) Goat, Cow and Elephant
37. In a food chain, the initial organism is usually:
   (a) Saprophytic (b) Photosynthetic (c) Parasitic (d) Herbivore
38. Which of the following is an artificial ecosystem?
   (a) Lake (b) forest (c) pond (d) crop field
39. The decomposers in an ecosystem:
   (a) Convert inorganic material into organic compounds
   (b) Do not break down organic compounds
   (c) Convert organic material to inorganic forms
   (d) Convert inorganic material to simpler forms
40. Which of the following is not a producer?
   (a) Zooplankton (b) Grass
   (c) Paddy (d) Phytoplankton
41. The harmful chemical which is accumulating in human beings through food chain is
   (a) Abscisic acid (b) Chlorofluorocarbon
   (c) Benzene hexachloride (d) Dichloropiphenyltrichloroethane
42. Which of the following gets the minimum energy through the food chain in an ecosystem?
   (a) Carnivore (b) Large carnivore (c) Producer (d) Herbivore

43. An agriculture cropland is
    (a) A biome
    (b) A natural ecosystem
    (c) An artificial ecosystem
    (d) A community of plants and animals only
44. The cleaners of nature are
    (a) Producers (b) Consumers (c) Decomposers (d) Mycorrhizae
45. Food levels in an ecosystem are called
    (a) Trophic levels (b) Consumer levels
    (c) Producer levels (d) Herbivore levels
46. Food web helps in
    (a) Ecosystem development
    (b) Providing alternative pathways of food availability
    (c) Checking the overpopulation
    (d) All of these
47. What provides the energy which then flows through a food chain?
    (a) Respiration (b) sunlight (c) glucose (d) oxygen
48. In the food chain given below, if the amount of energy available at fourth trophic level is 5 kJ, what was the energy available at the producer level?

    Grass → Grasshopper → Frog → Snake → Hawk

    (a) 5000 kJ (b) 500 kJ (c) 50 kJ (d) 5 kJ
49. Which of the following limits the number of trophic levels in a food chain?
    (a) Accumulation of harmful chemicals at higher trophic levels.
    (b) Increase in the number of organisms at higher trophic levels.
    (c) Insufficient food supply from producer level.
    (d) Decrease in energy at higher trophic levels.
50. The flow of energy in an ecosystem is always:
    (a) Bidirectional (b) cyclic
    (c) multidirectional (d) unidirectional
51. What percentage of sun's energy falling on the leaves of green plants is utilized by the plants in the process of photosynthesis and stored as chemical energy of food?
    (a) 1 percent (b) 20 percent (c) 99 percent (d) 10 percent

52. In an ecosystem, the ten per cent energy available for transfer from one trophic level to the next is in the form of:
    (a) Light energy (b) Chemical energy
    (c) Mechanical energy (d) Heat energy
53. If the energy transferred to a tertiary consumer in a food chain is 10 J, how much energy was available to the primary consumer?
    (a) 100 J (b) 500 J (c) 1000 J (d) 5000 J
54. The 10% Law is associated with
    (a) Transfer of ATP energy into muscular energy
    (b) Transfer of sun's energy to the organisms called producers.
    (c) Transfer of chemical energy from one organism to another
    (d) Transfer of energy from various trophic levels to decomposers in a food chain.
55. The depletion of ozone layer in the upper atmosphere is mainly due to the emission of:
    (a) Greenhouse gases (b) Chlorofluorocarbons
    (c) Ultraviolet radiations (d) Unburnt hydrocarbons
56. O_2 is converted into O_3 by the action of:
    (a) Gamma radiations (b) Infrared radiations
    (c) Ultraviolet radiations (d) Cosmic radiations
57. The excessive exposure of humans to ultraviolet rays results in:
    (i) Skin cancer (ii) Peptic ulcers
    (iii) Damage to lungs (iv) Damage to immune system
    (a) (i) and (ii) (b) (ii) and (iv) (c) (i) and (iii) (d) (iii) and (iv)
58. Which pollutant released into the air during refrigeration and air conditioning is the grated contributor to the depletion of ozone layer?
    (a) DDT (b) UNEP (c) BHC (d) CFC
59. Disposable plastic plates should not be used because:
    (a) They are made of biodegradable materials
    (b) They are made of lightweight materials
    (c) They are made of toxic materials
    (d) They are made of non-biodegradable materials
60. Accumulation of non-biodegradable pesticides in the food chain in increasing amount at each higher trophic level is known as:
    (a) Accumulation (b) Biomagnifications
    (c) Eutrophication (d) Pollution

61. What percentage of energy does a grasshopper gets after eating a green plant?
    (a) 10% (b) 20% (c) 30% (d) 40%
62. Kitchen wastes can be hygienically disposed off through
    (a) Dumping (b) Incineration (c) Recycling (d) Composting
63. Greenhouse effect is
    (a) A cooling effect (b) A warming effect
    (c) Carcinogenic effect (d) No effect
64. Which of the following is mainly responsible for the reduction of ozone of the atmosphere?
    (a) CFCs (b) CO_2 (c) NO_2 (d) SO_2
65. Ozone hole may cause
    (a) Mutations (b) Global warming (c) Skin cancer (d) All of these
66. Biodegradable wastes include
    (a) Aluminum cans (b) Glass bottles
    (c) Pesticides (d) Wood shavings
67. Excessive exposure to ultraviolet rays cause
    (a) Oral cancer (b) Skin Cancer (c) Lymphoma (d) Peptic ulcer
68. In the following groups of wastes, which group contains only non-biodegradable wastes?
    (a) Bakelite, Polythene, Wastes (b) Wood, Plastic, DDT
    (c) Wood, Polythene, Leaves (d) Plastic, DDT, Polythene
69. In the following group of wastes, which group/s contains only biodegradable wastes?
    (a) Aluminum cans, wood, synthetic fibres
    (b) Plastic, heavy metals, polyethylene
    (c) Paper, cloth, peels of fruits
    (d) Mercury, iron rods, DDT
70. Which one of the following would have a hazardous impact if they persist in the environment for a long time?
    (a) Pesticides (b) Vegetable peels
    (c) Newspapers (d) Livestock wastes
71. Which of the following could be recycled?
    (a) Paper (b) Torn cloths
    (c) Metallic wastes (d) All of these

72. Excessive growth of algae in water causes.
    (a) Biomagnifications (b) Eutrophication
    (c) Pollution (d) All of these
73. The main reason for excessive generation of waste is
    (a) Increased construction wastes
    (b) Increased availability of food
    (c) Use and throw policy
    (d) Non-utilization of all components of food
74. Use of disposable cups made of clay (Kulhads) is discouraged because
    (a) These are made of non-degradable substance.
    (b) These are unhygienic.
    (c) These are not liked by public.
    (d) Making them on large scale will result in loss of fertile top soil.
75. Full form of CFC
    (a) Chloro Fluro carbon (b) Chloro Fast carbon
    (c) Combined Flour carbon (d) Combined Fluorescent carbon

ANSWER KEY

1. (c)	2. (a)	3. (a)	4. (c)	5. (b)	6. (d)	7. (a)	8. (d)
9. (d)	10. (d)	11. (b)	12. (b)	13. (a)	14. (c)	15. (c)	16. (a)
17. (d)	18. (d)	19. (c)	20. (c)	21. (d)	22. (c)	23. (d)	24. (c)
25. (c)	26. (c)	27. (b)	28. (c)	29. (c)	30. (a)	31. (c)	32. (d)
33. (b)	34. (d)	35. (a)	36. (b)	37. (b)	38. (d)	39. (c)	40. (a)
41. (d)	42. (b)	43. (c)	44. (c)	45. (a)	46. (d)	47. (b)	48. (a)
49. (d)	50. (d)	51. (a)	52. (b)	53. (c)	54. (c)	55. (b)	56. (c)
57. (c)	58. (d)	59. (d)	60. (b)	61. (a)	62. (d)	63. (b)	64. (a)
65. (d)	66. (d)	67. (b)	68. (d)	69. (c)	70. (a)	71. (d)	72. (b)
73. (c)	74. (d)	75. (a)					

7 CHAPTER

MANAGEMENT OF NATURAL RESOURCES

Conservation and Judicious use of Natural Resources, Forest and Wild life, Coal and Petroleum conservation, Examples of People's Participation for Conservation of Natural Resources. Big Dams: Advantages and Limitations; Alternatives, if any, Water Harvesting, Sustainability of Natural Resources.

Anything which is useful to humans or anything which may be converted into useful products is called a resource. Natural resources refer to non-living and living components of nature which are used by humans for fulfilling their various requirements like food, fodder and shelter (Fig. 7.1). Since, natural resources are obtained from the earth; these are also called as earth resources (forests, wildlife, water, coal and petroleum). In earlier days, humans utilize the resources for their survival, but now they have started exploitation of natural resources for their luxurious lifestyle and mere pleasures also. This exploitation of natural resources has put an extra burden on our earth and disrupted the natural balance. There are three main reasons for the non-judicious alteration of natural resources such as

(A) Forests

(B) Wild Life

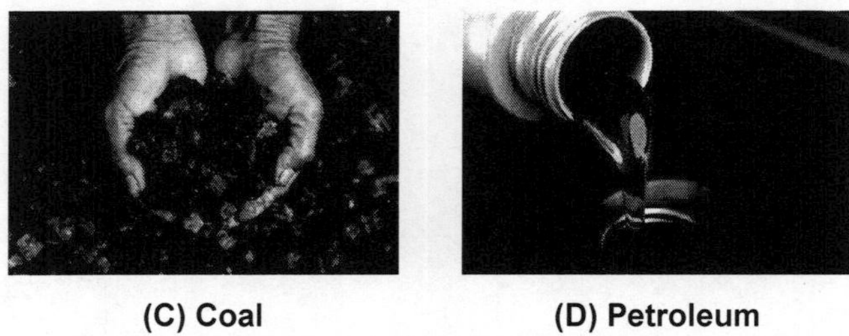

(C) Coal (D) Petroleum

Fig. 7.1 Our Natural Resources

- Increasing population
- Desire for the easy and luxurious lifestyle.

Development of modern technologies.

Natural resource management is a system to manage natural resources such as land, water, fossil fuels, plants and animals to meet the needs of present and future generations without harming the nature. The development which meets the need of the present generation as well as of the future generations without harming the environment is called sustainable development. Natural resources serve as the means of development of human beings by providing sources of energy. In this lesson, we will learn about various natural resources and their management and conservation. Some common examples of natural resources are **land resources** (e.g., mines and soil), **fresh water resources** (e.g., fish from aquaculture and river water for hydroelectricity), **marine resources** (e.g., edible algae, corals and pearls), **fossil fuels** (e.g., coal, natural gas, petroleum), **forests and wildlife** resources and **flora and fauna** (e.g., microorganisms, plants and animals). Natural resources are obtained from earth and environment, i.e., atmosphere, lithosphere and hydrosphere.

> Natural resource management is a system to manage natural resources such as land, water, fossil fuels, plants and animals to meet the needs of present and future generations without harming the nature. The development which meets the need of the present generation as well as of the future generations without harming the environment is called sustainable development. Some common examples of natural resources are land resources, fresh water resources, marine resources, fossil fuels, forests, wildlife, flora and fauna.

7.1 TYPES OF NATURAL RESOURCES

Natural resources can be classified on the basis of origin, availability and abundance, chemical nature and distribution.

7.1.1 On the basis of origin, natural resources can be classified into two types:

- **Biotic Resources**: The resources obtained from biosphere (living and organic materials) such as forests, animals and microorganisms are called biotic resources. The materials that are obtained from the fossil fuels (for example, coal and petroleum) are also included in this category.

- **Abiotic resources**: The resources obtained from non-living or inorganic materials, for example, land, fresh water, air, heavy metals including gold(Ai), iron(Fe), silver(Ag) are called abiotic resources. Biological activity is not involved in their formation but they are vital for the survival of organisms on the earth.

7.1.2 On the basis of availability and abundance, natural resources are of following types:

- **Inexhaustible resources**: The resources which occur in abundance and cannot be consumed completely are called inexhaustible resources. For example, air, sunlight, rain, tidal energy, etc.
- **Exhaustible resources**: The resources which are likely to be finished due to continuous human consumption (exploitation) are called exhaustible resources. For example, minerals, fossil fuels, etc. They are of two types such as **(i) Renewable Resources:** The exhaustible resources which can be replenished by natural cycling are called renewable resources, for example, forests, underground water, etc. **(ii) Non-renewable Resources**: The exhaustible resources which cannot be replenished or recycled again are called non-renewable resources. They are available in nature only in limited amount and cannot be regenerated once they are exhausted. For example, physical resources like coal, oil deposits, natural gas, minerals and soil.

7.1.3 On the basis of chemical nature, natural resources are of three types:

- **Inorganic Resources**: Inorganic natural resources generally include air, water and metallic minerals.
- **Organic Resources**: Organic natural resources include plants, animals, microorganisms and also their products such as fossil fuels.
- **Mixed Resources**: Mixed resources are those resources which generally consist organic as well as inorganic components. For example, soil.

7.1.4 On the basis of distribution, natural resources are of following three types:

- **National Resources**: Those resources which are generally restricted within the boundaries of a particular country are known as national resources. For example, minerals, land and forests, etc.
- **Multinational Resources**: Those resources which are not restricted within the boundaries of a particular country but are spread in two or more than two countries. For example, certain rivers and migratory birds.
- **International Resources:** These resources are shared uniformly by all the nations. For example, air and sunlight.

7.2 NECESSITY TO MANAGE OUR RESOURCES

All the things which we use such as food, water, clothes, furniture, fuels, vehicles etc. are obtained from the resources, which are available on the earth. There is a need to manage the natural resources due to following reasons:

(a) Limited resources are available on the earth due to rapid increasing the human population, the demand for resources is increasing day by day. The proper management can ensure that the natural resources are used judiciously, so that they fulfill the needs of present generation and also last for the generations to come.

(b) The proper management will take into consideration the damage caused to the environment during the extraction For example, if trees have to be cut for several purposes, and then the damage to the environment can be minimized by planting the new saplings (seedlings) in the place of cut down trees.

(c) The proper management of natural resources takes into consideration long-term viewpoint and prevents their exploitation to the hilt for short term gains.

(d) The proper management can ensure equitable distribution of natural resources so that all the people can benefit from the development of these resources.

The only solution of this problem is to save the life of mankind by the sustainable management of natural resources. The proper management of natural requires a long term view and a check on their exploitation to the hilt for short-term gains. With the increase in population growth, rapid urbanization and industrialization are also taking place. They require a vast input of natural resources. This gives rise to other serious problems like pollution (air, water and soil), soil erosion, deforestation, global warming and depletion of ozone layer. Therefore, there is an urgent need for management of natural resources.

7.3 CONSERVATION

It is the process of preservation of resources from loss, wastage and degradation or conservation as the controlled utilization of natural resources by humans in order to save them for present and future generations.

7.3.1 Aims of Natural Resource Conservation

Ensure safe disposal of wastes, for example, during metal extraction, several tones of slag is discarded which should be properly disposed to prevent various environmental hazards.

- Equitable distribution of resources among all sections of society
- Minimize wastage of resources

- To preserve a continuous yield of significant plants, animals and materials by establishing a balanced cycle of harvest and renewal of natural resources.
- To increase the preservation of a quality environment, this has aesthetic and recreational values.

7.3.2 Basic Practices of Conservation:

- Avoid wastage during their usage and keep long-term prospective in mind.
- Conservation and preservation of all kinds of natural resources by sustainable use through recycling and reuse.
- Dispose-off the wastes safely so as to protect the environment from different types of pollution.
- Educating the people regarding the outcome of overconsumption of natural resources.
- Judicious utilization of all natural resources especially exhaustible ones.
- Maintain a proper balance between economic development and utilization of resources.
- Prevent all types of pollution like soil, air, and water to conserve our natural resources.
- Search new resources to fulfill the requirements and also conservation of the exhaustible natural resources.

7.3.3 Conservation Strategies

There are two principal categories of conservation (i) In situ conservation, (ii) Ex situ conservation

(i) **In situ Conservation:** Conservation of natural resources in their original (natural) habitat is called in situ conservation. This conservation approach does not involve transfer of living organisms (animal/plant) to another suitable or protected place. For example, in situ approach is to protect the entire forest to save the tiger.

(ii) **Ex situ conservation**: Conservation by ex situ approach means that conservation of natural resources is done outside their habitats. Structures used in conservation by ex situ approach are botanical gardens, arboreta, aquaria, zoos, seed banks, pollen storage banks, tissue culture laboratories, etc.

7.4 THE THREE R'S TO SAVE RESOURCES/ENVIRONMENT:

Most of the natural resources belong to non-renewable resource category. Due to rapid development in human life, the resources are being depleted at a very fast rate. Thus, to save these resources as well as our environment, the 3R's principle, i.e., Reduce, Recycle and Reuse (Fig. 7.2) should be followed.

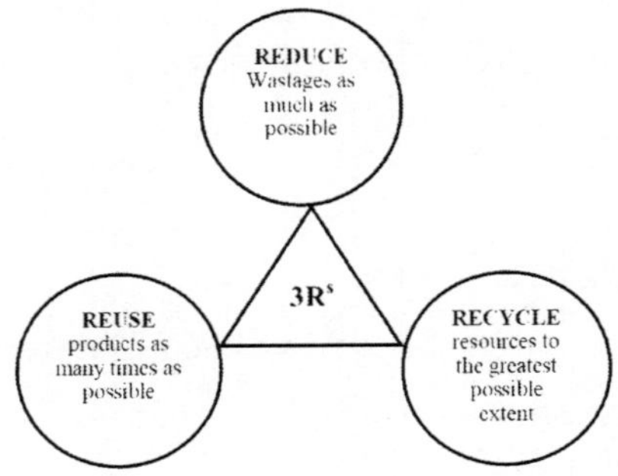

Fig. 7.2 3R^s: Reduce, Recycle and Reuse

Reduce:

It refers to the reduction in use of natural resources. The best way to save resources is that we use less of them by cutting down those practices which lead to their wastage. For example (a) Save water by turning off taps (b) Switch off lights and fans when they are not in use (c) Write on both sides of paper (d) Use public transport whenever possible (e) Buy only the things which are needed (f) Use solar cooker for cooking to reduce LPG consumption (g) Do not take extra food in your tiffin.

Recycle:

Recycling includes a series of steps that takes used materials, processes, re-manufactures and converts them into new products. During this process, discarded articles of plastic, glass, paper and metals are recycled for making fresh plastic, glass, paper and metallic items. New glass could be manufactured from old glass items. Paper, cotton, textiles, etc., are also recycled. Plastic are also recycled but inferior quality plastic is produced from that. In order to recycle, the garbage is segregated initially into recyclable materials and useless non-recyclable materials. Then, the recyclable materials are sent to separate recycling plants.

Reuse:

It refers to the repeated use of an article till it gets broken or is no more usable. Reuse makes economic and environmental sense to reuse products. For example (i) Plastic bottles and jars of pickles, jams, etc., can be used for storing other things in kitchen (ii) Plastic bottles can be used for planting. (iii) Carry bags should not be thrown but should be reused for shopping in future. (iv) Change the refill instead of throwing the old pen and buying a new one.

Natural resources can be classified on the basis of origin, availability and abundance, chemical nature and distribution. It is the process of preservation of resources from loss, wastage and degradation or conservation as the controlled utilization of natural resources by humans in order to save them for present and future generations. There are two principal categories of conservation (a) *In situ* (On spot) and (b) *Ex situ* (other than spot) conservation. To save the Resources/ Environment, the 3R's Principle (Reduce, Recycle and Reuse) is being used.

7.5 FORESTS

A forest is a large uncultivated piece of land which is dominated by trees forming a closed canopy which is self-maintained. Forests are important renewable natural resource. They are essential for maintaining ecological balance, preservation of biodiversity, soil erosion and floods. Forests play a vital role in maintaining life on the earth. About one third of the world's land surface is covered by forests which constitute 90% of global biomass. The forests are 'biodiversity hotspots' as they sustain a wide variety of life forms ranging from microorganisms to higher plants and animals. It includes bacteria, algae, fungi, gymnosperms, angiosperms and nematodes, insects, birds, reptiles, mammals, etc., which live together to maintain harmony of nature.

7.5.1 Forestry in India

Indian forest types include moist tropical, dry tropical, montane subtropical, montane temperate, sub-alpine and alpine forests. These are sub-divided into 16 major types. India has a geographical area of 31,61,414 sq km. Area of land utilized by agriculture is 306.05Mha and forest cover include 69.02Mha. India has a long history of traditional conservation and forest management practices. In 1950, the Government of India began the annual festival of tree planting called the Van Mahotsava, and the Gujarat was the first state to implement it. The FAO (Food and Agriculture Organization) of United Nations has defined forest as a land with tree crown cover of more than 10% and area of more than 0.5 hectare.

7.5.2 Forest Types of India

(A) Deciduous forest (B) Mangrove forest (C) Temperate forest

(D) Thorn forest (E) Wet Evergreen forest

Fig. 7.3 Various types of Forest in India

1. Several type of forests are found in India such as
2. Wet Evergreen forest
3. Semi Evergreen forest
4. Moist Deciduous forest
5. Dry Deciduous forest
6. Littoral and Swamp forest / Mangrove forest
7. Dry Evergreen forest
8. Thorn forest
9. Sub-tropical broad leaved forest
10. Subtropical Pine forest
11. Subtropical Dry Evergreen forest
12. Montane Wet Temperate forest
13. Montane Moist Temperate forest
14. Montane Dry Temperate forest
15. Sub Alpine forest

7.5.3 Significance of Forests:

Forests play a key role in the economy and prosperity of any country. Forestry is the second largest land use category after agriculture in India. Forests have very significant functions which are as follows (Fig. 7.4):

(A) Drug plants (B) Resin plants (C) Spices plants

(C) Woods (D) Walnut

Fig. 7.4 Some products obtained from Forests

(a) Economic or productive functions as wood and wood pulp, food and spices, nuts, tannins, gums, resins, dyes and drugs.

(b) Protective functions as protection of soil, anchorage, environmental pollution and shelter.

(c) Regulative functions as retention of water, control the amount of CO_2 in atmosphere and climate.

Forestry is the second largest land use category after agriculture in India. These are important renewable natural resource. They are essential for maintaining ecological balance, preservation of biodiversity, soil erosion and floods. Forests play a vital role in maintaining life on the earth. The forests are 'biodiversity hotspots' as they sustain a wide variety of life forms ranging from microorganisms to higher plants and animals. Forests are sub-divided into 14 major types include moist tropical, dry tropical, montane subtropical, montane temperate, sub-alpine and alpine forests. Forests have very significant functions which are as follows (a) Economic functions, (b) Protective functions and (c) Regulative functions.

7.5.4 Biodiversity Hotspots:

Biodiversity is the occurrence of diverse forms of organisms adapted to different environment of the earth. A biodiversity hotspot is a biogeographical region, where diverse forms of organisms are found or biodiversity hotspots are areas that are extremely rich in species, have high endemism and are under constant threat. The concept of biodiversity hotspot was given by Norman Myers (1988). Among the 25 hotspots of the world, two are found in India, extending into neighboring countries e.g., The Western Ghats and Indo-Myanmar region. Today, various international organizations are working in many ways to conserve the hotspots. Some of them are:

AZE	Alliance for Zero Extinction
BLI	Bird Life International
CEPF	Central Ecosystem Partnership Fund
NGS	National Geographical Society
PLI	Plant Life International
WWF	World Wildlife Fund

7.5.5 Critiques of Hotspots

The high profile biodiversity hotspot approach has resulted in considerable criticism. The two main criticisms involved are (i) biodiversity hotspots do not consider phylogenetic diversity, and (ii) biodiversity hotspots do not protect ecosystem services.

Biodiversity is the occurrence of diverse forms of organisms adapted to different environment of the earth. A biodiversity hotspot is a biogeographical region, where diverse forms of organisms are found. The concept of biodiversity hotspot was given by Norman Myers (1988). Among the 25 hotspots of the world, two i.e. The Western Ghats and Indo-Myanmar region are found in India, extending into neighbouring countries. Now a day's several national as well as international organizations are working to conserve the hotspots.

7.5.6 The Government Forest Department:

After independence, the employees of forest department took charge of the forests. They entirely ignored the needs of local people like leaves for fodder, herbs for medicines, fruits and nuts for food. Several timber industries largely depend on the forests for raw materials and in return, a handsome amount of money is obtained. This practice led to the destruction of a large amount of biodiversity. The development projects like building roads, dams, canals and reservoirs, etc., also damages the forests. Even the large inflows of tourists to the forests for observing wildlife create problems of pollution and destruction of habitats. The forests are now an important source of revenue for the forest department and the government.

7.5.7 The Nature and Wildlife Enthusiasts:

These are the people who are indirectly dependent on the forest products but they are interested in protecting the environment from damage. In the beginning, they worked for the protection and conservation of large wild animals but now they have realized the need for conserving forests in order to protect wildlife and also environment as a whole.

7.5.8 Deforestation:

Human population size has increased rapidly over the last hundred years. This means increase in demand for food, water, shelter and other requirements, resulting the pressure on natural resources of the country. Deforestation may be defined as the destruction, reduction or removal of forest cover or the conversion of foretasted areas to non-forested ones.

Consequences of Deforestation:

- Enhanced CO_2 concentration in atmosphere.
- Erosion of soil.
- Loss of biodiversity.
- Unbalancing of biogeochemical cycles.
- Global changes in climate.
- Scarcity of wood.

7.6 PARTICIPATION OF PEOPLE IN THE MANAGEMENT OF FORESTS

A person with an interest or concern in something is called a stakeholder. The people who live in village around the forests take firewood (fuel) from the forest trees. They usually lop (cut) the branches of the trees and pluck their leaves but do not cut down the whole trees. They take bamboo from the forest to make their huts and baskets for collecting and storing food materials. The local people take wood for making agricultural implements and gather fruits, nuts and medicinal herbs from the forest. They also collect green fodder and graze their cattle in the forest. On the whole, people living near the forests usually use the resources of the forests in a way that much damage is not done to the environment. In fact, the people living near forests had developed practices to ensure that the forest resources were used in a sustainable manner. So, the damage caused to forests cannot be attributed to only the local people living around the forests. Thus, it is our prime duty to protect the forests.

7.6.1 The Chipko Andolan

The Chipko Andolan (Hug the Trees Movement) originated from an incident in a remote village called 'Reni' in Garhwal, high up in the Himalayas in the early 1970s. A logging contractor has been allowed to cut down the trees in a forest close to a village. The villagers did not want to be cut down the forest because it would have spoiled their healthy environment. One day, when the men folk of the village were out for work, the contractor's workers came in the forest to cut down the trees. In the absence of men, the women of the village reached the forest quickly and clasped or gripped the tree trunks with their arms, preventing the workers from cutting down the trees. The forest trees were saved. Thus, the Chipko Movement quickly spread across all the communities and helped in the conservation of forests (Fig. 7.5).

Fig. 7.5 Chipko Andolan

7.6.2 A Case Study of Khejri Trees

In 1731, Amrita Devi Bishnoi led a group of 363 persons who sacrificed their lives for the protection of Khejri (*Prosopis cineraria*) trees (Fig. 7.6) in Khejrali village near Jodhpur in Rajasthan state of our country. It shows the determination of some people to work for the conservation of their natural environment. The Government has constituted an 'Amrita Devi Bishnoi National Award for Wildlife Conservation' in the memory of Amrita Devi Bishnoi.

Fig. 7.6 Khejri Tree

7.6.3 A Case Study of Regeneration of Sal Forests

In addition to conservationists, local people can also help in the management and regeneration of forests (Fig. 7.7). In one such incident, people led the revival of degraded forests in West Bengal when West Bengal Forest Department despite tremendous efforts could not revive the degraded Sal Forests of southwestern district of the state. In 1972, the forest officer at that time, AK Banerjee derived a strategy to involve local people (villagers) for reviving Arabari Forest range of Midnapore district covering 1272 hectares land. In return for their help, the villagers were given employment in silviculture and harvest operations of the forest, 25% of final harvest produce and were allowed to collect firewood and fodder from the forest at nominal charges. With the active and willing participation of local peoples and the government, degraded Sal forest of Arabari range became thick and green within 10 years.

Fig. 7.7 Regeneration of Sal forest

7.6.4 Joint Forest Management in India:

About half of the states in India have endorsed a strategy of joint forest management (JFM), in which forestry departments and communities jointly manage forests and share rights and responsibilities. The idea of JFM originated from the management of Sal (*Shorea robusta*) forests in West Bengal. Here, community involvement had a remarkable effect on the rehabilitation of degraded Sal forests. Landsat images have shown that the area of closed Sal forest increased from 11% to 20% in Midnapore District alone, and that many square kilometres of degraded scrub forest have been restored to open forest. Encouraged by this success, the Indian government expanded the programme during the 1990s. Under JFM, the ownership of the land remains with the government. Village committees, who are the co-managers, are entitled to the benefits from forest products. Forest protection committees control access to jointly managed forests. These local institutions are demonstrably more effective in protecting forests than the state forest departments. The JFM strategy has required a change of attitude from both forest departments and rural communities. Rural communities have had to organize themselves in new ways, overcome village and inter-village conflicts, and work together with forestry officials. Foresters have had to communicate with local people and share the responsibility for decision-making. To facilitate this process, the Indian government has provided legal and institutional backing, including land reforms, social forestry programmes, sharing of user rights with local people and educating foresters in participatory processes.

Several moments or case studies are noted for the conservation of forests and protection of environment

The Chipko Andolan (Hug the Trees Movement) originated from an incident in a remote village called 'Reni' in Garhwal, high up in the Himalayas in the early 1970s

Amrita Devi Bishnoi (In 1731) led a group of 363 persons who sacrificed their lives for the protection of Khejri (*Prosopis cineraria*) trees in Khejrali village near Jodhpur in Rajasthan state of India.

Sh A. K. Banerjee (Forest officer) in 1972, the derived a strategy to involve villagers for reviving Arabari Forest range of Midnapore district covering 1272 hectares land.

JFM (Joint Forest Management) originated from the management of Sal (*Shorea robusta*) forests in West Bengal.

7.7 CONSERVATION STRATEGIES:

Conservation strategies for forests involve two principal aspects such as **(i)** prevention of deforestation and **(ii)** expansion of country's forest areas. For establishment of ecological stability (ecological balance), the forest area of a country should remain stable and self-sustained. Some important measures necessary for conservation of forests are:

- Afforestation should be done immediately in fields unfit (sterile) for agricultural purposes
- Controlling the expansion of agriculture at the expense of forests
- Enhancing active participation of local people near forest area
- Fertility should be enhanced in forest soil by adding extra fertilizers and water
- Preventing cattle ranching in grasslands and forest areas
- Preventing destruction of green cover
- Reducing timber harvesting and use of fuel wood

7.7.1 Afforestation:

The conversion of free, bare or uncultivated land into forest by planting, trees for commercial purposes as well as to conserve habitat is called afforestation. Afforestation is usually done on lands that are unprotected and barren, for example, grasslands (Fig. 7.8).

Fig. 7.8 Afforestation by Plantation

7.7.2 Reforestation:

The re-establishment of forest cover, either naturally or artificially, that have been usually depleted through deforestation is called reforestation. Natural means of reforestation include natural seeding and root suckering while artificial means of reforestation include the direct seeding or planting through machine or hand (Fig. 7.9).

Fig. 7.9 Reforestation

7.7.3 Benefits of Afforestation:

(a) It creates new forests.

(b) It increases carbon capture and sequestration (sequestration is the process of capture and long term storage of atmospheric CO_2.)

(c) It helps in the improvement of biodiversity anthropogenically.

(d) It helps to increase the soil fertility by preventing soil erosion and providing water and minerals.

Afforestation in degraded areas/soil: in some places, forests need help to re-establish themselves between environmental factors. For example, in arid zones forest cover is destroyed and land may dry and become unsuitable for new growth.

7.7.4 Programmes of Afforestation necessary for Conservation of Forest:

(i) Urban forestry: Raising and proper management of trees on public and private owned lands.

(ii) Social forestry: Forestry of the people, by the people and for the people. It is associated with raising the forests on government land for food and wood, etc.

(iii) Agroforestry: It is the raising of trees and agricultural crops, either on the same land or in a close association with the waste patches of land for better use.

Conservation strategies for forests involve two principal aspects such as (a) prevention of deforestation, and (b) expansion of country's forest areas. The forest area of a country should remain stable and self-sustained for establishment of ecological stability (ecological balance),

The conversion of free, bare or uncultivated land into forest by planting, trees for commercial purposes as well as to conserve habitat is called afforestation. While re-establishment of forest cover, either naturally or artificially, that have been usually depleted through deforestation is called reforestation. Several programmes as (a) Urban forestry (b) Social forestry and (c) Agroforestry are going on for Conservation of Forests.

7.8 WILDLIFE

Any living organism in its natural habitat is called wildlife. It may be a plant or an animal or a microbe which is neither domesticated nor cultivated. The wildlife helps to maintain the balance of nature by maintaining food chain and food web. It acts as a gene bank in relation to breeding of plants and animals. India with its unique geographical location and various climatic conditions harbor and protect a large number of wild plants and animals. The wildlife is not only productive but also acts as an important natural resource.

7.8.1 Importance of Wildlife

- Wildlife along-with forests maintains the ecological balance of nature.
- A huge number of species live in a particular area and thus, the wildlife provide a great bio-diversity in a particular area.

- It also serves as a source of animal domestication.
- It is also a source for cultivation of crop plants. Most of the food plants are originated from wild ancestors.
- Lifesaving drugs, honey, feathers, musk, ivory, leather, silk, lac and other valuable products are obtained from wildlife.

7.8.2 Threats to Wildlife

(i) **Hunting:** killing of wild animals

(ii) **Destruction of habitats:** Population of wild animals is under threat due to destruction of their natural habitats, i.e., forests for the construction of dams, reservoirs, filling of wetlands

(iii) **Overgrazing:** Wildlife is threatened due to destruction of forests. Forests are also destroyed and converted into desert due to overgrazing by domestic animals.

(iv) **Economic value:** The skin, fur, wool, ivory, etc., obtained from wild animals have great demand because various valuable, high-priced articles are made from these. Even wild animals are in demand for keeping them in zoos for the people. These factors have also put the wild animals under threat.

7.9 ENDANGERED SPECIES

The first decade of 20th century has witnessed an alarming increase in the number of species threatened with danger of extinction. All species of plants (flora) and animals (fauna) which are threatened to become extinct in near future are called endangered species. IUCN, i.e., International Union for Conservation of Nature and Natural Resources (Headquarters: Morges, Switzerland) maintains a Red Data Book to (i) Create awareness about importance of biodiversity, (ii) Maintain a record of species which are on the verge of extinction, and (iii) Highlight the conservation priorities

A. **Examples of Endangered Animals** (Fig 7.10) **species such as** (i) Indian wild ass (ii) Great Indian rhinoceros (iii) Indian wolf (iv) Lion-tail macaque (v) Indian antelope (vi) Indian salamander (vii) Swans

(A) A great Indian Rhinoceros

(B) Lion-tail Macaque

(C) Indian Antelope

(D) Swans

Fig. 7.10 Some Endangered Animals

B. Examples of endangered plant species (Fig 7.11) **such as** (i) *Drosera indica* (an insectivorous plant) (ii) *Nepenthes khasiana* (iii) *Aldrovanda vesiculosa* (iv) *Rhus hookeri* (v) Snow orchid

(A) *Drosera indica* (An Insectivorous Plant)

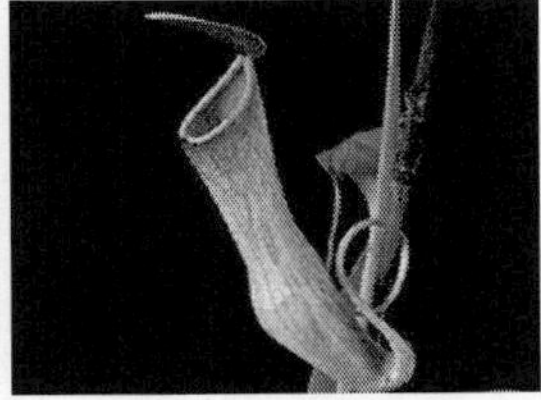

(B) *Nepenthes khasiana*

(C) *Aldrovanda vesiculosa*

(D) *Snow orchid*

Fig. 7.11 Some endangered plant species

7.10 CONSERVATION OF WILDLIFE

The large-scale killing of wild flora and fauna is a serious threat for the survival of species and it also disturbs the food chains and food web and thus have undesirable consequences on the whole ecosystem. We must take following steps to conserve our wildlife:

(a) Strict laws should be made in impose ban on poaching or capturing plant and animal species especially the ones which are endangered. These laws should be enforced strictly.

(b) Indiscriminate killing of all wild animals and birds should be completely banned.

(c) Natural habitats of wild animals must be protected.

(d) National parks, sanctuaries and biospheres should be established in order to preserve the natural habitats of flora and fauna.

- Sanctuaries are tracts of land where wild animals are not hunted, neither their habitats are disturbed.
- National parks are areas where wildlife is protected from all types of exploitation, only tourism is permitted.
- Biosphere reserves are multipurpose protected areas where selected human activities are allowed.
- The unauthorized cutting of forest trees should be banned in order to save the natural habitat of wildlife, i.e., the forests.
- For every cut tree, a sapling should be planted to meet the loss.
- Medical facilities should be provided to wildlife in case of epidemics, drought and flood.
- Special care should be given to specific species by collecting research information from data bank.
- Various awareness programmes should be conducted to educate more and more people about the importance of wildlife.

The legal perspective in conservation refers to the development of effective laws and their enforcement to prevent damaging practices, destruction of environment and natural resources. A number of wildlife acts are made by the state and central governments. Some are listed below:

- Assam Rhinoceros Preservation Act (1954)
- Convention on International Trade in Endangered Species (CITES), 1976
- Fisheries Act (1897)
- Forest Conservation Act (1980)
- Indian Board of Wild Life (IBWL), 1952
- Madras Wildlife Preservation Act (1873)
- Preservation of Trees Act, 1975 (Amendment in 1996)
- The Indian Forest Act (1927)
- The Man and Biosphere (MAB) Programme of UNESCO, 1971
- Wild Life (Protection) Act, 1972

Wildlife Protection Act (1972) is the most important act regarding wildlife conservation in India. Various projects are running in India under this act. Some special conservation projects for endangered species are:

- Cir Lion project (1972)
- Crocodile breeding project (1974)
- Project elephant (1992)

- Project Tiger (1973)
- Rhinos conservation (1987)
- Snow leopard project (1992)

Any living organism in its natural habitat is called wildlife. The wildlife helps to maintain the balance of nature by maintaining food chain and food web. The wildlife is not only productive but also acts as an important natural resource. India with its unique geographical location and various climatic conditions harbour and protect a large number of wild plants and animals.

All species of plants (flora) and animals (fauna) which are threatened to become extinct in near future are called endangered species. IUCN (International Union for Conservation of Nature and Natural Resources) maintains a Red Data Book to (a) Create awareness about importance of biodiversity, (b) Maintain a record of species which are on the verge of extinction, and (c) Highlight the conservation priorities Few examples of endangered animals species are (a) Indian Wild Ass (b) Great Indian Rhinoceros as well as endangered plant species are *Drosera indica* and (b) Snow orchid.

Wildlife Protection Act (1972) is the most important act regarding wildlife conservation in India. Various projects are running in India under this act. Some special conservation projects for endangered species are such as (a) Cir Lion Project (1972), (b) Project Elephant (1992), (c) Rhinos Conservation (1987) and Snow Leopard Project (1992) etc.

7.11 WATER

Water is a basic necessity of life. It is essential for plants, animals as well as human beings. Human beings use water for drinking, bathing, washing, agricultural activities and other industrial activities. Agriculture consumes maximum amount of fresh water, i.e., about 70% of total water. 26% of total consumable water is used in industries and only 1.1% of water is taken for domestic use. With the increasing population, the demand for fresh water is also increasing day by day. So, there is a need of proper management and distribution of available water to ensure equitable distribution for all the people.

7.11.1 Water Resources

Hydrosphere constitutes a major portion of globe (about 2/3). Out of this, 97.5% of water on the earth is found in the oceans and seas (saline water), which cannot be consumed directly. Terrestrial life can't use this saline water because their bodies can neither tolerate salt content of sea water nor eliminate the salts from the body. Only 2.5% of the total water available on the earth is fresh water. About 80% of the fresh water is found at the polar region in the form of ice caps and glaciers. Antarctica is the only continent, completely covered with ice, as a form of fresh water. The remaining 20% of the fresh water is found in the form of (a) Underground water

(b) Surface water (Ponds, lakes and rivers). Apart from these sources, rain is also responsible for providing the fresh water. Water is unevenly distributed on the earth, so the availability of water varies from one region to another region/area. The management of river water is done by constructing dams on rivers. Despite good rains, we are not able to meet the demand of water for all the people because of following reasons:

- Cultivation of high water-demanding varieties of crops
- Discharge of untreated sewage and industrial wastes into nearby rivers or any other water body
- Lack of vegetation requirement of urban population due to changing lifestyle
- More water requirement of urban population due to changing lifestyle
- Rapid increase in human population

7.11.2 Ground Water

A major part of fresh water (90%) is present inside the earth known as ground water and a small fraction (10%) found on the earth is called as surface water which is found in lakes, ponds, streams and rivers etc. When the rain falls, some water is lost as run off. This water enters into the rivers and seas and some of it lost through evaporation into the atmosphere. The remaining water that percolates into the soil through the soil particles with the effect of gravity is called groundwater. On the basis of medium and source of water can be classified as:

- **Capillary Water:** The water moved against the pull of gravity in the small pores is known as capillary water. Plant can draw only capillary water from the soil. This water is known as available water.
- **Hygroscopic water:** The water that forms is a light film around the soil particles is known as hygroscopic water.
- **Combined Water:** When the water in small quantity is held chemically with soil particles is known as combined water.
- **Gravitational Water:** The water which moves down by the action of gravity and finally reaches the water table is known as Gravitational water.

7.11.3 Benefits of Underground Water

The main purpose of water harvesting is to conserve rainwater for various uses and also to recharge the ground water level or underground water. Underground water has various advantages such as:

- It provides moisture for plant growth over a wide area.
- It does not evaporate and recharges wells.
- It is protected from contamination by human beings and animals.
- It is available throughout the year for irrigation, industry and human activities.

- Unlike stagnant water reservoirs like ponds or artificial lakes, ground water does not promote breeding of mosquitoes and flies.

Water is a basic necessity for plants, animals as well as human beings. Only 2.5% of the total water available on the earth is fresh water. About 80% of the fresh water is found at the polar region in the form of ice caps and glaciers. Antarctica is the only continent, completely covered with ice, as a form of fresh water. The remaining 20% of the fresh water is found in the form of (a) Underground water (b) Surface water (Ponds, lakes and rivers).

A major part of fresh water (90%) is present inside the earth known as ground water and a small fraction (10%) found on the earth is called as surface water which is found in lakes, ponds, streams and rivers etc. On the basis of medium and source of water can be classified as Capillary Water, Hygroscopic water, Combined Water and Gravitational Water. Underground water is available throughout the year for irrigation, industry and human activities. It provides moisture for plant growth over a wide area, protected from contamination by human beings and animals.

7.12 WATER POLLUTION

The water in most water bodies is highly polluted. Human beings have been abusing the water bodies around the world. This pollution is caused due to dumping of untreated sewage and industrial wastes into it. Realizing the importance of maintaining the cleanliness of the water bodies, the government of India has passed the Water Prevention and Control of Pollution Act in 1974 which aims to safeguard the country's water resources. Materials which pollute water bodies are categorized into following three groups such as (a) **Colloidal materials as** Cloth, paper fibres, faecal and bacteria (b) **Dissolved materials** as compounds and salts of Ca^{2+}, Na^{+}, PO_4^{3-}, Mg^{2+}, NO^-_3 and (c) **Suspended materials** as clay, silt and sand, etc.

7.12.1 GAP (Ganga Action Plan)

Ganga is the largest river system of India. It originates as Bhagirathi in North Uttarakhand from Gangotri glaciers and ends up at Ganga Sagar in Bay of Bengal (West Bengal). Downstream it connects Alaknanda and Mandakini in Devprayag to form Ganga. In this way, it passes through four states, Uttarakhand, Uttar Pradesh, Bihar and West Bengal. Ganga supplies water to more than 50 million people. Every day a large amount of untreated sewage is dumped into the Ganga river. Bathing of animals and washing clothes are other pollution causing activities. The result was the BOD of the river had risen to 9.7 mg/L instead of the maximum of 2mg/L. Pollution load and toxicity had started killing fishes in large sections of the river. A multicore Ganga Action Plan (GAP) project (Fig. 7.12) was launched in 1985 to clean the river Ganga and make it pollution free and to reduce pollution load of Ganga by more than 75%. This plant has been carried out in two phases since 1985.

Fig. 7.12 Ganga Action Plan (GAP)

7.12.2 Measurement factors to check the quality of Ganga water:

The contamination of water can be checked by any of the following methods:

(i) **Biological Oxygen Demand (BOD):** It is the amount (in mg) of oxygen required by the microorganisms present in a water body to decompose organic wastes in one litre of water. In polluted water, value of BOD is more. Pollution is low if BOD is below 1500 mg/L, medium pollution represents BOD value between 1500-4000 mg/L and in the highly water body. BOD is above 4000 mg/L.

(ii) **Presence of Coliform bacteria:** Coliform is a group of bacteria found in human intestine. Coliform count measures the MPN (most probable number) of coliform bacteria in 100 ml of water.

(iii) **pH of water:** It is the easiest method for finding the contamination of water. It is done by dipping pH indicator paper in the water sample and comparing the reading with the standard chart of pH. The pH of pure water is = 7. If the pH comes to be more or less than 7, then the water is polluted.

(iv) **Total Dissolved Substances (TDS):** TDS indicated hardness or softness of water. The TDS of hard water is above 120 ppm which is not suitable for drinking, cooking, washing, bathing and industrial purposes.

7.13 MANAGEMENT AND CONSERVATION OF WATER RESOURCES

A. **Reduce water wastage:** Water can be conserved through reducing its wastage (i) Reduction consumption in domestic works, irrigation, and industrial wastage

B. **Economical use of water:** Water should be used economically. Water harvesting involves the collection and storage of rain water for future use.

C. **Increasing water storage capacity:** Every year, when it rains a large amount of rainwater flows into the oceans. So, we lose a huge quantity of water. This water can be stored in tanks, reservoirs, etc., and must be used later on, in dry seasons.

D. Afforestation: Large scale plantation of trees helps in the prevention of water loss and also helps in bringing rains.

E. Efficient water distribution system: There are places where ample water is available while some other places where there is scarcity of water. There must be an efficient water distribution system so that areas with scarcity of water, ensuring proper water management.

> Human beings have been abusing the water bodies around the world. Water pollution is caused due to dumping of untreated sewage and industrial wastes into it. Materials which pollute water bodies are categorized into following three groups such as (a) Colloidal materials (b) Dissolved materials and (c) Suspended materials
>
> GAP (Ganga Action Plan) was launched in 1985 to clean the river Ganga and make it pollution free and to reduce pollution load of Ganga by more than 75%. This plant has been carried out in two phases since 1985. The contamination of water can be checked by any of the following methods such as (i) Biological Oxygen Demand (BOD) (ii) Presence of Coliform bacteria (iii) pH of water and (iv) Total Dissolved Substances (TDS).

7.14 DAMS

Dams are barriers constructed across a river or stream to hold back water in a reservoir and regulate the flow of water. Dams store enormous amount of water in the reservoirs. This water is then allowed to flow according to the requirement and desire. In India, dams have been built across many rivers like Ganga, Brahmaputra, Godavari, Krishna, Mahanadi, Narmada and Satluj Major dams in India (Fig. 7.13).

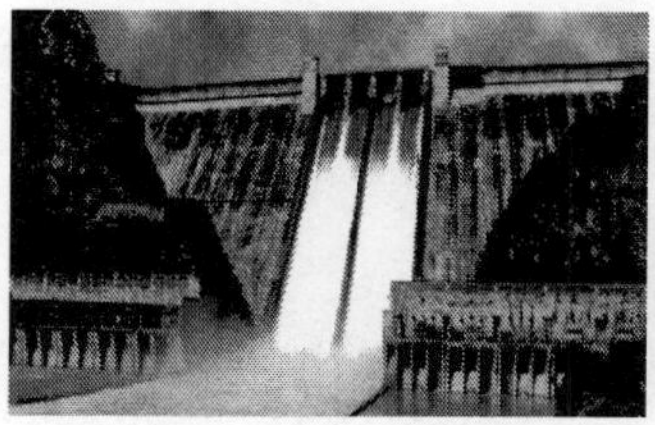

(A) Bhakra Nangal Dam **(B) Hiradund Dam**

Fig. 7.13 Dams

S.No.	Name of the Dam	Name of the River	State
1.	Bhakra Nangal	Satluj	Punjab
2.	Sardar Sarovar	Narmada	Gujarat
3.	Tawa	Tawa	Madhya Pradesh
4.	Mettur	Kaveri	Tamil Nadu

5.	Tehri	Ganga	Uttrakhan
6.	Srisailam	Krishna	Andhra Pradesh
7.	Banasura Sagar	Kabini	Kerala
8.	Hirakud	Mahanadi	Odisha
9.	Nagarjuna Sagar	Krishna	Andhra Pardesh
10.	Indira Sagar	Narmada	Madhya Pradesh
11.	Tungabhadra	Tungabhadra	Karnataka
12.	Bhavani Sagar (Lower Bhawani)	Bhawani	Tamil Nadu
13.	Idukki	Periyar	Kerala
14.	Krishna Raja Sagara (KRS)	Kaveri	Karnataka

7.14.1 Advantages of Dams

- Water from a dam is used for irrigation in fields through a network of canals. Dam ensures round the year water supply to the crop fields and help raise agricultural production. For example, Indira Gandhi Canal originating from Bhakra Dam has brought greenery to considerable areas of Rajasthan.
- Water from a dam is supplied to the people in towns and cities through pipelines after suitable treatment. In this way, construction of dams ensures continuous water supply in the region.
- The falling water (or flowing water) from the dam is used for generating electricity. The water rushing down the dam turns turbines which run electric generators. The electricity thus produced is called hydroelectricity.

The construction of high-rise dams for the management of river water and generation of electricity has certain problems associated with it.

7.14.2 Problems facing by the Public due to DAMs

The public opposition to the construction of large dams on rivers is mainly due to the following problems likely to be created by them.

(a) **Environmental Problems:** The construction of high-rise dams on the rivers contributes to deforestation and loss of biodiversity. This is because a vast variety of flora and fauna (plants and animals) get submerged in the water of large reservoir formed by the dam and disturb the ecological balance.

(b) **Social Problems:** Due to the construction of high-rise dams, a large number of human settlements (villages) are submerged in the water of large reservoir formed by the dam and many people are rendered homeless. It creates a social problem. It is, therefore, necessary that all the people who are displaced from the dam site are given adequate compensation by the Government for rehabilitation so as to start their life afresh.

(c) **Economic Problems:** Some people say that the construction of high-rise dams involves the spending of huge amount of public money without the generation of proportionate benefits. On the other hand, others say that there can be no real progress without building dams because they allow us to manage our water resources properly and at the same time give us much needed electricity (without causing any air pollution). So, whether the construction of dams on rivers is an economic problem or not is a debatable question.

The opposition to the construction of Tehri Dam on the river Ganga and raising the height of Sardar Sarover Dam on the river Narmada are due to such problems.

Dams are barriers constructed across a river or stream to hold back water in a reservoir and regulate the flow of water. In India, dams have been built across many rivers like Ganga (Tehri Dam), Narmada (Sardar Sarovar Dam) and Satluj (Bhakra Nangal Dam) etc. Water of dam supplied to the people in towns and cities, used for irrigation in fields and for generating electricity. The public opposition to the construction of large dams on rivers is mainly due to environmental, social and economic problems

7.15 WATER HARVESTING

The availability and management of fresh water has become a gigantic task for the government. So, it is our duty to participate in the cause and help the government in increasing water resources. Water harvesting is an age old practice in our country. It depends on the area where it is to be used.

7.15.1 Need of Water Harvesting

Water is a basic need for human society. We know that water is required for drinking, bathing, washing, irrigation, industry, cooling, and construction work, disposal of sewage and industrial effluents. Agriculture consumes the maximum amount of water, approximately 70% of total water. Rainwater is a primary source of irrigation of crop plants around the world. In India, 80% of farmers are totally dependent on rainwater for agriculture because they lack irrigation facilities. Various industries required approximately 26% of water while domestic and municipal consumption of water account for only 1.1%. The construction of pavements, roads and pukka drains prevent the percolation of water into the ground. Because of rising population, increasing urbanization, rapid industrialization and the rising standards of urban areas to meet various agricultural, industrial and domestic needs.

7.16 RAIN AND RAIN HARVESTING

Rain is a very important source of water. In India, Rains are largely due to monsoon which lasts for a few months. Rainwater is stored in lakes for use over a long period of time. There are some natural lakes in our country. Some artificial lakes have also been made at various places to store rainwater to meet the increasing demand for water. Some rainwater also seeps into the ground and becomes available as ground water or rather underground water. Thought most of the parts of our country get a good rainfall during monsoon but due to the loss of vegetation cover, much water does not seep into the ground, it rather flows into rivers. In fact, many cities of our country depend on such lakes for their water supply during the year. Despite good rains, we are unable to meet the demand for water of all the people due to following reasons:

(a) Changing life-style of people is consuming more water especially in urban areas.

(b) Lack of sufficient vegetation cover on ground, only a little rain water seeps into the ground and gets stored as ground water.

(c) Rapid increasing the population of our country.

(d) Reduces the availability of usable water due to discharge of untreated sewage and industrial wastes into rivers and lakes.

(e) Requirement of more water by high yielding varieties of crops for irrigation.

Rivers are another important source of water. Rivers get their water supply form the melting of snow lying on the speaks of snow mountains (for glaciers) as well as from rains. The management of river-water is done by constructing dams on rivers.

7.16.1 Rainwater Harvesting

Developing special water harvesting structures to capture and store rainwater is called as rainwater harvesting. It recharges the underground water level. It can be efficiently achieved by the drought wells, percolation pits, dams and lagoons. It not only benefits the poor but is also helpful for all the people who basically reside in the areas of low rainfall. In India, although the annual average rainfall is around 120mm but major rainfall is mostly confined to the months of monsoons. In non-monsoons seasons, there is a severe shortage of water due to unmanaged water loss. Since, we don't have any control on the flowing water or the runoff water, the places with maximum of rainfalls like Cherrapunji also suffer from the lack of water (Fig 7.14).

Fig. 7.14 Cherapunji Rainfalls

Rainwater harvesting is a necessity in areas having water shortage where there is less rain. Also at places with ample rainfall, the rainfall does not percolate in the soil and flows to the stream, river and sea without recharging the underground water level. This causes loss of fresh water and lowering of groundwater level day by day.

The harvested rainwater is used for watering the plants in the garden, toilet flushing and washing clothes. This practice can save upto 50% of the household water.

7.16.2 Methods of Rainwater harvesting

Water harvesting methods are employed according to the concerned area. In rural areas, i.e., the villages, water harvesting is a community-based programme. The people in rural India use a large number of water collecting methods to capture as much rainwater as possible which had fallen on their land. Some of the methods applied by rural people for water harvesting are (i) Digging of small pits and lakes (ii) Construction of small-sized earthen dams (embankments) (iii) Development of sand and limestone reservoirs (iv) Setting up of roof-top water collecting units, and (v) Establishment of a wall to regulate water levels

In addition to these methods, water is also harvested from flooded rivers and through building of community wells and bore-well with hand-operated pumps. In rural areas, water is also conserved through water shed managements. All these methods recharge the depleted ground water levels. All kinds of storage systems vary in complexity and their complexity depends on requirement of water. To fight water scarcity various states of India have been using old age water harvesting practices.

Traditional Methods of Rainwater Harvesting:

In earlier times, rainwater was harvested in ponds and tanks in villages and this water was used for animals and irrigation purposes throughout the year. In foot hills, the water from springs was collected into the small embankments and then the water was diverted and transported by small hollow bamboo sticks. Other traditional methods were Khadins, Kulhs, Hauze, Baolis and Johads etc. (Fig 7.15). Two important traditional ways of water harvesting are as:

(A) Khadin **(B) Kulhs**

(C) Eris and Ooranis

Fig. 7.15 Traditional Methods of Rainwater Harvesting

(a) Khadins: These are traditional rainwater harvesting structures functional in Rajasthan. These consist of long (100m-300m) earthen embankments called 'bund' build across the slope of catchment area used to collect runoff water. A shallow well is often dug off behind the embankment to store extra run off, if it overflows the embankment. Slowly and gradually, the stored water percolates down into the deeper layers of ground. This water saturates land and is subsequently used for growing crops.

(b) Kulhs: About four hundred years ago, Kulhs were built in various regions of Himachal Pradesh. These were local canal irrigation systems having man-made channels by which water flowing in the streams was diverted to numerous villages down the hill side. The entire system was managed by villagers through a common agreement. It is interesting to know that the water was first used by villages farthest away from the source and then the villages progressively higher up. In addition to irrigation, this system also helped in percolation of water into soil and fed springs at various places. Later kulhs were taken over by irrigation department and most of them became defunct.

Table 7.1 List of the ancient water harvesting structures of India

Sr. No.	State	Harvesting structure	Description
1.	Bihar	Ahars and Pynes	Embankment, catchment basin and channels
2.	Maharashtra	Bandharas, Tals	Check dams
3.	Madhya Pradesh	Bhundhis	Reservoirs
4.	Uttar Pradesh	Bhundhis	Reservoirs
5.	Kerala	Surangams	Horizontal wells
6.	Karnataka	Kattas	Embankment across a drainage line
7.	Jammu and Kashmir	Ponds and Kul	Water channels
8.	Tamil Nadu	Eris and Ooranis	Tanks and ponds
9.	Rajasthan	Khadin	Embankments across lower hill slopes
10.	Himachal Pradesh	Kulhs	Headwall across a river to divert water from the natural stream for irrigation

New Methods of Rainwater Harvesting:

Now days, various techniques of water harvesting are used e.g. percolating tanks (Artificial ground water recharging tanks). Water is diverted from the catchment areas with the help of check dams made by indigenous materials such as stone, sacks etc.

The best new technique is the rooftop water harvesting. It is cheap and effective harvesting method. According to this method, the rooftops of building are cleaned and the rainwater collected on the rooftop is directed through pipes to the large underground storage tanks. This water is used for several purposes throughout the year.

In urban areas, i.e., cities is scarcity of space and most of the ground is covered with buildings, pavements made up of concrete and roads, unlike the rural where open ground has soil. So, unlike rural areas where rainwater seeps into the soil and increases the ground water level, in urban areas most of the rainwater goes into the drains and becomes useless. So, rainwater harvesting is necessary in urban areas.

The principal elements of rainwater harvesting include (i) rainwater collection, (ii) storage, (iii) distribution and (iv) system maintenance.

Generally, rainwater is collected from roof top area with the help of drain pipes. This collected amount of water is stored at least three feet away from the foundation of the building. The storage system varies according to the need.

The rainwater collected from roof top or from other sources is drained to specially made percolation pits. Percolation pits are constructed in the concrete slabs in open spaces. These percolation pits are connected to the recharge wells. The recharge well is generally 1 metres in diameter and 3 metres deep. The rainwater goes to these recharge wells through the outlet pipes from percolation pit. This water can be taken out according to the requirement through tube wells. Before storage, debris and leaves are filtered from the water with the help of a screen over the gutter. The storage tank is placed above or below the ground level. The rainwater is distributed by the regular drip irrigation system or according to the need. Due to deposition of waste materials or debris over the gutters, the water harvesting system requires occasional maintenance. The debris is removed and the storage tank is cleaned and dried as per convenience. Rainwater harvesting thus helps in meeting the growing water requirements and increases the groundwater availability.

7.16.3 Benefits of Rainwater Harvesting

Rainwater harvesting has several advantages. These are as follows:

- It provides year round supply of drinking and irrigation water.
- It increases the ground water recharge.
- It decreases sea water loss in coastal areas and dependence on ground water for domestic uses.
- It reduces urban floods and stagnation of water during the monsoon season.
- It reduces storm-water discharge and thereby prevents soil erosion.

Rainwater harvesting is a necessity in areas having water shortage where there is less rain. Developing special water harvesting structures to capture and store rainwater is called as rainwater harvesting. Traditional Methods of Rainwater Harvesting are Khadins (Rajasthan), Kulhs (Himachal Pradesh), Ahars and Pynes (Bihar) etc. Now days, various techniques of water harvesting are used e.g. percolating tanks. Rainwater harvesting provides year round supply of drinking and irrigation water, reduces urban floods and stagnation of water and storm-water discharge.

Some of the measures may be used for conservation and management of water resources such as (a) Awareness Programmes (b) Control of flood and storage of flood water (c) Economical use for water (d) Interlinking of Indian rivers (d) Recharge of groundwater by artificial means etc.

7.17 CONSERVATION AND MANAGEMENT OF WATER RESOURCES

In India, often summer seasons the groundwater level goes down. Our wells and hand-pumps dry and the water supply by Municipal Corporation (Nagar Palika) become minimal. Demand of water is increasing day by day due to rapid increase in human population. There is a limited amount of water available for use. Therefore, there is an urgent need for conservation and proper management of water resources. Following measures are necessity for checking the wastage of water.

- Awareness programmes among peoples.
- Control of flood and storage of flood water for use in dry periods
- Economical use for water domestic as well as industrial purposes
- Interlinking of Indian rivers to transfer surplus water to the water deficient areas
- Recharge of groundwater by artificial means
- Use of drip and sprinkler techniques of irrigation to save water.

7.18 WATERSHED MANAGEMENT

Watershed management is a process of managing various human activities in an area defined by watershed boundaries to protect the land water resources. A watershed is a high raised area varying in size (few hectares to several hundred hectares) jointly managed by a large number of farmers for the purpose of collecting and storing rainwater or water from rivers, streams and ground water aquifers to use in future for various purposes like drinking and irrigation. The area from where the rainfall flows into a reservoir river system is called catchment area. The vegetation in watershed area increases retention of rainwater and protects the soil from erosion. Thus, watershed management develops primary resources of land and water and increases the development of secondary resources of plants and animals. They also supply water for irrigation, hydroelectricity, transportation and vegetative growth. Himalayas are one of the most critical water sheds in the world.

7.18.1 Advantages of Watershed management

- It improves economy of the region.
- It reduces the occurrence of drought and floods in the downstream area.
- It reduces silting in downstream dams and reservoirs.
- It increases the production and income of watershed community.

7.19 FOSSIL FUELS

Fossil fuels are carbon containing energy yielding substances formed by degradation of remains of plants and animals buried deep under the earth crust millions of years ago. They are formed due to high pressure, heat and anaerobic condition present inside the earth. Most of our energy needs are met through fossil fuels. Coal, petroleum and natural gas are called fossil fuels. With rapid industrialization and urbanization, the demand for energy is also rising. To fulfill the growing demands of energy, exploitation of fossil fuel is done at such a rapid rate that they are likely to be exhausted very soon. The formation of coal and petroleum is continuing in different parts of the world, however, their formation is a very slow process. Once consumed, they further cannot be replaced in the near future and so, they are considered as non-renewable resources. It has been estimated that at the present rate of consumption, the known petroleum reserves of the earth will last for about 40 years more and the coal will last for about another 200 years. It is, therefore, necessary to conserve these resources for present as well as for future generations. Apart from this, burning of coal and petroleum produce different gases which are poisonous at high concentration.

7.19.1 Coal

Coal is a blackish, solid fuel that originates inside the earth and is extracted from the coal mines dug into the earth. Once exhausted, coal will not be available for use in future because they are formed through a very slow and time-consuming process. Therefore, it is necessary to conserve coal resources of the earth by reducing their consumption so that they may last for long.

Uses of Coal

- It is a fuel for domestic and commercial establishments.
- It is used in thermal power plants to produce electricity.
- It is used for the manufacture of coke.
- It is used in preparation of synthetic petrol and fuel gas by liquefaction and gasification.
- It is used in the production of various organic compounds like benzene, toluene, aniline and naphthalene.

7.19.2 Petroleum

Petroleum is a dark-coloured, liquid fuel that is taken out from various depths of earth on land and sea shore. It is commonly known as mineral oil or rock oil. The petroleum which is extracted from the earth is called crude oil. It is also considered as a non-renewable source of energy. The consumption of petroleum is several times faster than its formation. India's contribution to global oil production is quite low. Petroleum oil is depleting very fast. Disappearance of it will not only cause energy crisis, but it will also adversely affect our agricultural productivity, industrial production and finally, the economy of the county. Till independence, Assam was the only petroleum producing state of India. After independence, other oil sources were also discovered in Gujarat plains at Digboi and in Bombay High. Bombay High is an offshore oil field, 160 km (99 miles) off the coast of Mumbai (India) in Arabian Sea. It was discovered by a Russian and Indian oil exploration team. A number of products are obtained through the fractional distillation of petroleum. These are petroleum gas, petrol, diesel, kerosene, bitumen, paraffin wax, dyes and organic compounds.

Uses of Petroleum derivatives

(a) LPG (Liquefied Petroleum Gas) is a derived product from petroleum which is used for cooking and lighting purposes.

(b) Diesel, petrol and kerosene are used as a fuel for internal combustion engines in transport vehicles.

(c) Lubricating oils provide lubrication to machinery.

(d) Paraffin wax is used in candles, Vaseline and water proofing.

(e) Asphalt (semi-solid form of petroleum) is a source of several chemicals and dyes used in carpeting of the roads.

Fossil fuels are carbon containing energy yielding substances formed by degradation of remains of plants and animals buried deep under the earth crust millions of years ago. Fossil fuels are formed due to high pressure, heat and anaerobic condition present inside the earth. Coal, petroleum and natural gas are called fossil fuels. Coal is a blackish, solid fuel that originates inside the earth and is extracted from the coal mines dug into the earth. Petroleum is a dark-coloured, liquid fuel that is taken out from various depths of earth on land and sea shore. It is commonly known as mineral oil or rock oil. It is also considered as a non-renewable source of energy.

7.20 FOSSIL FUELS AS A SOURCE OF POLLUTION

Fossil fuels, on combustion, produce gases like carbon dioxide (CO_2), carbon monoxide (CO), sulphur dioxide (SO_2), oxides of nitrogen (NO_x) and unburnt carbon particles. These gases disturb the natural composition of air and cause air pollution. Pollution causes harm to both biotic and abiotic components. The substances which contaminate air are called air pollutant. The main air pollutants are –

(a) Carbon dioxide (CO_2): This gas is produced during the combustion of fossil fuels. This is a greenhouse gas which is responsible for the rise of temperature on earth (global warming).

(b) Carbon monoxide (CO): This is produced as a result of incomplete combustion of fossil fuels. It combines with haemoglobin (blood protein) forming carboxyhaemoglobin and it replaced oxygen, thereby reducing the oxygen carrying capacity of blood. At high concentration, it causes headache and suffocation which ultimately cause death.

(c) **Fly ash:** Burning of coal produces fly ash which contains various toxic particulate matters.

(d) **Nitrogen oxides (NO_x):** Nitrogen oxides are the products obtained by incomplete burning of fuels in automobiles. They attack the breathing system and also cause acid rain. They cause necrosis and killing of plant parts and produce various diseases in humans and animals.

(e) **Sulphur dioxide (SO_2):** This is produced during burning of coal in power houses and other places. It absorbs water vapour and enters into the soil as sulphuric acid (H_2SO_4). It also gets dissolved in rainwater, making it acidic and results in acid rain H_2SO_4(70%) + HNO_3 (30%). The acid rain, thus produced, has damaging effects on vegetation, aquatic life, buildings and metal structures. It causes lung diseases, asthma, bronchitis and emphysema in humans.

(f) **Smog:** Many of the products of incomplete combustion of petrol and diesel generate photochemical smog. It has toxic effects on humans, animals and vegetation's and also cause the cracking of rubber.

7.21 STEPS TO REDUCE THE CONSUMPTION OF FOSSIL FUELS

Fossil fuels like coal are extensively used in thermal power plants for producing electricity. Petroleum products like petrol and diesel are used in automobiles and LPG and kerosene are used for domestic purposes. We should take following steps to reduce the consumption of fossil fuels and conserved them (Fig. 7.16):

(A) Cycling

(B) CFL

(C) Solar device

Fig. 7.16 Some ways to consumpt the fuels

- Educating peoples through various awareness programmes.
- Fuel efficient engines of motor vehicles should be designed to reduce the consumption.
- In kitchens, food should be cooked in pressure cooker or covered utensils to save LPG.
- Judicious utilization of petroleum is the most important way of oil conservation.
- Public transport system should be improved in the cities to minimize the use of personal vehicles.

- Solar cooker can be used for cooking the food.
- Switch off the electrical appliances when they are not use.
- Switch off the engines of vehicles at red light.
- The use of biogas should be encouraged in rural areas.
- Use CFLs instead of traditional lamps, and save electricity.
- Use of alternate sources of energy i.e. solar energy, wind energy, hydro energy etc. should be encouraged.
- Walking or cycling should be preferred for covering short distances instead of vehicles. We should try to minimize the use of automobiles.

Fossil fuels, on combustion, produce gases like carbon dioxide (CO_2), carbon monoxide (CO), sulphur dioxide (SO_2), oxides of nitrogen (NOx) and unburnt carbon particles. These gases disturb the natural composition of air and cause air pollution. Pollution causes harm to both biotic and abiotic components. The substances which contaminate air are called air pollutant. i.e. CO_2, CO, Fly ash, NOx, SO_2 and Smog. Some of the steps are to be taken to reduce the consumption of fossil fuels and conserved them as (a) awareness programmes, (b) use of Solar systems and CFLs, (c) Caring and willingness to save the electricity (d) use of Biogas (e) Walking or cycling etc.

7.22 BIOGAS

Biogas is a mixture of gases produced by anaerobic decomposition of biological wastes by the activities of microbes. Biogas predominantly contains methane (CH_4) and is used as fuel. It is mainly composed of methane (about 50-65%), carbon dioxide (25-35%) and of other gases such as nitrogen (N_2), hydrogen (H_2) and hydrogen sulphide (H_2S) in traces. Among these, methane is highly inflammable. Micro-organisms produce different types of gaseous end-products during their growth metabolism. The type of the gas produced depends upon the microbes and the organic substances they utilize. For example, during formation of dough and production of beverages, the main gas produced is CO_2.

It is a good alternative energy source in rural areas. In this technology, the organic wastes are anaerobically fermented in fermentation tank by microorganisms and produce biogas. In India, biogas is mainly produced by the use of cattle dung. There are two types of bacteria involved in the production of biogas such as non-methanogenic and methanogenic. The non-methanogenic bacteria (*Propionibacterium, Peptostreptococcus* and *Clostridium*) hydrolyze and ferment the cattle dung along with other wastes into acetic acid. The methanogenic bacteria or methanogens (*Methanobacterium, Methanosarcina, Methanococcus,* and *Methanospirillum)* convert citric acid into biogas (i.e., methane and carbon dioxide.) For biogas production, water hyacinth, *Hydrilla,* Duckweeds and algae can act as supplement along with cattle dung as a chief source. The production of biogas requires 80-90% water content and a temperature range of 20-35°C. Methanogens are also found in the rumen (a part of stomach) of cattle where cellulosic materials are present. These bacteria are helpful in degradation of cellulose and play in important role in the nutrition of cattle. Thus, the excreta (dung) of cattle are rich in these bacteria, and can be used for biogas production.

7.23 BIOGAS PLANT

Biogas is prepared in biogas plants. The biogas plants are of two types such as (a) Fix dome type and (b) Floating gas holder type. The biogas plant consists of a concrete tank (10-15 feet deep) in which bio-wastes are collected. The biogas plant has an outlet which is connected to a pipe to supply biogas to nearby houses. Cattle dung is available in large quantities in rural areas where cattle are used for a variety of purposes. So, biogas plants are more often built in rural areas. The biogas, thus produces, is used for cooking and lighting. Methane is a pollution-free gas and cheap source of energy (Fig. 7.17). The slurry which is a residual matter after fermentation of dung is used as biofertilizer (manure). The technology of biogas production was developed in India mainly by the efforts of India Agricultural Research Institute (IARI) and Khadi and Village Industries (KVI).

Fig. 7.17 Biogas Plant

Biogas is a mixture of gases produced by anaerobic decomposition of biological wastes by the activities of microbes. It is mainly composed of methane (about 50-65%), carbon dioxide (25-35%) and of other gases such as nitrogen (N_2), hydrogen (H_2) and hydrogen sulphide (H_2S) in traces. Two types of bacteria are involved in the production of biogas as non-methanogenic bacteria (*Propionibacterium*), which hydrolyze and ferment the cattle dung along with other wastes into acetic and methanogenic bacteria or methanogens (*Methanobacterium*), which convert citric acid into biogas.

Biogas plants are of two types such as (a) Fix dome type and (b) Floating gas holder type. The technology of biogas production was developed in India mainly by the efforts of India Agricultural Research Institute (IARI) and Khadi and Village Industries (KVI).

7.24 SUSTAINABILITY OF NATURAL RESOURCES

Resources happen to be the determination of every economy. With the help of these resources judiciously, a country shows that it is concerned about the modern and forthcoming generations. However, looking at current positions, the indiscriminate use of our modern resources are such that the possibilities of succeeding generations and the developing countries to have access to their fair share of rare resources which are endangered. Furthermore, the consequences are dreadful and the result on the environment will induce severe damages which go beyond the carrying capacity of the environment. We all have heard numerous tales about the world's natural ecosystems_being under increasing pressure to provide an expanding population with a sustainable supply of food, fiber, fuel, etc. Preservation or management of natural resources is essential for maintaining community sustainability.

- **Water:** Water is an important resource which we use for every daily activity like drinking, cooking, cleaning, washing, etc. It is the precious natural resource that is largely getting wasted. To avoid the scarcity of Water both government and people must work together to assure that water is used judiciously and there is adequate water supply to meet future needs.
- **Energy:** Energy happens to be a nonrenewable source for power generation which is mainly used for home, workplace, and transportation, causing pollution. The need for energy conservation and the use of renewable fuels providing cost-effective and more sustainable alternatives are the need of the hour.
- **Biodiversity:** It plays a vital role in satisfying our need for different types of food, raw materials, fuel, etc. Over 4 billion years ago, the earth was crowded with a variety of species and every single species played a specialized role in maintaining ecological balance. As time passed, more than 80% of species got extinct. Communities are involved in their preservation by supporting, managing, promoting, protecting and enhancing healthy wildlife by establishing areas like the zoo, sanctuaries, national parks, biosphere reserves, wildlife safaris, and sanctuaries.
- **Air and Climate:** The human health and ecosystem can be dramatically affected by declining air quality and change in climate. The communities can contribute in controlling the sources of air pollution by limiting the use of harmful chemicals in the air.
- **Land, Forests, and Ecosystems:** An ecosystem consists of both living things and non-living things which interact with each other. Light, air, soil, water, energy, atmosphere, forests are renewable sources which need to be protected. Communities have to develop judicious land use practices for both conserving ecosystems and enhance local economies as they maintain the equilibrium in the environment.

Preservation or management of natural resources (water, energy, biodiversity, air and climate, land, forests, and ecosystems) is essential for maintaining community sustainability

REMEMBER SOCKETS

- Natural resources are those living or non-living substances available in the normal environmental which are being exploited for supporting life and meeting human requirements.
- Examples are natural resources are: land and soil resources; fresh water resources; marine or ocean resources; fossil fuel and other energy resources; forest and wild life resources; etc.
- Main reasons for the non-judicious alteration of natural resources are such as (a) Increasing population (b) Desire for the easy and luxurious lifestyle, and (c) Development of modern technologies.
- Aims of natural resource conservation are (a) ensure safe disposal of wastes, (b) equitable distribution of resources among all sections of society, (c) Minimize wastage of resources, (d) to preserve the continuous yield of significant plants, animals and materials and to increase the preservation of a quality environment.
- The major natural resources are either inexhaustible or exhaustible. The exhaustible resources are either renewable or non-renewable.
- Renewable natural resources can be replenished if managed wisely. Non-renewable natural resources get exhausted with use and cannot be replenished.
- Conservation is controlled utilization of natural resources for the benefit of all life.
- Forests are important renewable natural resources. Destruction, reduction or removal of forest cover is called deforestation.
- Conservation of forest is done by afforestation, reserve forests, social forestry, agroforestry and urban forestry.
- Active participation of public and their support must be generated in order to conserve our forests and to achieve the real goal of eco-development. One such public support is the Chipko Movement of Bishnoi Women in U.P.
- JFM (Joint Forest Management) was originated from the management of Sal (*Shorea robusta*) forests in West Bengal.
- In 1731, Amrita Devi Bishnoi led a group of 363 persons who sacrificed their lives for the protection of Khejri (*Prosopis cineraria*) trees in Khejrali village near Jodhpur in Rajasthan state of India.

- The Three R's Principles (Reduce, Recycle and Reuse) is used to save resources/environment:
- The concept of biodiversity hotspot was given by Norman Myers (1988).
- A biodiversity hotspot is a biogeographical region where diverse forms of organisms and their variations are found.
- Among the 25 hotspots of the world, two e.g., The Western Ghats and Indo-Myanmar region are found in India.
- Wild life includes all naturally occurring animals, colourful birds, plants and other life forms in the forests, which are neither domesticated nor lamed.
- Conservation of wild life is the management of wild flora and fauna. It is done by protection of natural habitats, maintenance of wild life in protected areas and protection through legislation.
- The wildlife is not only productive but also acts as an important natural resource.
- All species of plants (flora) and animals (fauna) which are threatened to become extinct in near future are called endangered species.
- IUCN (International Union for Conservation of Nature and Natural Resources) maintains a Red Data Book to (a) Create awareness about importance of biodiversity, (b) Maintain a record of species which are on the verge of extinction, and (c) Highlight the conservation priorities
- Examples of endangered animals species such as (a) Indian Wild Ass (b) Great Indian Rhinoceros (c) Indian Wolf (d) Lion-tail Macaque (e) Indian Antelope (f) Indian Salamander and (g) Swans.
- Examples of endangered plant species such as (a) *Drosera indica* (an insectivorous plant) (b) *Nepenthes khasiana* (c) *Aldrovanda vesiculosa* (d) *Rhus hookeri* (e) Snow orchid.
- Sanctuaries are tracts of land where wild animals are not hunted, neither their habitats are disturbed.
- National parks are areas where wildlife is protected from all types of exploitation, only tourism is permitted.
- Biosphere reserves are multipurpose protected areas where selected human activities are allowed.
- Wildlife Protection Act (1972) is the most important act regarding wildlife conservation in India. Various projects are running in India under this act. Some special conservation projects for endangered species.
- Any living organism like animals, birds, plants and microbes in its natural habitat is called wildlife.
- Conservation of wildlife involves management of wild flora and fauna.

- Wildlife is protected in natural habitats by forming national parks, biosphere reserves and sanctuaries.
- Wildlife is under continuous threat from hunting, destruction of habitat, overgrazing and economic reasons.
- All species of flora and fauna which are threatened to become extinct in near future are called endangered species.
- The species that have completely vanished from the biosphere are called extinct species.
- Depletion of quantity and deterioration in quality of water are two major problems.
- The quality of water could be checked by BOD, presence of coliform bacteria, pH of water and TDS.
- Dams store enormous amount of water in reservoirs.
- Dams provide water (drinking and irrigation), generate electricity and prevent floods.
- Dams have some harmful effects also. Dams cause people to leave their houses, loss of money of people and deforestation and loss of biodiversity.
- Non-living and living components of nature used by humans for fulfilling their various requirements like food, fodder and shelter are called natural resources.
- Natural resource management involves the management of natural resources to meet the needs of present and future generations without harming the nature.
- The development which meets the need of the present generation as well as of the future generation without harming the environment is called sustainable development.
- Some examples of natural resources are land and soil resources, fresh water and marine resources, fossil fuels, forest and wildlife resources.
- The natural resources are categorized as inexhaustible (renewable and non-renewable) and exhaustible; inorganic, organic and mixed; biotic and abiotic; national, multinational and international.
- The stakeholders of forests include local people, forest department, industrialists and wildlife enthusiasts.
- There are two types of conservation, *in situ* and *ex situ*. The conservation done in the natural habitat is the in situ conservation (on spot) whereas outside the natural habitat is ex situ conservation (other than spot).
- Destruction, reduction or removal of forest cover is called deforestation.
- Establishment of forest at a place where there was no forest previously is called afforestation.

- Re-establishment of forest cover either naturally or artificially is called reforestation.
- The water moved against the pull of gravity in the small pores is known as capillary water. Plant can draw only capillary water from the soil. This water is known as available water.
- The water that forms is a light film around the soil particles is known as hygroscopic water.
- When the water in small quantity is held chemically with soil particles is known as combined water.
- The water which moves down by the action of gravity and finally reaches the water table is known as gravitational water.
- Rainwater harvesting is the process of capturing, collecting and storage of rainwater and surface run off for filling either small water bodies or recharging groundwater.
- Traditional methods of rainwater harvesting are khadins, kulhs, hauze, baolis and johads etc.
- New Methods of Rainwater Harvesting is used of percolating tanks (Artificial ground water recharging tanks).
- Khadin is a traditional water harvesting structure in Rajasthan.
- Water harvesting methods are employed according to the area of concern, i.e., rural or urban.
- A watershed is a specific bounded area varying in size jointly managed by people for the purpose of collecting and storing rainwater.
- Watershed management increases the production and income of watershed community. It reduces occurrence of droughts and floods and silting in downstream dams.
- Watershed management is a process of managing various human activities in an area defined by watershed boundaries to protect the land water resources.
- A watershed is a high raised area varying in size (few hectares to several hundred hectares) jointly managed by a large number of farmers for the purpose of collecting and storing rainwater or water from rivers, streams and ground water aquifers to use in future for various purposes like drinking and irrigation.
- Water quality can be checked by Biological Oxygen Demand, Presence of Coliform bacteria, pH of water and Total Dissolved Substances
- Biological Oxygen Demand (BOD) is the amount (in mg) of oxygen required by the microorganisms present in a water body to decompose organic wastes in one litre of water. In polluted water, value of BOD is more. Pollution is low if BOD is below 1500 mg/L, medium pollution represents BOD value between 1500-4000 mg/L and in the highly water body. BOD is above 4000 mg/L.

- Coliform is a group of bacteria found in human intestine. Coliform count measures the MPN (most probable number) of coliform bacteria in 100 ml of water.
- Total Dissolved Substances (TDS) indicated hardness or softness of water. The TDS of hard water is above 120 ppm which is not suitable for drinking, cooking, washing, bathing and industrial purposes.
- Fractional distillation of petroleum gives a number of products like petroleum gas, petrol, diesel, kerosene, bitumen, paraffin wax, dyes and organic compounds.
- On combustion, fossil fuels produce gases like CO_2, CO, SO_2 and oxides of nitrogen. These cause air pollution.
- Fossil fuels should be used in a judicious way so as to minimize its wastage.
- Fossil fuels are formed by degradation of remains of plants and animals beneath the earth's surface.
- Coal is extensively used in thermal power plants.
- Conservation is the controlled utilisation of natural resources by humans in order to save them for present and future generations.
- Forests are biodiversity hotspots as they sustain a wide variety of life forms.
- Forests have very significant functions in terms of economic, protective and regulative form
- Coal and petroleum are non-renewable and exhaustible natural resources found in earth's crust. They must be conserved to get their advantage in present and future.
- Biogas is a mixture of gases (mainly methane and CO_2) produced by anaerobic decomposition of biological wastes by the activities of microbes. It is used as a fuel in several areas.
- Biogas is produced from plant material and animal waste, garbage, waste from house-holds and some types of industrial wastes, such as fish processing, dairies, and sewage treatment plants.
- Two types of bacteria are involved in the production of biogas such as non-methanogenic and methanogenic.
- The non-methanogenic bacteria i.e. *Propionibacterium, Peptostreptococcus* and *Clostridium* hydrolyse and ferment the cattle dung along with other wastes into acetic acid.
- The methanogenic bacteria or methanogens i.e. *Methanobacterium, Methanosarcina, Methanococcus* and *Methanospirillum* convert citric acid into biogas (i.e., methane and carbon dioxide.)

VERY SHORT ANSWER TYPE QUESTIONS:

1. Define resource.
2. Define natural resources.
3. Name any two common natural resources.
4. Name any two renewable resources.
5. Name any two non-renewable resources.
6. Name any three inexhaustible resources.
7. Name any two exhaustible resources.
8. What does the presence of coliform bacteria in water body indicate?
9. Define conservation.
10. What are forests?
11. What is afforestation?
12. Define deforestation.
13. Write the expanded form of IUCN.
14. Name any one important act related with wildlife conservation.
15. What are endangered species?
16. Name any two mammalian endangered species in India.
17. What is 'Chipko Andolan'?
18. Who started 'Chipko Movement'?
19. Where was Chipko Movement started?
20. What is the slogan of Chipko Movement?
21. What is the contribution of Amrita DevWhat is meant by 3-R strategy?
22. What are 'hot spots of biodiversity'?
23. Give three uses of wood.
24. Name at least one resin and one dye that we obtain from forests.
25. Give two major functions of forests.
26. Name one method of replenishing forests.
27. Which fossil fuel is conserved when we save electricity?
28. Name one major factor behind destructions of habitat.
29. Name any two environmentalists of India.
30. When is Van Mahotsava practiced?
31. Name a pollution free source of energy.
32. Name a process which causes ecological imbalance in the biosphere.

33. When was project tiger started in India?
34. Name an animal which has become extinct in India.
35. When was the Forest Conservation Act enacted?
36. Which is the major coal producing state in India?
37. Which is the biggest reservoir of water on the earth?
38. Name the most common practice of recharging ground water.
39. Name the person who is most remembered for the protection of Khejri trees in Rajasthan.
40. Name any two major dams in India.
41. Name the river with which Tehri Dam is associated.
42. Give the expanded form of CFFL and CFC.
43. What are national parks?
44. What are biosphere reserves?
45. What are different sources of water?
46. Name a greenhouse gas produced during burning of fossil fuels.
47. Write two advantages of reforestation or replenishment of forests.
48. State any two advantages of constructing a dam on a river.
49. What is water harvesting?
50. Name any two traditional systems of water harvesting.
51. What are biodiversity hotspots?
52. What is coal?
53. Which is cleanest fossil fuel?
54. What is BOD?
55. What is TDS?
56. What percentage of solid waste is formed of packing material?
57. Why should we use metallic, glass and ceramic containers?
58. What is the meaning of garbage management?
59. Name the group of bacteria the presence of which confirms the water pollution.
60. What are national parks?
61. What do you mean by wildlife?
62. What are the advantages of dams?
63. Name two endangered plant species.
64. What do you understand by sustainable development?

65. What is in situ conservation?
66. What is ex situ conservation?
67. What is urban forestry?
68. What is social forestry?
69. What is agroforestry?
70. Give an example of endangered animal species.
71. Give an example of endangered plant species.
72. Define the term capillary water.
73. Define the term hygroscopic water.
74. Define the term combined water.
75. Define the term gravitational water.
76. Give the examples of onenon-methanogenic bacteria.
77. Give the examples of one methanogenic bacteria.

SHORT ANSWER TYPE QUESTIONS

1. What is a natural resource? Name three important natural resources.
2. Why do we need to manage our natural resources?
3. Why should fossil fuels like coal and petroleum be used judiciously?
4. What are the three R's to save the environment? Explain with one example of each.
5. What are the main uses of coal and petroleum products?
6. Name two factors which can be used to find whether river water has been contaminated.
7. What is the main purpose of rainwater harvesting?
8. Write on the process in which rainwater falling on the earth is stopped form flowing and made to percolate into the ground?
9. Name the most common practice of recharging ground water.
10. The pH of a river water sample as measured by pH paper is found to be 6. What does it tell us about water?
11. Why is the process of 'reuse' better than that of 'recycling'?
12. Name a clean gaseous fuel other than LPG and natural gas.
13. Write a short not on "sustainable development"?
14. Write a short not on 'Chipko Andolan' (Hug the Trees Movement).
15. Write down on Joint Forest Management in India.

16. Why should we conserve forests and wildlife?
17. Describe briefly the 'Khadin' system of rainwater harvesting practiced in Rajasthan.
18. What measures would you take to conserve electricity in your house?
19. Is water conservation necessary? Give reasons.
20. Name the products of combustion of fossil fuels like coal and petroleum products. How do they affect us and our environment?
21. State any five steps to reduce the consumption of coal and petroleum products.
22. Explain why, despite good rains, we are not able to meet the demand for water of all the people in our country.
23. Give one example to show how the participation of local people can lead to the efficient management of forests.
24. State the advantages of constructing dams across the rivers.
25. Describe some of the problems associated with the construction of dams.
26. Name any five source of water (other than rivers).
27. Describe how, the water of river Ganga has been highly polluted.
28. State the main aim of the management of forests and wildlife.
29. Name the three main stakeholders in the management of forest resources.
30. What is meant by rainwater harvesting? Name some of the ancient structures used for rainwater harvesting by the rural people.
31. Prepare a list of items that you use daily in the school and also can be recycled.
32. List two advantages associated with water harvesting at community level.
33. Suggest few measures for controlling carbon dioxide levels in the atmosphere.
34. How sustainable development can be obtained?
35. Write down the methods of water harvesting.
36. Explain the effects of deforestation.
37. Why do we need to manage our natural resources?
38. Justify the statement 'The growth of industries has led to destruction of forests'.
39. Suggest four ways to manage the garbage.
40. What are forests considered as biodiversity hotspots?
41. What are different harmful gases produced on combustion of coal and petroleum?
42. Give two measures for checking the quality of water.
43. Why is fuel oil better than coal for industrial use?

44. If forests get depleted, what effect do they have on rainfall?
45. What is deforestation? Explain the reasons of deforestation.
46. Explain the measures to manage and conserve water resources.
47. Explain the importance of forests.
48. Give the types of forests in India
49. Write down on conservation strategies
50. What are the reasons for depletion of wildlife?
51. What are the aims to conserve natural resources?
52. What is meant by recycling of wastes?
53. Give two examples to emphasize the concept of reuse.
54. Mention one negative effect of our affluent life style on the environment.
55. Write down the aims of natural resource conservation
56. Write down the basic practices of conservation
57. Write differences between in situ conservation and ex situ conservation.
58. Define sustainable development. Give one way to achieve sustainable development.

LONG ANSWER TYPE QUESTIONS

1. What is conservation? Explain about the major categories of conservation.
2. What is rainwater harvesting? Describe roof top rainwater harvesting.
3. Discuss the significance of forests in detail.
4. What are fossil fuels? Give two examples of fossil fuels.
5. Why should we conserve forests? Suggest any two ways to conserve forests.
6. What are dams? What are different problems associate with the construction of dams?
7. How can we reduce the consumption of fossil fuels?
8. Describe the methods of rainwater harvesting and their benefits.
9. Describe the conservation of wildlife in detail.
10. Discuss the three R's to save natural resources in detail.
11. What is the role of forest in conserving the environment?
12. Describe briefly the Ganga Action Plan.
13. Define the term rain water harvesting and explain the methods.
14. Why are coal and petroleum considered to be non-renewable sources of energy?

15. Water is a valuable resource. And list two ways that you would suggest every family member to save this resource.
16. Write two advantages of classifying energy sources as renewable and non-renewable.
17. Industrialization is one main cause of deterioration of environment. Discuss.
18. Explain the role of forests in conserving the environment. How do the forests get depleted? What are its consequences?
19. Dams are constructed on the river for multipurpose use. Give its advantages and disadvantages.
20. What is 'Chipko movement'? Who provided leadership for this movement in Tehri region?
21. Name the movement similar to Chipko movement that happened in Karnataka. Who provided leadership for this movement?
22. Name any major dam constructed in northern India. Across which river it is constructed? And List any one category of problem associated with the construction of large dams.
23. Explain briefly, how rainwater harvesting is done from open spaces around the buildings in city areas.
24. What is sustainable development? Explain it with respect to forest, wildlife and water resources.

MULTIPLE CHOICE QUESTIONS:

1. A successful forests conservation strategy involve
   (a) Protection of only consumers
   (b) Protection of only herbivores
   (c) Protection of animals at the highest tropic level
   (d) Comprehensive programmes to protect all the physical and biological components
2. Which of the following R's is not used to save the environment?
   (a) Remove (b) Reduce (c) Reuse (d) Recycle
3. The main aim of conservation of preservation is to
   (a) Preserve the plants (b) Preserve the animals
   (c) Preserve the biodiversity (d) Preserve the fuels
4. Climate change may impact.
   (a) Air quality, oceans, and coastal zones
   (b) Energy and human health
   (c) Agriculture, natural terrestrial ecosystems and water resources
   (d) All of the these

5. The most rapidly dwindling natural resource in the world is
   (a) Forest (b) Sunlight (c) Water (d) Wind
6. Which of the following fuels are formed by the degradation of biomass?
   (a) CNG (b) Biogas (c) Nuclear fuel (d) Fossil Fuel
7. The major role of forest is
   (a) Productive (b) Regulative (c) Protective (d) All of these
8. Deforestation causes
   (a) Soil erosion and degradation (b) Loss of wildlife
   (c) Environmental pollution (d) All of these
9. Conservation is
   (a) Proper use of natural resources
   (b) Protection of natural resources
   (c) Management of natural resources
   (d) All of these
10. Tracts of land with or without a lake where wild life is not hunted but other activities are allowed.
    (a) National Parks (b) Zoos
    (c) Sanctuaries (d) Biosphere reserves
11. Ecological imbalance in the biosphere is created by
    (a) Cutting of forest (b) Conservation of forest
    (c) Production of more paper (d) Both (a) and (c)
12. Measurement of water pollution is made by
    (a) pH (b) BOD
    (c) Coliform count (d) All of these
13. Animals get extinct mainly due to
    (a) Predation (b) Pollution
    (c) Habitat destruction (d) Afforestation
14. Soil erosion can be checked by
    (a) Afforestation (b) Deforestation
    (c) Removal of vegetation (d) Overgrazing
15. Water scarcity is not caused due to
    (a) Increasing urbanization (b) Increased demands for crops
    (c) High living standards of people (d) Afforestation

16. The carbon dioxide emission can be regulated by
    (a) Increasing vegetation
    (b) Reduced burning of organic matter
    (c) Use of improved technology in automobiles
    (d) All of these
17. Water from dam is used for irrigation in fields through
    (a) Canals (b) Ocean (c) River (d) None of these
18. Select the eco-friendly activity among the following:
    (a) Using polybags for shopping
    (b) Using car for transportation
    (c) Using dyes for colouring clothes
    (d) Using windmills to generate power for irrigation
19. Which of the following are the uses of forest?
    (a) Provide raw material for paper
    (b) Prevent soil erosion, floods and dust storms.
    (c) Help in maintaining a balance of CO_2 and O_2 in the atmosphere
    (d) All of the above
20. The bacteria present in contaminated water is
    (a) *Mycobacterium* (b) *Streptococcus*
    (c) *Clostridium* (d) *Coliform*
21. Water is
    (a) National resource (b) Organic resource
    (c) Exhaustible resource (d) Inexhaustible resource
22. The Indian Forests Act was started in
    (a) 1927 (b) 1995 (c) 1997 (d) 1999
23. Biodiversity hotspots are
    (a) Oceans (b) Rivers (c) Glaciers (d) Forests
24. Bhundhis system of water harvesting is found in
    (a) Bihar (b) Kerala (c) Tamil Nadu (d) Uttar Pradesh
25. Among the following animals which one has become extinct in India?
    (a) Bison (b) Cheetah (c) Tiger (d) Rhinoceros

26. Forests is an
    (a) Exhaustible resource
    (b) Renewable resource
    (c) Inexhaustible resource
    (d) Both a and b
27. Opposition to the construction of large dams is due to
    (a) Social reasons
    (b) Economic reasons
    (c) Environmental reasons
    (d) All of these
28. Petroleum is
    (a) Inexhaustible resource
    (b) Exhaustible resource
    (c) Non-renewable resource
    (d) Both b and c
29. Which of the following can be recycled?
    (a) Broken glass (b) Waste food (c) Cattle dung (d) fruit peels
30. The three R's that will help us to conserve natural resources for long term use are
    (a) Recycle, regenerate, reuse
    (b) Reduce, reuse, redistribute
    (c) Reduce, recycle, reuse
    (d) Reduce, regenerate, reuse
31. Extensive plantation of trees to increase forest cover is
    (a) Deforestation
    (b) Afforestation
    (c) Agro-forestry
    (d) Social forestry
32. The poisonous gas which reduces the oxygen-carrying capacity of blood to a large extent is:
    (a) SO_2 (b) NO (c) CO (d) CO_2
33. Which one of the following is not a natural resource?
    (a) Soil (b) Water (c) Air (d) Electricity
34. Which of the following gas is released during combustion of coal and petroleum?
    (a) Carbon dioxide
    (b) Carbon monoxide
    (c) Nitrogen oxides
    (d) All of these
35. Which factor is mainly responsible for increase in demand of natural resources?
    (a) Scientific advancement
    (b) Increased human population
    (c) Environmental pollution
    (d) Use of Biodegradable chemicals
36. Which one of the following is a multinational resource?
    (a) Forest
    (b) Minerals
    (c) Air
    (d) Migratory birds

37. The pollution in River Ganga is caused by
    (a) Bathing
    (b) Washing clothes
    (c) Dumping of unburnt corpses
    (d) All of these
38. The pH range most conducive for life of fresh water plants and animals is
    (a) 6.5-7.5 (b) 2.0-3.5 (c) 3.5-5.0 (d) 9.0-10.5
39. Expand the abbreviation GAP.
    (a) Ganga Action Plan
    (b) Gross Assimilation by Photosynthesis
    (c) Governmental Agency for Animal Protection
    (d) Governmental Agency for Pollution Control
40. Water with TDS 140ppm is not suitable for
    (a) Drinking (b) Bathing (c) Washing (d) All of these
41. If PH of water is acidic, it indicates that water is
    (a) Hard (b) Pure
    (c) Polluted (d) Suitable for drinking
42. Khadins, Bundhis, Ahars and Kattas are ancient structures that are examples for
    (a) Grain storage (b) Wood storage
    (c) Water harvesting (d) Soil conservation
43. Kulhs are irrigation canals of
    (a) Rajasthan (b) Bihar
    (c) Himachal Pradesh (d) Karnataka
44. Sardar Sarovar Dam was constructed on
    (a) Yamuna (b) Narmada (c) Godavari (d) Ganga
45. The famous movement that started by women of Advani village in Tehri-Garhwal against felling of trees
    (a) Appiko Movement (b) Bishnoi Movement
    (c) Bahuguna Movement (d) Chipko Movement
46. Arabari forest of Bengal is dominated by:
    (a) Teak (b) Sal (c) Bambooo (d) Mangrove

47. With which tree Amrita Devi Bishnoi is associated?
    (a) Khajoor (b) Khejrali (c) Khejri (d) Keekar
48. The Bishnoi community of Rajasthan is associated with the conservation of:
    (a) Coal and petroleum (b) Forests and wildlife
    (c) Water resources (d) Abiotic environment
49. The Chipko Andolan is associated with:
    (a) Tigers (b) Turtles (c) Trees (d) Tomatoes
50. Amrita Devi Bishnoi was associated with
    (a) Preventing the custom of child marriage in Rajasthan
    (b) Campaign to save the girl child
    (c) Conservation of cultural heritage of Rajasthan
    (d) Conservation of forests and wildlife
51. The important message conveyed by the 'Chipko Movement' is
    (a) To ignore the community in forest conservation efforts
    (b) To involve the community in forest conservation efforts
    (c) To cut down forest trees for developmental activities
    (d) Government agencies have the unquestionable right to order destruction of trees in forests
52. Which of the following communities in Rajasthan has been a religious tenet of conservation of forest and wild life?
    (a) Bishnoi (b) Agargwal (c) Jaiswal (d) Jain
53. Who protected Sal forest in Midnapore?
    (a) H.K. Banerjee (b) D.K. Banerjee
    (c) M.K. Banerjee (d) A.K. Banerjee
54. A national award for wildlife conservation has been named after
    (a) Amrita Devi Bishnoi (b) Chandi Prasad Bhatt
    (c) Gaura Devi (d) Sunder Lal Bahuguna
55. Red Data Book
    (a) Maintain record of dead animals
    (b) Maintains record of human population
    (c) Maintains the record of those creatures which are threatened with extinction
    (d) Recommends names of biodegradable substances

ANSWER KEY

1. (d)	2. (a)	3. (c)	4. (d)	5. (d)	6. (d)	7. (d)	8. (d)
9. (d)	10. (c)	11. (a)	12. (d)	13. (c)	14. (a)	15. (c)	16. (d)
17. (c)	18. (d)	19. (d)	20. (d)	21. (d)	22. (a)	23. (d)	24. (d)
25. (d)	26. (b)	27. (d)	28. (c)	29. (a)	30. (c)	31. (b)	32. (c)
33. (d)	34. (d)	35. (b)	36. (d)	37. (d)	38. (a)	39. (a)	40. (d)
41. (c)	42. (c)	43. (c)	44. (b)	45. (d)	46. (b)	47. (c)	48. (a)
49. (c)	50. (d)	51. (b)	52. (a)	53. (d)	54. (a)	55. (c)	

CONCEPT MAPS

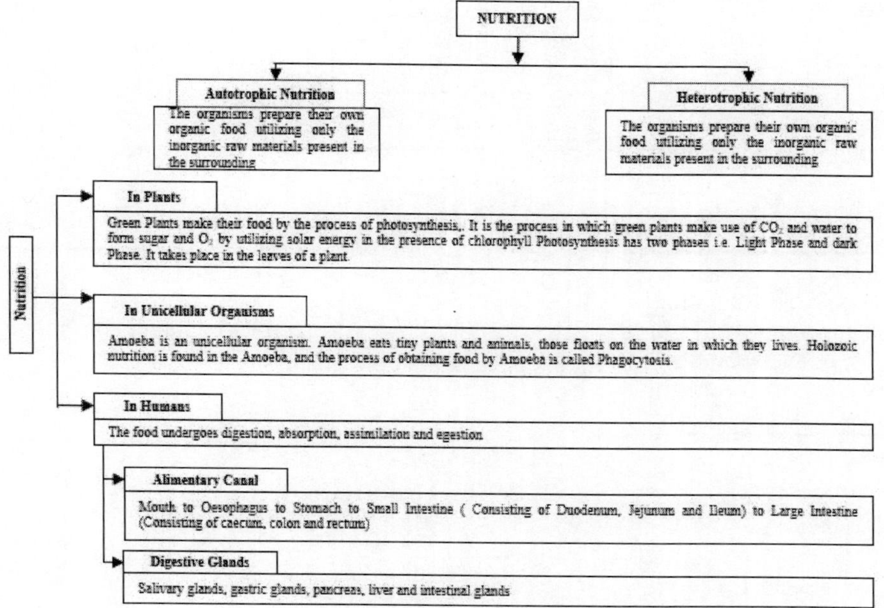

Concept Map 1

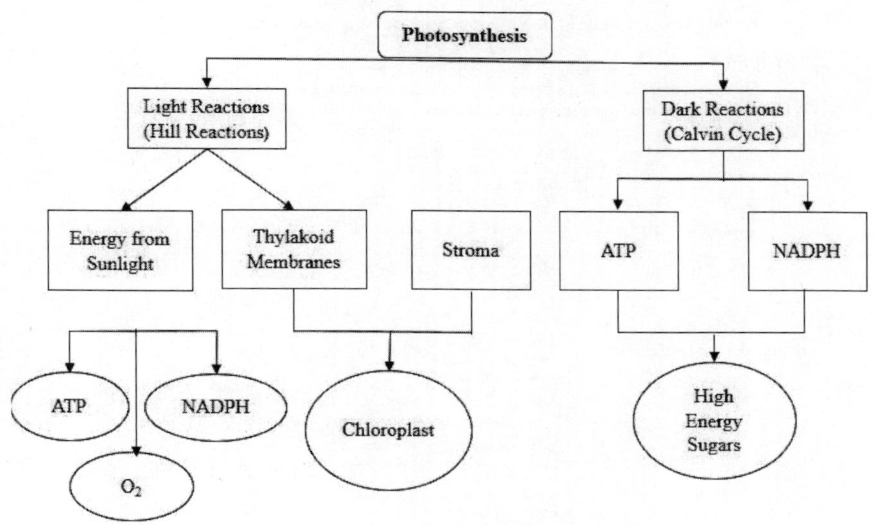

Concept Map 2

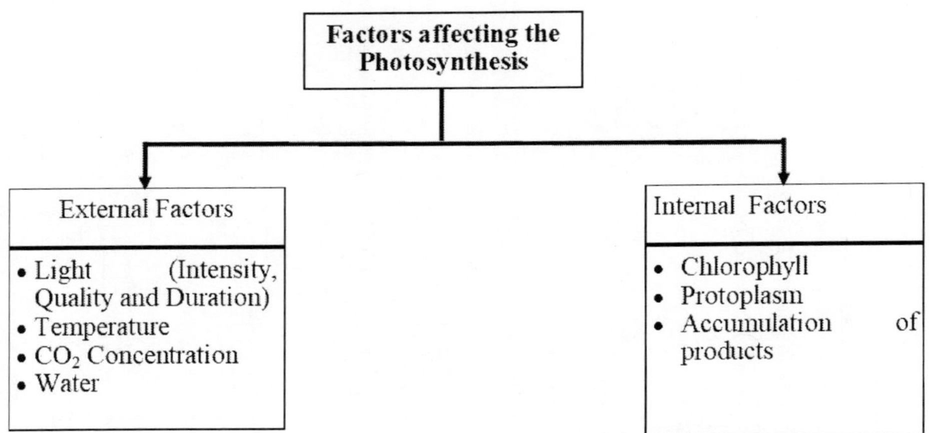

Concept Map 3

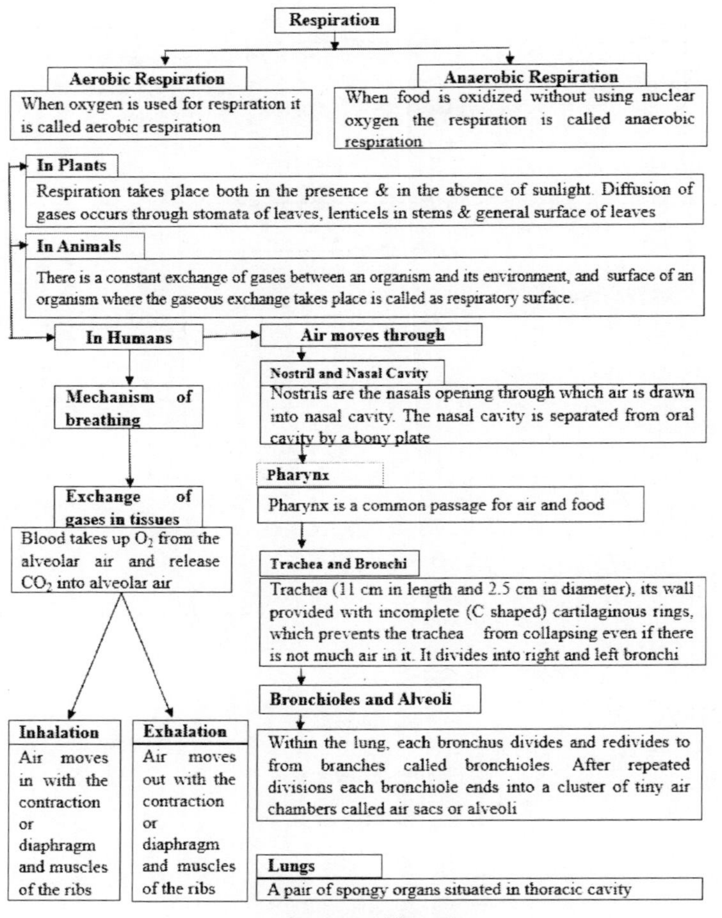

Concept Map 4

Transport of Material in Body

Transport

The transport of substances like food, water, oxygen, hormones and enzymes etc, inside the body of an organism is called internal transport and the organs which carry out this transport within the body together constitute the internal transport system

In Plants

Plants absorb water and mineral salts from the soil by their roots. Absorbed water and minerals are transported through xylem tissue, while food prepared by the leaves is transported through the phloem

Ascent of Sap

Upward movement of minerals and water from root to the aerial part of plants against the gravitational force is called ascent of sap.

Transpiration

The loss of water in the form of vapours from the aerial parts of a plant (specially leaves) into the atmosphere is called transpiration.

Translocation

Movement of food from leaves to different parts of plants is termed as translocation

Humans

The circulatory system is a transport system in humans, carrying O_2, nutrients, hormones and other substances to the tissues, CO_2 to the fungus and other waste products to the kidney. Blood carries the food materials, O_2, waste materials etc., is forced by a pump, the heart, through a system tubes, the blood vessels (arteries and veins) to the close vicinity of every cell and then back to the heart to be pumped round again and again. Lymph is another fluid that transports some materials through the blood to the heart.

Concept Map 5

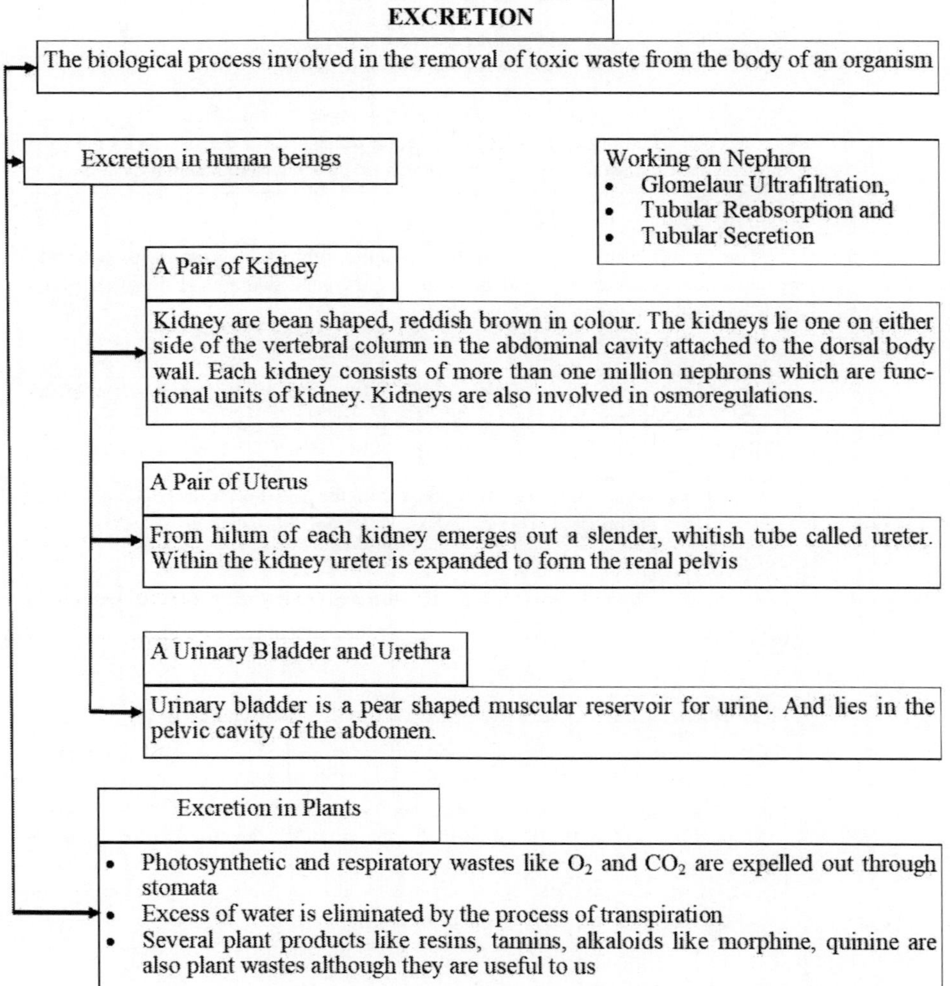

Concept Map 6

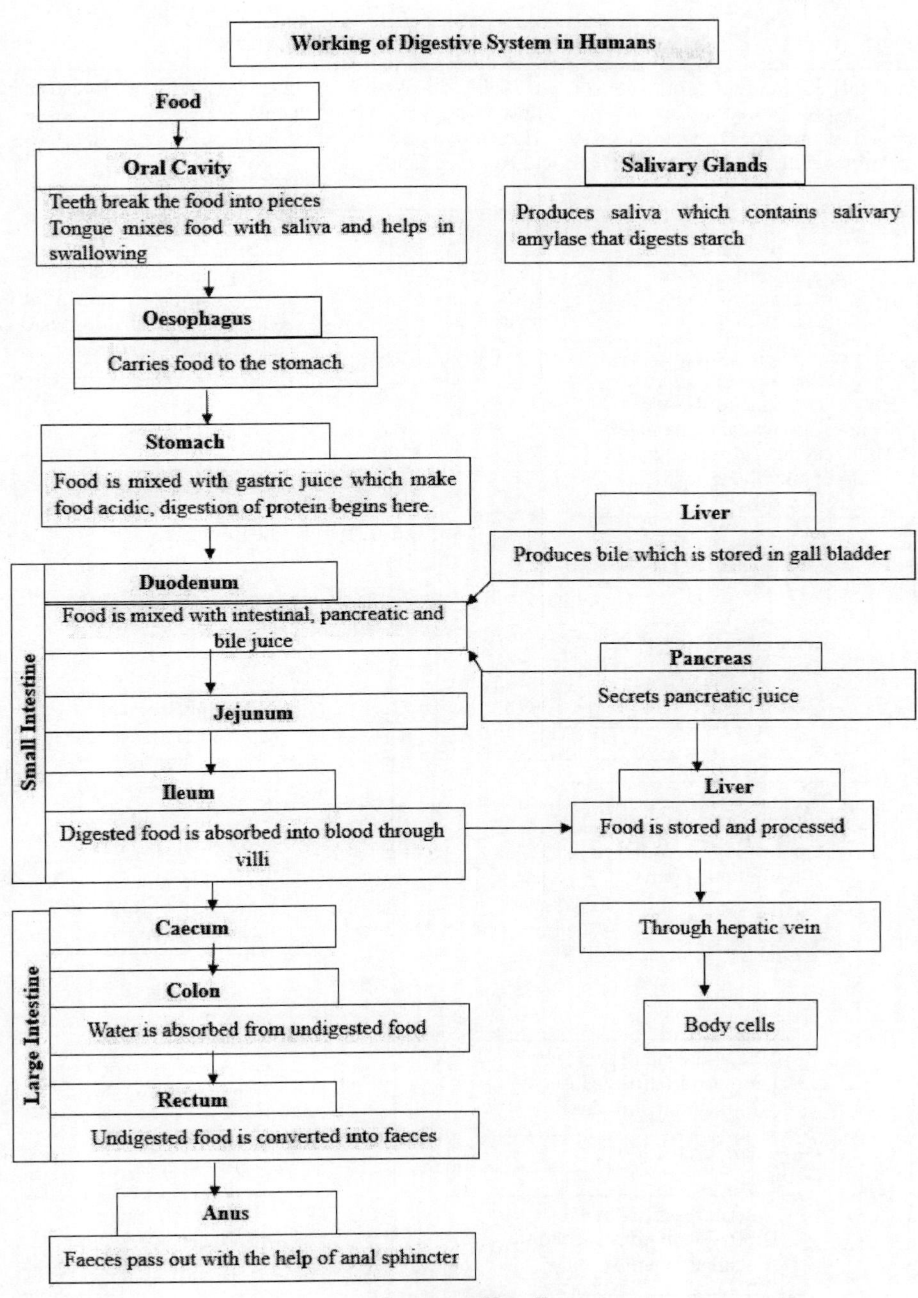

Concept Map 7

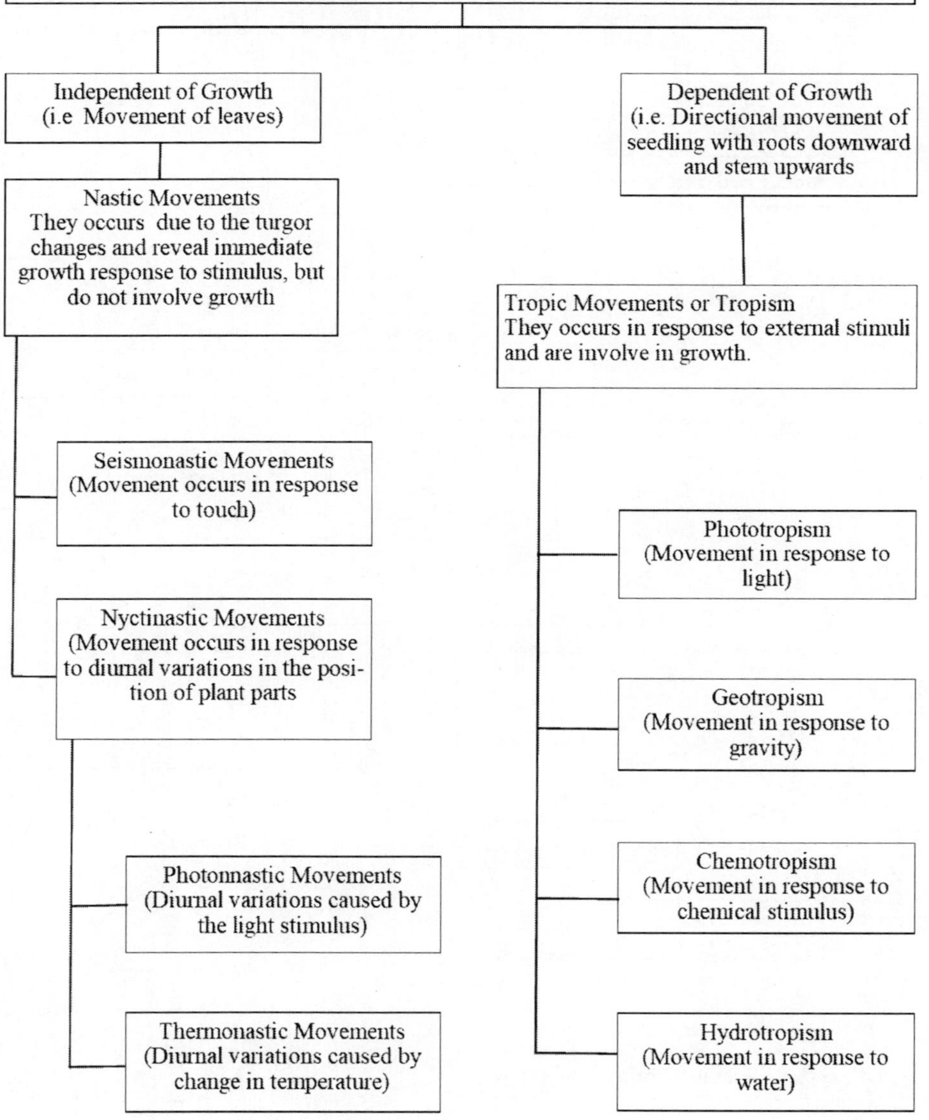

Concept Map 8

Working of Sensory Receptors

The receptors collect information from stimuli, i.e. the environmental changes around the animal

↓

Electrical impulses are set by stimuli in the sensory receptors. These sensory receptors pass the information as electrical impulses to brain or spinal cord through nerves.

↓

These impulses which reach the brain from sensory neurons are called sensations

↓

Sensations, which reach the brain, are analyzed and interpreted as perceptions. Some of the reactions are also produced in the spiral cord without the involvement of brain.

↓

These motor impulses are transmitted to appropriate effectors via motor neurons from brain or spinal cord

↓

Suitable responses are produced by effectors

Concept Map 9

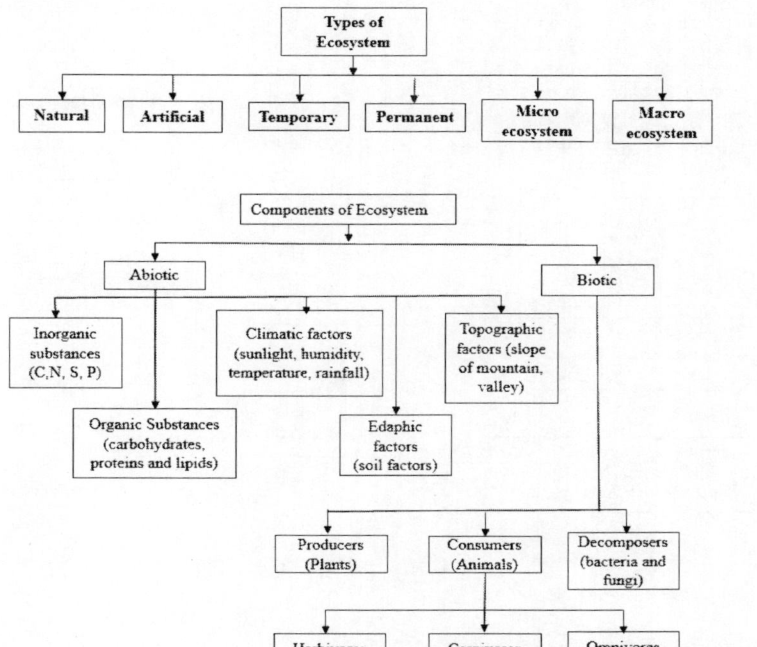

Concept Map 10

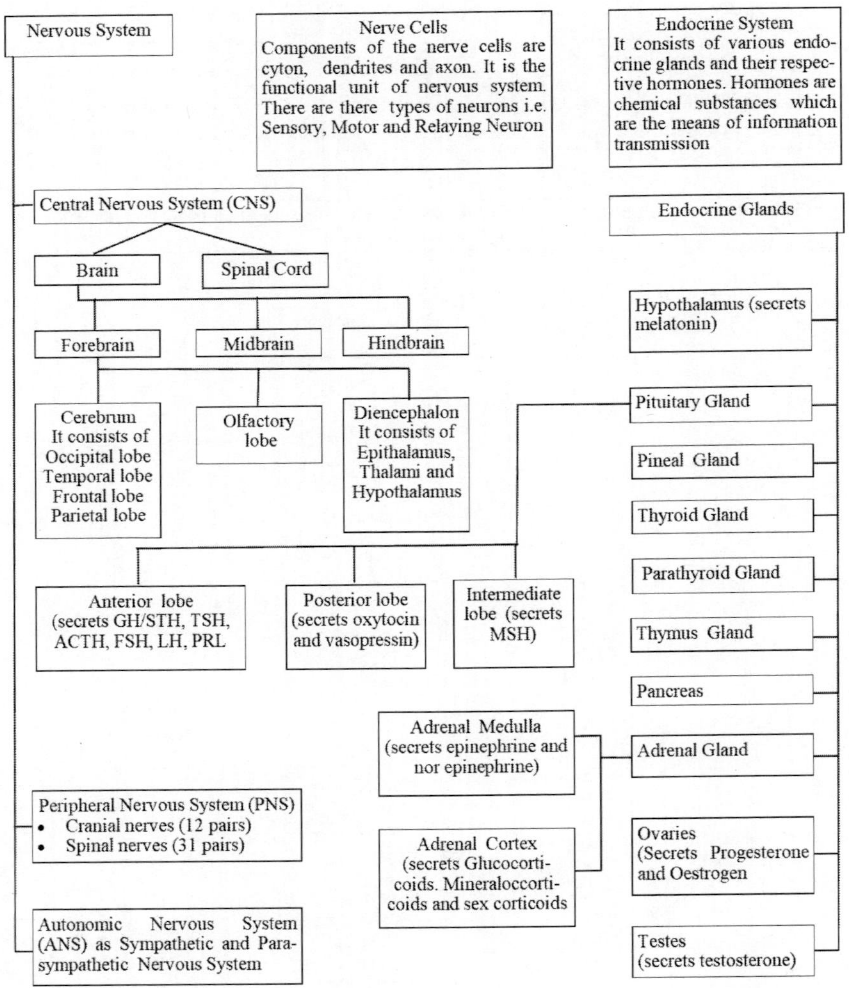

Concept Map 11

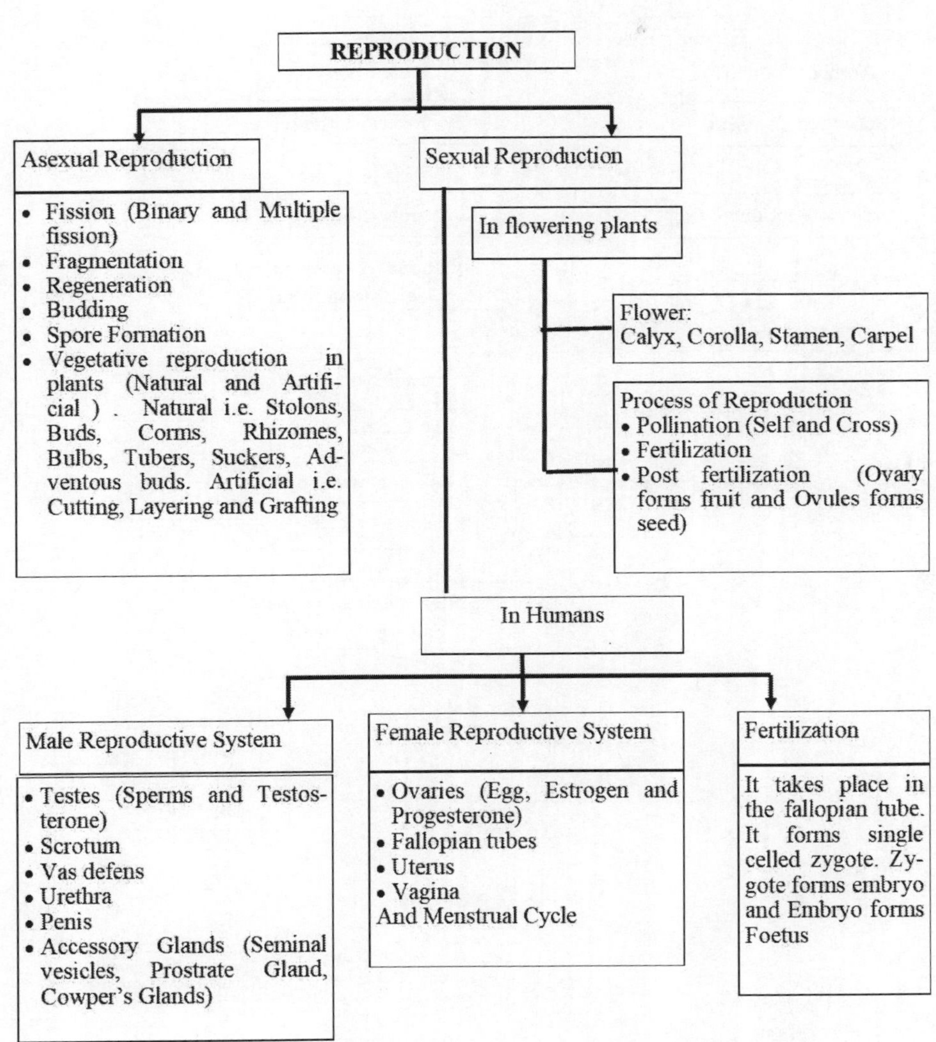

Concept Map 12

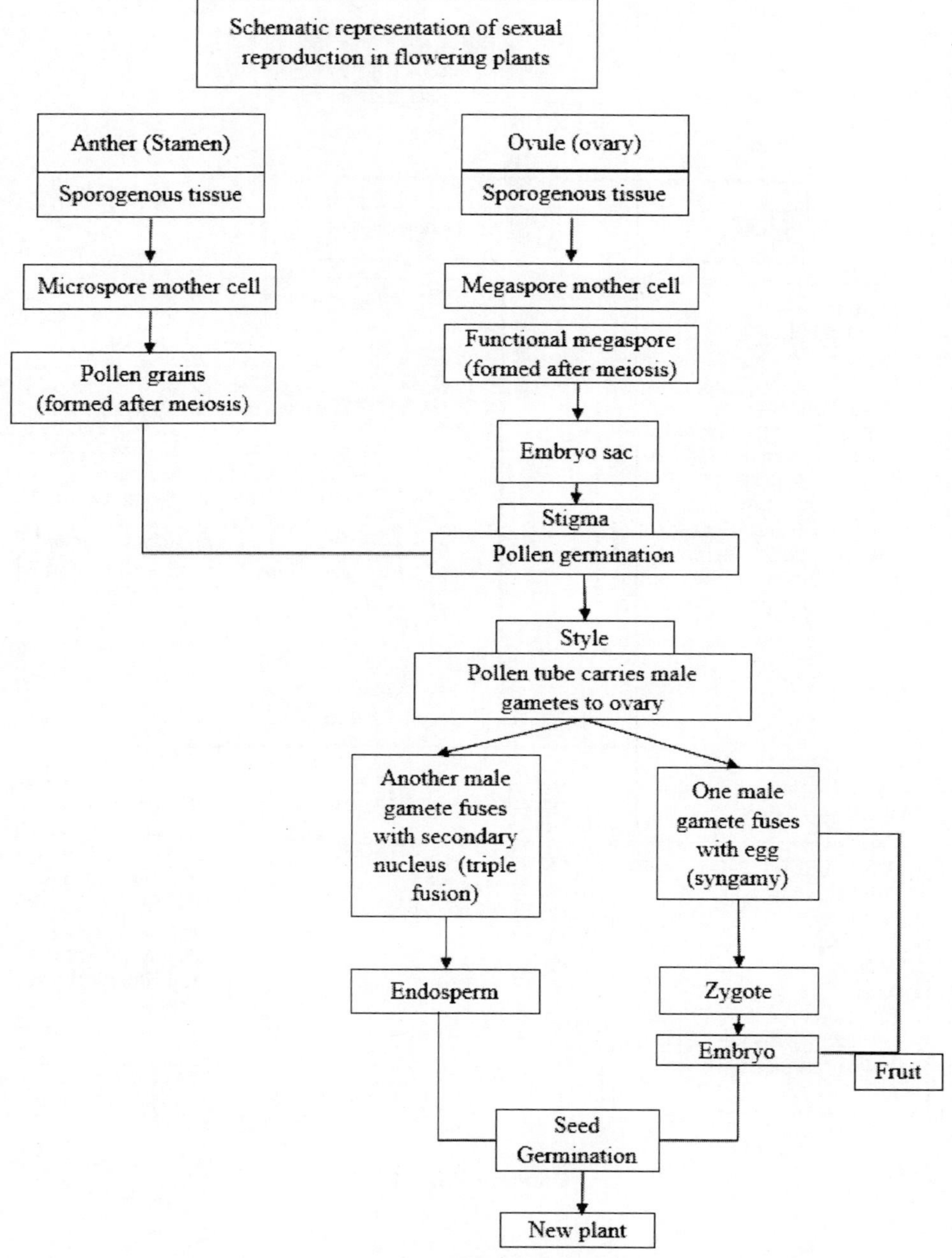

Concept Map 13

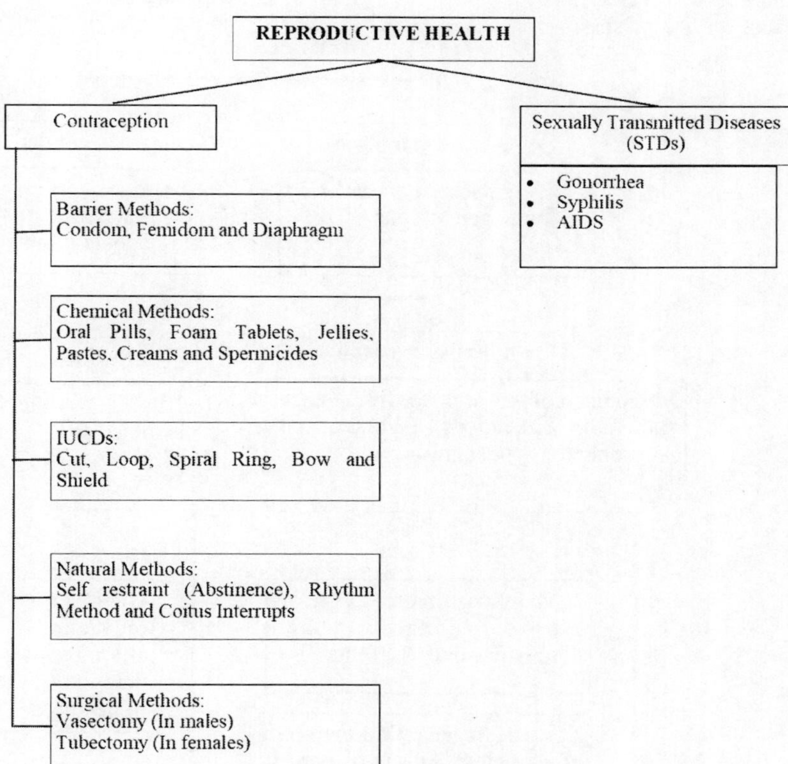

Concept Map 14

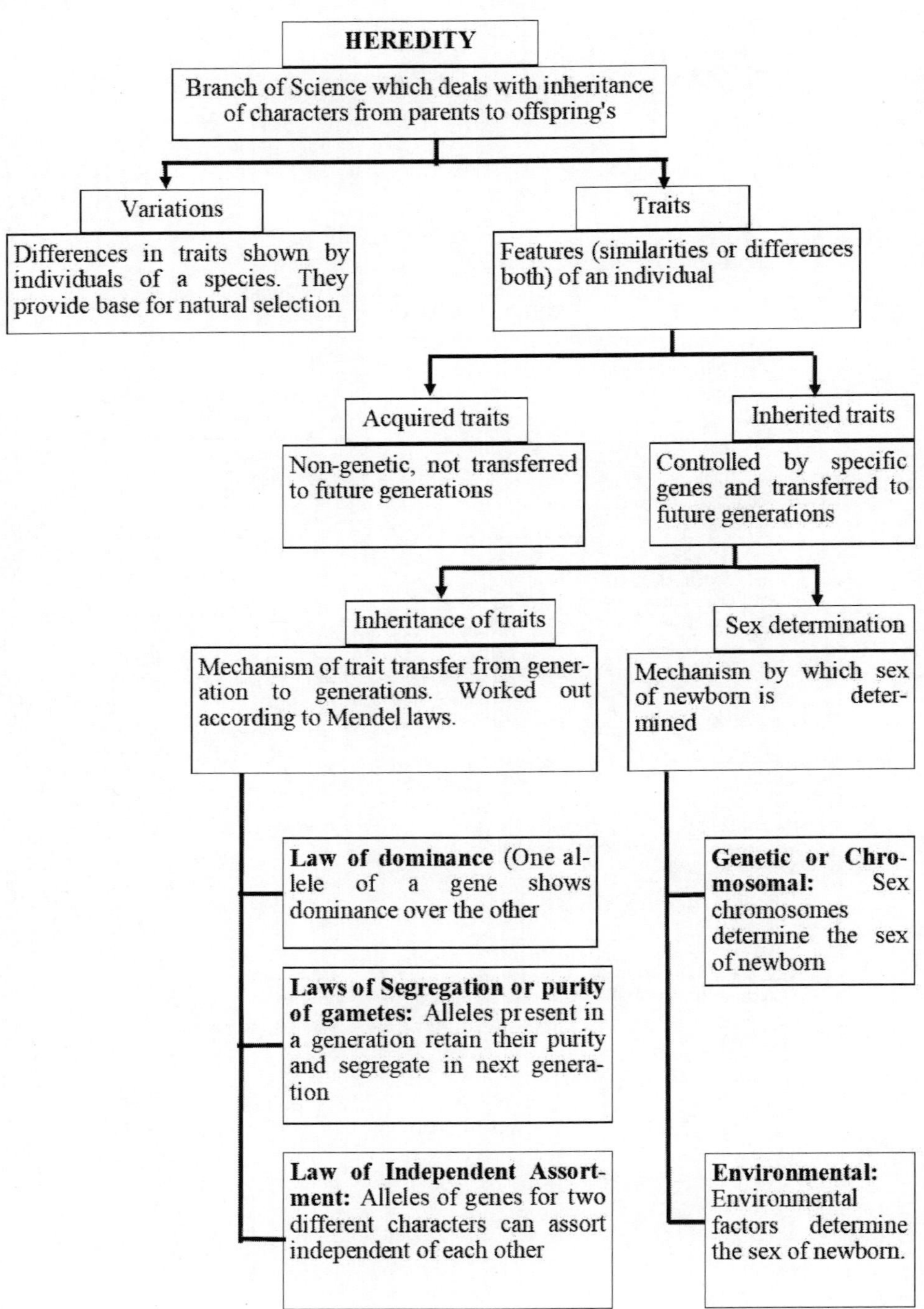

Concept Map 15

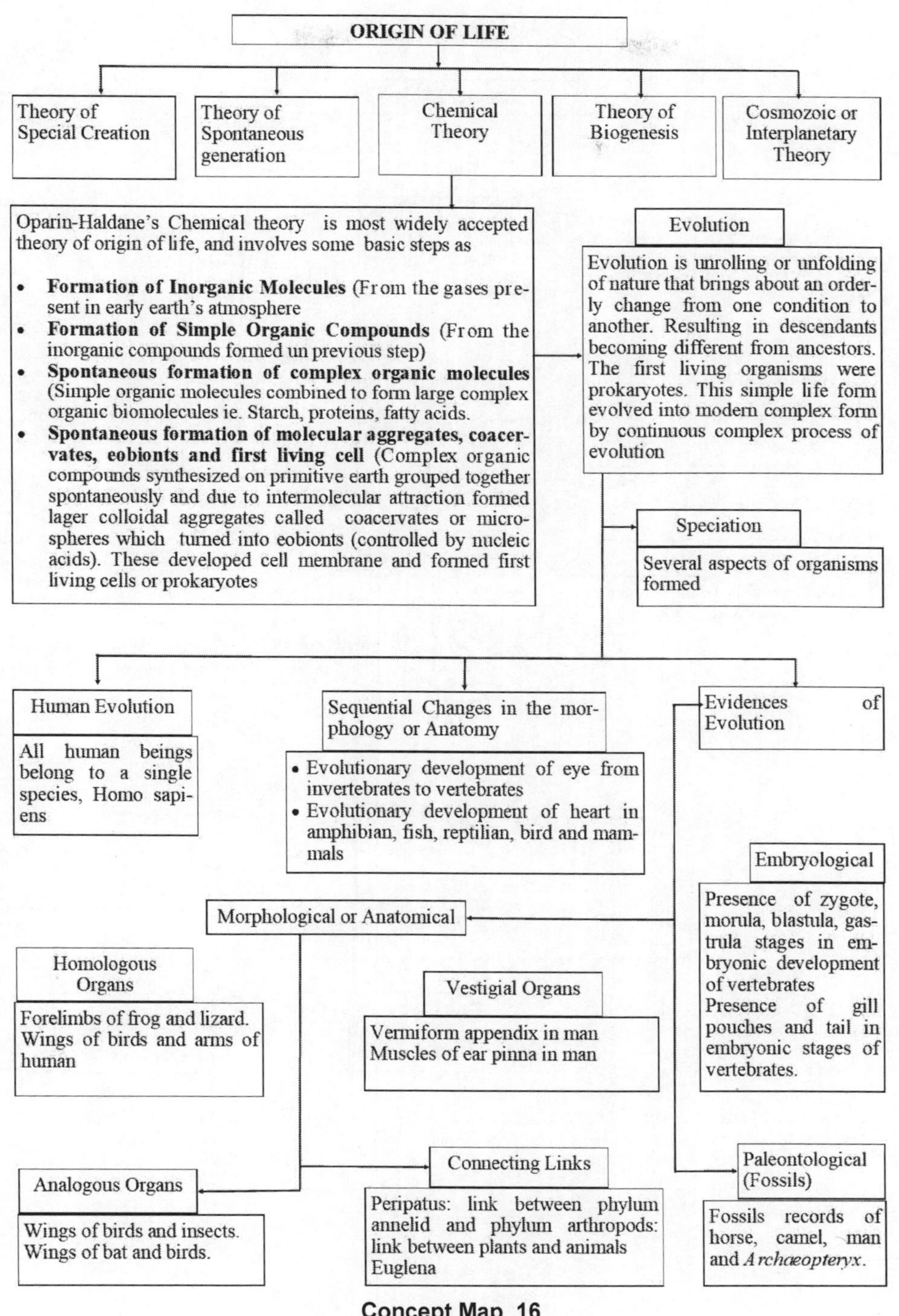

Concept Map 16

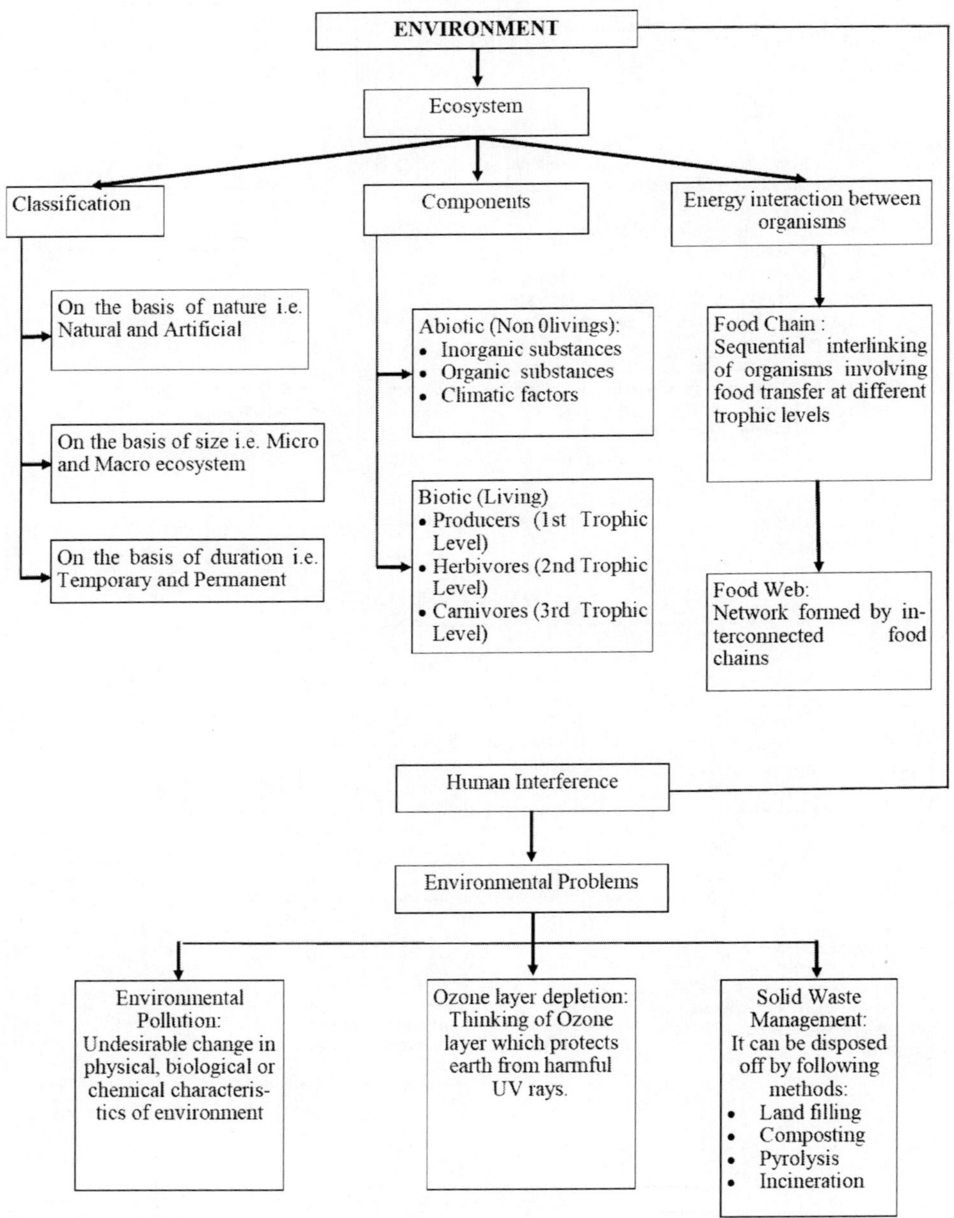

Concept Map 17

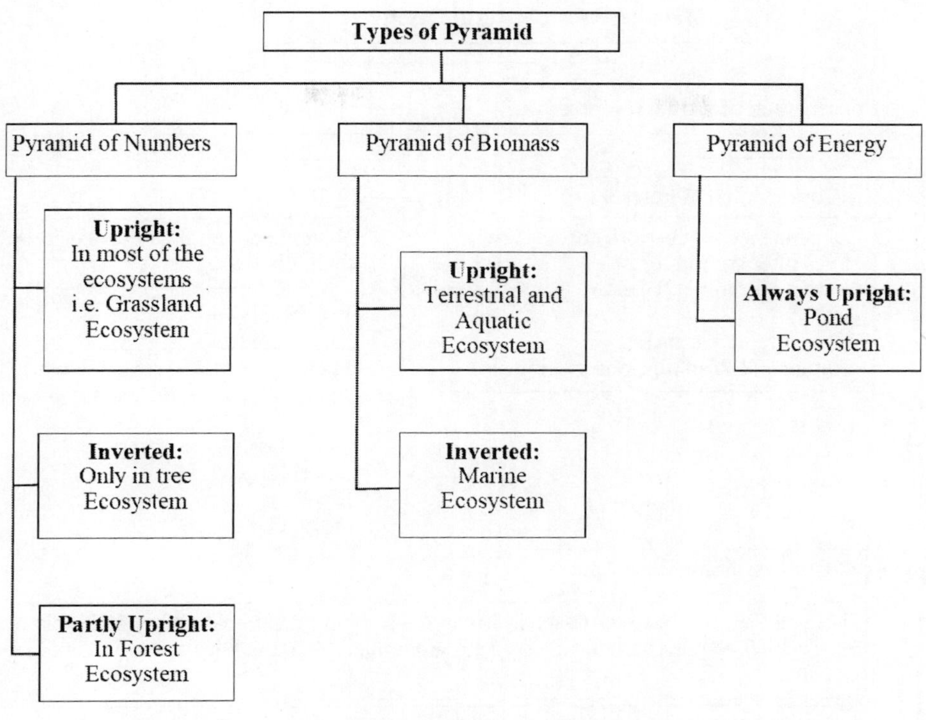

Concept Maps 18

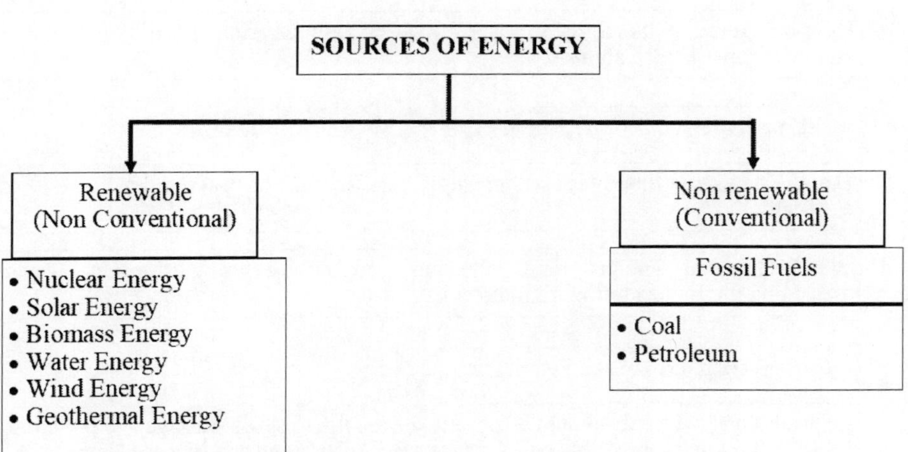

Concept Map 19

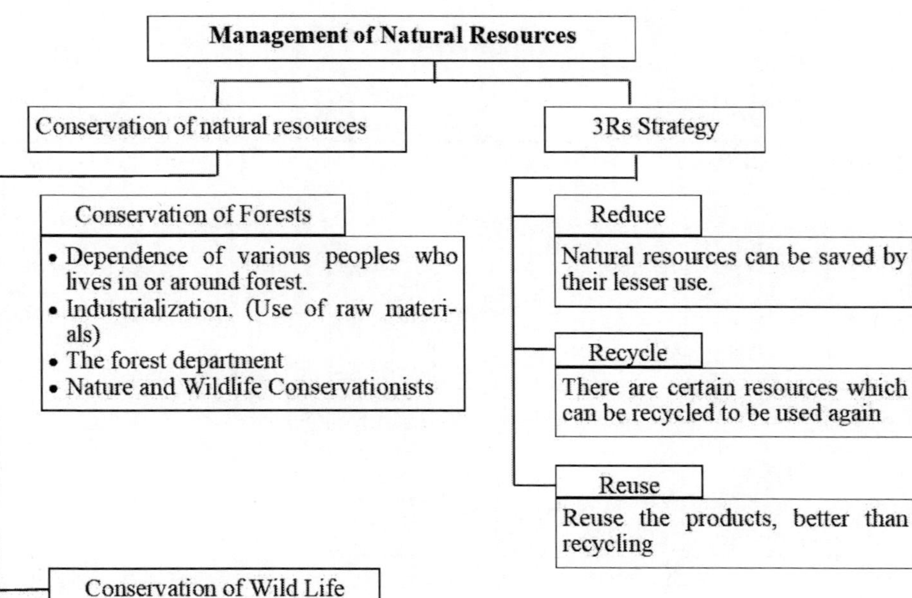

Conservation of Wild Life

National Park: It is an area which is strictly reserved for the welfare of wildlife only & where other activities are not allowed e.g. Kaziranga (Assam), Gir Forest (Gujarat)

Sanctuary: It is an area which is preserved for the conservation of wildlife and operation, such as collection of minor forest products, cutting of timber trees and private ownership are allowed, so long as they don't affect the animals adversely i.e. Bha-

Biosphere Reserve: It is an area meant for conservation of wildlife, life style of tribals and their domesticated animals.

Conservation of Water

Dams: They serve dual purpose-storage of huge amount of water and generation of electricity.

Water Harvesting: The process of collecting rainwater and storing it for direct use or recharging it into the ground for indirect use.

Conservation of Coal and Petroleum:

Oil wells and coal mines should be protected from fire to avoid wastage and pollution
Use of alternative sources of energy i.e. nuclear, solar, wind and biogas plants
Use of CNG and less consumption of oil automobiles.

Concept Map 20

INDEX

R

S

T

U

V

W

X

Y

Z